Lecture Notes in Artificial Intelligence 13506

Subseries of Lecture Notes in Computer Science

Series Editors

Randy Goebel
University of Alberta, Edmonton, Canada

Wolfgang Wahlster
DFKI, Berlin, Germany

Zhi-Hua Zhou
Nanjing University, Nanjing, China

Founding Editor

Jörg Siekmann
DFKI and Saarland University, Saarbrücken, Germany

More information about this subseries at https://link.springer.com/bookseries/1244

Sylvie Le Hégarat-Mascle · Isabelle Bloch ·
Emanuel Aldea (Eds.)

Belief Functions:
Theory and Applications

7th International Conference, BELIEF 2022
Paris, France, October 26–28, 2022
Proceedings

 Springer

Editors
Sylvie Le Hégarat-Mascle ⓘ
University of Paris-Saclay
Gif sur Yvette, France

Isabelle Bloch ⓘ
Sorbonne University
Paris, France

Emanuel Aldea ⓘ
University of Paris-Saclay
Gif sur Yvette, France

ISSN 0302-9743   ISSN 1611-3349 (electronic)
Lecture Notes in Artificial Intelligence
ISBN 978-3-031-17800-9   ISBN 978-3-031-17801-6 (eBook)
https://doi.org/10.1007/978-3-031-17801-6

LNCS Sublibrary: SL7 – Artificial Intelligence

This Springer imprint is published by the registered company Springer Nature Switzerland AG
The registered company address is: Gewerbestrasse 11, 6330 Cham, Switzerland

Preface

The theory of belief functions, also referred to as evidence theory or Dempster-Shafer theory, was first introduced by Arthur P. Dempster in the context of statistical inference, and was later developed by Glenn Shafer as a general framework for modeling epistemic uncertainty. These early contributions have been the starting points of many important developments not only in statistics but also in computer science and engineering. The theory of belief functions is now well established as a general framework for reasoning with uncertainty, and has well understood connections to other frameworks such as probability, possibility, and imprecise probability theories. It has been applied in diverse areas such as machine learning, information fusion, and pattern recognition.

The series of biennial International Conferences on Belief Functions (BELIEF), sponsored by the Belief Functions and Applications Society (BFAS), is dedicated to the confrontation of ideas, the reporting of recent achievements, and the presentation of the wide range of applications of this theory. The first edition of this conference series was held in Brest, France, in 2010. Later editions were held in Compiègne, France, in 2012, Oxford, UK, in 2014, Prague, Czech Republic, in 2016, again in Compiègne, France, in 2018, and in Shanghai, China, in 2021 together with the 1st International Conference on Cognitive Analytics, Granular Computing, and Three-way Decisions (CCGT 2021).

The 7th International Conference on Belief Functions (BELIEF 2022) was held in Paris, France, during October 26–28, 2022. It was held both onsite and online due to the COVID-19 situation. This volume represents the proceedings of BELIEF 2022, and it contains 29 accepted submissions, each reviewed by either two or three peers in a single-blind review process. Original contributions were solicited on theoretical aspects (including, for example, mathematical foundations, links with other uncertainty theories) as well as on methods for various problems including classification, clustering, data fusion, and on applications in various areas including medical data processing, environmental studies, and so on.

We would like to thank all the people who made this volume and this conference possible: all contributing authors, the organizers, and the Program Committee members who helped to build such an attractive program. We are especially grateful to our four invited speakers, Stéphane Canu (INSA Rouen Normandie, France) for his talk "Robustness of neural networks and adversarial attacks", Rémi Bardenet (CNRS and Lille University, France) for his talk "Topics in Monte Carlo computation and Bayesian learning", Ozgur Erdinc (Raytheon Technologies Research Center, USA) for his talk "Challenges in Automating Mission-Critical Decision Making Systems: A Practitioner's Perspective", and Philippe Xu (Université de Technologie de Compiègne, France) for his talk "Fusion of heterogeneous deep neural networks with belief functions". We would also like to thank all our generous sponsors: the Belief Functions and Applications Society (BFAS), the DATAIA Institute, the SATIE Laboratory, the Sorbonne Center of Artificial Intelligence (SCAI), the International Journal of Approximate Reasoning, and Elsevier. Furthermore, we would like to thank the editors of the Springer series Lecture

Notes in Artificial Intelligence (LNCS/LNAI) and Springer for their dedication to the production of this volume.

August 2022   Sylvie Le Hégarat-Mascle
                                                        Isabelle Bloch
                                                        Emanuel Aldea

Organization

General Chair

Sylvie Le Hégarat-Mascle — Paris-Saclay University, France

Program Committee Chairs

Isabelle Bloch — Sorbonne Université, France
Emanuel Aldea — Paris-Saclay University, France

Steering Committee

Thierry Denœux — Université de Technologie de Compiègne, France
Sylvie Le Hégarat-Mascle — Paris-Saclay University, France
Isabelle Bloch — Sorbonne Université, France
Emanuel Aldea — Paris-Saclay University, France
Frédéric Pichon — Université d'Artois, France

Program Committee

Emanuel Aldea — Paris-Saclay University, France
Alessandro Antonucci — Dalle Molle Institute for Artificial Intelligence, Switzerland
Isabelle Bloch — Sorbonne Université, France
Ines Couso — University of Oviedo, Spain
Fabio Cuzzolin — Oxford Brookes University, UK
Thierry Denœux — Université de Technologie de Compiègne, France
Sébastien Destercke — Université de Technologie de Compiègne, France
Jean Dezert — ONERA, France
Didier Dubois — Toulouse Institute of Computer Science Research, France
Zied Elouedi — Institut Supérieur de Gestion de Tunis, Tunisia
Sabine Frittella — Université d'Orléans, France
Radim Jiroušek — Prague University of Economics and Business, Czech Republic
Anne-Laure Jousselme — Centre for Maritime Research and Experimentation, Italy
John Klein — Université de Lille, France

Václav Kratochvíl	Institute of Information Theory and Automation, CAS, Czech Republic
Sylvie Le Hégarat-Mascle	Paris-Saclay University, France
Éric Lefèvre	Université d'Artois, France
Xinde Li	Southeast University, China
Zhunga Liu	Northwestern Polytechnical University, China
Arnaud Martin	Université de Rennes 1, France
Ryan Martin	North Carolina State University, USA
David Mercier	Université d'Artois, France
Enrique Miranda	University of Oviedo, Spain
Serafín Moral	University of Granada, Spain
Frédéric Pichon	Université d'Artois, France
Benjamin Quost	Université de Technologie de Compiègne, France
Emmanuel Ramasso	École nationale supérieure de mécanique et des microtechniques, France
Roger Reynaud	Paris-Saclay University, France
Johan Schubert	Swedish Defence Research Agency, Sweden
Prakash P. Shenoy	University of Kansas, USA

Additional Reviewers

Constance Thierry
Sajad Nazari

Contents

Links with Other Uncertainty Theories

Applications

.

Evidential Clustering

A Distributional Approach for Soft Clustering Comparison and Evaluation

Andrea Campagner[1]([✉]), Davide Ciucci[1], and Thierry Denœux[2,3]

[1] University of Milano-Bicocca, viale Sarca 336, 20126 Milan, Italy
a.campagner@campus.unimib.it
[2] Université de technologie de Compiègne, CNRS,
UMR 7253 Heudiasyc, Compiègne, France
[3] Institut universitaire de France, Paris, France

Abstract. The development of external evaluation criteria for soft clustering (SC) has received limited attention: existing methods do not provide a general approach to extend comparison measures to SC, and are unable to account for the uncertainty represented in the results of SC algorithms. In this article, we propose a general method to address these limitations, grounding on a novel interpretation of SC as distributions over hard clusterings, which we call *distributional measures*. We provide an in-depth study of complexity- and metric-theoretic properties of the proposed approach, and we describe approximation techniques that can make the calculations tractable. Finally, we illustrate our approach through a simple but illustrative experiment.

Keywords: Soft clustering · Evidential clustering · External validation

1 Introduction

External clustering evaluation, defined as the act of objectively assessing the quality of a clustering result by means of a comparison between two or more clusterings (one of which is usually assumed to be the *correct one*), is one of the most relevant steps in clustering analysis [19]. In the case of hard clustering (HC), where each object is *unambiguously* assigned to a single cluster, several criteria have been considered in the literature [5,17,18].

By contrast, how to properly evaluate the results of a clustering analysis is much less clear in the case of soft clustering (SC) methods. Several SC methods have been proposed, including rough clustering (RC) [14], fuzzy clustering (FC) [2], possibilistic clustering (PC) [13] and evidential clustering (EC) [7,9]. The development of evaluation measures for SC has largely focused on the extension of common measures [1], notably the Rand index, to the setting of FC [4,11,12], while only recently a formulation of this approach has been introduced for the more general case of EC [8]. Nonetheless, a general approach to extend other comparison measures to SC is still lacking. Furthermore, existing measures fail to properly distinguish and quantify different types of uncertainty that can arise

S. Le Hégarat-Mascle et al. (Eds.): BELIEF 2022, LNAI 13506, pp. 3–12, 2022.
https://doi.org/10.1007/978-3-031-17801-6_1

in SC [8], namely: *ambiguity*, i.e., the inability to uniquely assign an object to a single clustering (typical of RC); and *partial assignment*, i.e., the assignment of objects to multiple clusters (typical of FC and PC).

In this article, we propose a general method to address these limitations, which makes it possible to extend any clustering evaluation to the case of SC. This approach allows us providing a full account of the uncertainty in the two clusterings to be compared. It relies on a novel interpretation of SC as representing distributions over HCs, referred to as *distributional measures*. We provide an in-depth study of the proposed approach, with respect to both computational complexity and metric properties[1] Furthermore, we describe approximation techniques that can make the approach tractable. Finally, we illustrate the application of the proposed method through a simple but illustrative example involving commonly adopted SC algorithms.

2   Background and Related Work

In the following section we provide basic background on clustering (Sect. 2.1) and evaluation measures for SC (Sect. 2.2).

2.1   Background on Clustering

Let $X = \{x_1, ..., x_n\}$ be a set of objects. A *HC* is an unique assignment of objects in X to *clusters*. Formally, a HC can be represented as a mapping $C : X \mapsto \Omega$, where $\Omega = \{\omega_1, ..., \omega_k\}$ is a set of clusters. This representation is called *object-based*. An equivalent representation, called *relational representation*, can be obtained by defining a clustering as an equivalence relation $[C] \subseteq X \times X$. Clearly, for the case of HC the two representations are equivalent. Given a clustering C, $[C]$ denotes its relational representation. Given two clusterings C_1, C_2 we say that they are equivalent iff $[C_1] = [C_2]$; we then write $C_1 \sim C_2$.

As mentioned in the introduction, in *SC* the unique assignment assumption is relaxed: the intuition is that we allow uncertainty in the cluster assignments. In the most general framework of EC, the uncertainty about cluster assignment is represented as a Dempster-Shafer mass function. Formally, using the object-based representation, an EC is a set $M = \{m_x\}_{x \in X}$, where each m_x is a *mass function*: i.e., a function $m_x : 2^\Omega \mapsto [0,1]$ such that $\sum_{A \subseteq \Omega} m_x(A) = 1$. If the mass functions m_x are *logical*, then the collection $R = \{m_x\}_{x \in X}$ is said to be a *RC*. A RC can be seen equivalently as a set of HCs [3]. Namely, a HC C is *compatible* with R if $\forall x \in X$, $C(x) \in R(x)$. Then, R can be represented by $C(R) = \{C : C$ is compatible with $R\}$. If all m_x are Bayesian, then the collection $F = \{m_x\}_{x \in X}$ is a FC. Finally, if all m_x are consonant, then the collection $P = \{m_x\}_{x \in X}$ is a PC. Both FC and PC can be alternatively represented as a collection of cluster membership vectors $F = \{\mu_x\}_{x \in X}$. In PC it is assumed that $\forall x \in X$, $\max_{\omega \in \Omega} \mu_x(\omega) \leq 1$, while in FC that $\forall x \in X$, $\sum_{\omega \in \Omega} \mu_x(\omega) = 1$.

---

[1] Due to space constraints, the complete version of all proofs appears online at https://arxiv.org/abs/2206.09827.

A relational representation can also be defined for the case of EC. Let $\Theta = \{s, \neg s\}$ be the frame where s denotes that two objects are in the same cluster, and $\neg s$ the opposite event. Given an EC M, the corresponding relational representation can be obtained, for any two objects $x, y \in X$, by combining m_x, m_y using Dempster's rule of combination and then computing the restriction m^{xy} of the resulting mass function to Θ [8].

Finally, we note that, if we interpret a SC as describing the uncertainty in regard to an underlying (unknown) HC, then two types of uncertainty can be distinguished. First, *partial assignment*, i.e., the fact that for $\omega_1, \omega_2 \in \Omega$ it may happen that $m_x(\omega_1), m_x(\omega_2) > 0$. Second, *ambiguity*, i.e., the assignment of some mass to non-singleton events, describing our inability to exactly determine to which cluster an object belongs. It is easy to observe that in a FC only partial assignment is relevant, since all the mass is assigned to the singletons, while in the case of RC, only ambiguity is present. By contrast, the EC formalism is flexible enough to represent both types of uncertainty.

2.2 Clustering Comparison Measures

Several measures have been defined to compare clusterings. Given two HCs C_1, C_2, a commonly adopted approach to compare them is to evaluate the number of object pairs $x, y \in X$ on which they agree. Formally, the Rand index can be defined as:

$$\mathsf{Rand}(C_1, C_2) = \frac{|\{(x, y) \in X^2 : (x, y) \in ([C_1] \cap [C_2]) \cup ([C_1]^c \cap [C_2]^c)\}|}{|X|^2}. \quad (1)$$

It is easy to show that the Rand index is a similarity on HCs.

Several extensions of the Rand index to the setting of SC have been considered. In the case of FC and PC, Campello et al. [4] and Frigui et al. [11] proposed to use a t-norm $\wedge$ and a t-conorm $\vee$ in place of classical set operators. This setting was later generalized to the setting of RC [10], by means of a transformation from RC into FC. A further generalization of the Rand index was proposed by Hüllermeier et al. [12], in the setting of FC, and Denœux et al. [8], in the more general setting of EC. This approach is based on the use of a normalized metric d_M on mass functions, that is then extended to compare ECs as

$$\mathsf{Rand}_E(M_1, M_2) = \frac{2}{n(n-1)} \sum_{x \neq y \in X} 1 - d_M(m_1^{xy}, m_2^{xy}). \quad (2)$$

The authors of [8,12] note that, while the measure they propose is a pseudo-similarity, it is not completely satisfactory for the comparison of ECs as it does not distinguish between *partial assignment* and *ambiguity*.

Another commonly adopted approach for the definition of clustering evaluation measure grounds on information theory. Let C_1, C_2 two HCs. The mutual information is defined as

$$MI(C_1, C_2) = \sum_{\omega_i^1 \in \Omega_1} \sum_{\omega_j^2 \in \Omega_2} p_{ij} \log \frac{p_{ij}}{p_i^1 \cdot p_j^2}, \quad (3)$$

where $p_i^1 = |\{x \in X : C_1(x) = \omega_i^1\}|/|X|$, and similarly for each $\omega_i^2 \in \Omega_2$, while $p_{ij} = |\{x \in X : C_1(x) = \omega_i^1 \text{ and } C_2(x) = \omega_j^2\}|/|X|$. The mutual information is a similarity over HCs. This measure has been extended to the case of RC in [3], by representing RCs as collections of compatible HCs. To our knowledge, no proposal to extend these metrics to the more general case of EC can be found in the literature.

Finally, another comparison approach grounds on the notion of *edit distance* between partitions. The *partition distance* [5] which, for two HCs C_1, C_2, is defined as the minimum number of objects to be moved to transform C_1 into C_2 (or, equivalently, C_2 into C_1), can be computed as

$$d_\pi(C_1, C_2) = \frac{1}{|X| - 1} \min_w \frac{1}{2} \sum_i |\omega_i^1 \Delta \omega_{w(i)}^2|, \tag{4}$$

where w is a permutation function, and Δ is the symmetric difference operator. It is easy to note that d_π is a normalized metric. An extension of the partition distance to the case of FC was proposed by Zhou [20]. The obtained measure is a proper generalization of the partition distance and is a metric. As for the mutual information, to our knowledge, the extension of the partition distance to rough and EC has not been considered in the literature.

3   A General Framework for Soft Clustering Evaluation Measures

As shown in the previous section, most of the research on clustering comparison measures for SC has focused on the analysis of some specific indices, while a general methodology for obtaining such measures is still missing. Furthermore, as noted in [8,12], most of the existing methods fail to satisfy reasonable metric properties and can thus hardly be used for the objective comparison of SCs. Notably, also the more principled approaches [8] can have some drawbacks, such as the inability to properly distinguish between different types of uncertainty arising in SC. In this section, we propose an approach that attempts to address these limitations, based on the representation of a SC as a distribution over HCs.

3.1   Distribution-Based Representation of Soft Clustering

As shown in Sect. 2.1, a RC R can be represented as a set $C(R)$ of HCs C. Based on this observation, we extend this representation to general SCs. Formally, given an EC M, we consider the following probability distribution over RCs:

$$m_M(R) = \prod_{x \in X} m_x(R(x)), \tag{5}$$

which can also be seen as a Dempster-Shafer mass function over HCs. Given an EC M and its distribution-based representation m_M, we denote with $\mathcal{F}(M)$ the collection of focal RCs of m_M, that is $\mathcal{F}(M) = \{R : m_M(R) > 0\}$.

The distribution-based representation for RC, FC and PC can then be obtained as a special case of Eq (5). Indeed, in the case of RC, m_M is logical (i.e. $|\mathcal{F}(m_M)| = 1$). In the case of a FC $F = \{\mu_x\}_x$, where $\mu_x : \mathcal{C} \mapsto [0, 1]$ is a probability distribution, the focal RCs are all singletons (i.e., HCs), thus we can define $Pr_F(C) = \prod_{x \in X} \mu_x(C(x))$ Finally, given a PC P and a t-norm $\wedge$, we can view P as a possibility distribution over HCs $\mathsf{Poss}_P(C) = \bigwedge_{x \in X} \mu_x(C(x))$. The possibility distribution Poss_P can equivalently be represented as a consonant mass function, i.e., a mass function for which the focal RCs are nested. Note that, when $\wedge$ is the product t-norm, we recover the case of FC.

3.2 Distributional Measures

Let d be a normalized metric on HCs. Since, as shown in the previous section, any SC can be seen as a distribution over HCs, an intuitive approach would be to extend d to a distribution-valued function, providing a quantification of the belief about the real value of the evaluation measure. The intuition behind this idea is based on the definition of SC as representing a clustering with some uncertainty affecting our knowledge with respect to the assignment of objects to clusters. Thus, it is natural to require that an evaluation measure for SC should transfer this uncertainty to the possible outcomes of the evaluation.

Therefore, intuitively, a measure over RCs would provide, given two RCs R_1, R_2, a set of values, representing all possible distances between HCs compatible with R_1, R_2. More generally, a measure over ECs would provide a mass function over possible values of d. Formally, we define the *distributional measure*, based on d, between two RCs (resp., PCs, ECs) as, respectively:

$$d_R(R_1, R_2) = \{d(C_1, C_2) : C_1 \in C(R_1) \text{ and } C_2 \in C(R_2)\} \quad (6)$$

$$\forall v \in \mathbb{R}, \ d_P(P_1, P_2)(v) = \bigvee_{C_1, C_2 : d(C_1, C_2) = v} \mathsf{Poss}_{P_1}(C_1) \wedge \mathsf{Poss}_{P_2}(C_2) \quad (7)$$

$$\forall V \in 2^{\mathbb{R}}, \ d_E(M_1, M_2)(V) = \sum_{R_1, R_2 : d_R(R_1, R_2) = V} m_{M_1}(R_1) \cdot m_{M_2}(R_2) \quad (8)$$

where $\wedge, \vee$ are a t-norm and the corresponding dual s-conorm. It is easy to observe that d_E is a generalization of d_R and d_P. For ease of notation, we distinguish the case of d_P where $\wedge = \otimes_P, \vee = \oplus_P$ (resp., the product t-norm and the corresponding t-conorm) and we denote it as d_F, since it can be applied directly to the case of FC. Intuitively, $d_E(M_1, M_2) = V$ can be interpreted as the evidence supporting the statement "The distance between the two real HCs underlying M_1, M_2 is within V". Therefore, d_E provides a complete representation of the possible distance values that arise when comparing HCs compatible with M_1, M_2. Though, clearly, d_E is not a metric, we can see that it satisfies the properties stated in the following theorem.

Theorem 1. *Function d_E satisfies (M3) (see Appendix A[2]). $d_E(M_1, M_2)(0) = 1$ iff there exists a HC C s.t. for $i = 1, 2$, $R \in \mathcal{F}(M_i) \implies \forall C' \in C(R), C' \sim C$.*

---

[2] https://arxiv.org/abs/2206.09827.

As a consequence of the previous result, the value $d_E(M_1, M_2)(0)$ can be interpreted as the evidence that the unknown HCs corresponding to M_1 and M_2 are the same (have a distance equal to 0). Indeed, d_E assigns full evidence to value 0, if and only if M_1, M_2 are totally compatible. In particular, simple equality between M_1, M_2 does not suffice to obtain $d_E(M_1, M_2)(0) = 1$, unless M_1, M_2 are HCs.

In regard to computational complexity, it is easy to show that computing d_R (resp., d_F, d_P, d_E) is computationally easy w.r.t. the size of the distribution-based representation introduced in the previous section, while it is intractable w.r.t. the size of the object-based representation:

Theorem 2. *The problem of computing d_R (resp., d_F, d_P, d_E) has complexity $O(k^m)$, where $m = |\{x \in X : |R(x)| \neq 1\}|$ and $k = |\Omega|$ is the number of clusters. More precisely, d_R can be computed in constant amortized time, while d_F, d_P, d_E can be computed in at most linear amortized time.*

Interval Representation. A possible solution to the intractability of computing the distributional measures would be to consider a compact representation of the latter. For the case of RC, d_R could be summarized as the interval:

$$\langle d_R^l, d_R^u \rangle (R_1, R_2) = \langle \min\{v \in \mathbb{R} : v \in d_R(R_1, R_2)\}, \max\{v \in \mathbb{R} : v \in d_R(R_1, R_2)\}\rangle.$$

We note that this definition satisfies the following properties:

Proposition 1. *Let R_1, R_2 be two RCs. Then, $1 - d_R^l$ is a consistency: in particular $d_R^l = 0$ iff $C(R_1) \cap C(R_2) \neq \emptyset$. By contrast, d_R^u is a meta-metric: in particular, $d_R^u = 0$ iff $R_1 = R_2$ and $|C(R_1)| = 1$, that is iff $R_1 = R_2$ is a HC.*

Corollary 1. *d_R^u is a metric iff either R_1, R_2 is a HC.*

As a result of the previous corollary, in the special case where the aim is to evaluate a RC R with respect to a HC C representing the ground truth, then d_R^u is guaranteed to be a metric. Nonetheless, it is easy to observe that computing $\langle d_R^l, d_R^u \rangle$ is still computationally hard:

Theorem 3. *Let R_1, R_2 be two RCs, represented through the object-based representation. Then, the problem of computing $\langle d_R^l, d_R^u \rangle$ is NP-HARD[3].*

For the case of FC, PC and, more generally, EC, a possible approach to obtain a similar summarization would be to apply a decision rule to transform the distribution-valued d_F, d_P, d_E into simpler indices [6]. An example of this approach would be to compute the following lower and upper expectations:

$$\underline{E}(d_E)(M_1, M_2) = \sum_{V \subseteq 2^{\mathbb{R}}} d_E(M_1, M_2)(V) \min_{d \in V} d = E(d_R^l), \tag{9}$$

$$\overline{E}(d_E)(M_1, M_2) = \sum_{V \subseteq 2^{\mathbb{R}}} d_E(M_1, M_2)(V) \max_{d \in V} d = E(d_R^u). \tag{10}$$

---

[3] The problem is trivially in P w.r.t. the distribution-based representation of R_1, R_2.

If M_1, M_2 are two FCs we obtain that $\underline{E}(d_E) = \overline{E}(d_E) = E(d_F)$. Similarly to the case of RC, it is easy to show that the following properties hold:

Theorem 4. *Let M_1, M_2 be two ECs. Then $\overline{E}(d_E)$ is a meta-metric. Furthermore, $\underline{E}(d_E)$ satisfies only (M1b) and (M2) (see Appendix A^4). In particular:*

- *If M_1, M_2 is a HC, then $1 - \underline{E}(d_E)$ is a consistency and $\overline{E}(d_E)$ is a metric;*
- *If F_1, F_2 are two FCs, then $E(d_F)$ is a meta-metric.*

From the computational complexity point of view, computing $\underline{E}(d_E), \overline{E}(d_E)$ is at least as hard as computing $\langle d_R^l, d_R^u \rangle$. However, for the case of FC, it is easy to show that for certain base distances d, $E(d_F)$ can be computed efficiently:

Proposition 2. *Let $d = 1 - Rand$. Then, $E(d_F)$ can be computed in time $O(n^2)$.*

We leave the problem of characterizing the general complexity of computing $\underline{E}(d_E), \overline{E}(d_E)$ as open problem.

3.3   Approximation Methods

In the previous section we proposed distributional measures as a general approach to extend any HC comparison measure to a SC comparison measure. Nonetheless, the computation of these distributional measures is, in general, intractable. For this reason, in this section, we introduce some approximation methods and algorithms, based on a sampling approach, which can be applied to any base distance between HCs.

We start with the case of the summarized representation of d_R, that is with d_R^l, d_R^u. Given two RCs R_1, R_2, we draw s samples $(C_1^1, C_2^1), \ldots, (C_1^s, C_2^s)$ uniformly from $C(R_1), C(R_2)$. Then, we can approximate d_R^l and d_R^u as, respectively, $\hat{d}_R^l = \min_{i \in \{1,\ldots,s\}} d(C_1^i, C_2^i)$ and $\hat{d}_R^u = \max_{i \in \{1,\ldots,s\}} d(C_1^i, C_2^i)$. Clearly, the following result holds:

Proposition 3. *The following bounds hold for any $\epsilon > 0$:*

$$Pr(d_R^u - \hat{d}_R^u > \epsilon) \leq F(d_R^u - \epsilon)^s, \quad Pr(\hat{d}_R^l - d_R^l > \epsilon) \leq 1 - \left(1 - F(\epsilon - d_R^l)\right)^s \tag{11}$$

where F is the cumulative distribution function (CDF) of the probability distribution p_R defined as $p_R(t) = \frac{|\{C_1 \in C(R_1), C_2 \in C(R_2) : d(C_1, C_2) = t\}|}{|d_R(R_1, R_2)|}$.

Since for each ϵ, the quantity $F(d_R^u - \epsilon)$ (resp., $F(\epsilon - d_R^u)$) is strictly less than 1, it holds that $Pr(d_R^u - \hat{d}_R^u > \epsilon)$ (resp., $Pr(\hat{d}_R^l - d_R^l > \epsilon)$) decays exponentially w.r.t. the size of the sample s. However, we note that the quality of the previously described approximation method largely depends on d_R. In particular, the convergence in Eq (11) is influenced by the *tailedness* of p_R: the heavier the tails of p_R, the lower the approximation error.

For the case of FC, if we use the expected value $E(d_F)$ to summarize d_F and we use a sampling procedure to estimate $E(d_F)$ as $\hat{d}_F$ then we can obtain a tail bound by applying Hoeffding's inequality:

---

[4] https://arxiv.org/abs/2206.09827.

Proposition 4. *Assume that d is a normalized metric on HCs. Then:*

$$Pr(|\hat{d}_F - E(d_F)| \geq \epsilon) \leq 2e^{-2s\epsilon^2} \tag{12}$$

Hence, the deviation between the empirical mean $\hat{d}_F$ and $E(d_F)$ has exponential decay in the size of the sample s.

Combining Eqs (11) and (12), a similar result holds also for d_E:

Proposition 5. *Assume that d is a normalized metric on HCs. Let $\hat{d}_E^l, \hat{d}_E^u$ be the sample estimates of $\underline{E}(d_E), \overline{E}(d_E)$. Then:*

$$Pr(|\hat{d}_E^l - \underline{E}(d_E)| \geq \epsilon) \leq 2e^{-2s\epsilon^2}, \quad Pr(|\hat{d}_E^u - \overline{E}(d_E)| \geq \epsilon) \quad \leq 2e^{-2s\epsilon^2} \tag{13}$$

Given two ECs M_1, M_2, the previous estimate requires that $\hat{d}_E^l, \hat{d}_E^u$ are computed by sampling pairs R_1, R_2 of RCs from the distributions m_{M_1}, m_{M_2} and then computing the exact values of $d_R^l(R_1, R_2), d_R^u(R_1, R_2)$. As a consequence of Proposition 3, this may not be feasible when $|X|$ is large. In such cases, a possible solution would be to compute $\hat{d}_E^l, \hat{d}_E^u$ by means of nested sampling (i.e., first we sample a RC R from m_M, then we sample a HC C from $C(R)$). In this case, however, one should expect a larger approximation error. Finally, we note that all the above mentioned sampling-based approximation methods can easily be implemented in time complexity $O(n^2 s)$.

4   Illustrative Experiment

In this section, we illustrate the use of the proposed metrics using the Anderson's Iris dataset. This latter is a small-scale benchmark problem comprising 150 objects, four numerical features and three perfectly balanced classes. We selected this dataset as, the three classes being approximately linearly separable, it can be expected that any SC algorithm would give as output a clustering close to being an HC. As a consequence of Theorem 3, this is a necessary condition for efficient computation of the exact versions of the distributional measures.

We considered five different clustering algorithms, namely: k-means (KM), rough k-means (RKM) [16], fuzzy c-means (FCM) [2], possibilistic c-means (PCM) [13] and evidential c-means (ECM) [15]. In order to reduce the complexity of computing the distributional measures, we set the algorithm hyper-parameters to obtain clusterings close to being hard. In particular, for RKM we set $\epsilon = 1.1$, for FCM and PCM we set $m = 5$, and for ECM we set $\delta = 10, \beta = 5, \alpha = 5$. The output of each algorithm was compared with the ground truth labeling of the iris dataset. We considered, in particular, the distributional generalizations of the Rand index (D-RI) and the partition distance (D-PD), as well as their sampling-based approximations (S-RI, S-PD). Code was implemented in Python (v. 3.8.8), using scikit-learn (v. 0.24.1), numpy (v. 1.20.1) and scipy (v. 1.6.2).

The results of the experiment are reported in Table 1, in terms of the metrics values as well as running time (in seconds). As far as running time is concerned, we can observe that the cost of computing the exact versions of the proposed measures sharply increases when considering more general SC algorithms.

Table 1. Results of the experiment. For the Rand index higher is better, while for the partition distance lower is better.

Metric	KM	RKM	FCM	PCM	ECM
D-RI	0.877 (0.034 s)	(0.874, 0.886) (0.802 s)	0.876 (784.388 s)	(0.839, 0.941) (979.053 s)	(0.781, 0.944) (1394.69 s)
S-RI	-	(0.874, 0.886) (0.429 s)	0.876 (11.266 s)	(0.860, 0.927) (19.848 s)	(0.681, 0.819) (19.845 s)
D-PD	0.111 (0.031 s)	(0.099, 0.113) (0.803 s)	0.112 (184.707 s)	(0.033, 0.122) (224.31 s)	(0.041, 0.229) (431.57 s)
S-PD	-	(0.100, 0.113) (0.202 s)	0.112 (11.391 s)	(0.072, 0.103) (13.739)	(0.154, 0.209) (16.424 s)

Indeed, the running times of D-RI and D-PD for ECM were approximately twice those of FCM and PCM. On the other hand, the differences in running time for the approximation algorithms were much smaller, and indeed the running times for FCM, PCM and ECM were similar.

In terms of approximation quality, even though for RKM and FCM there were no differences between the approximated and exact results, this was not the case for PCM and ECM. In particular, we note that the sampling-based approximation algorithm systematically underestimated the uncertainty in clustering comparison results, by producing intervals that were narrower than the exact ones. Nonetheless, we note that the approximation methods provided results that were aligned with the exact ones, with smaller values according to one method associated to smaller values according to the other one.

5   Conclusion

In this article we proposed a general framework for extending clustering comparison measures from HC to EC (hence, as special cases, also to RC, FC and PC), that we called *distributional measures*. We studied the metric- and complexity-theoretic properties of the proposed measures and, since a major limitation of the proposed approaches lies in their high computational complexity, we also proposed some strategies for approximation based on sampling . Finally, we illustrated the application of the proposed methods through a simple experiment.

We believe that this article could provide a first step toward the development of general and principled approaches for the comparison of SC algorithms. For this reason, we deem the following problems to be worthy of further investigation: 1) Generalizing Proposition 2 to other base distance measures, and determining whether this result can be extended to EC; 2) Designing more refined sampling approaches that can be used to correct the uncertainty underestimation that we observed in the experiments.

References

1. Anderson, D.T., Bezdek, J.C., Popescu, M., et al.: Comparing fuzzy, probabilistic, and possibilistic partitions. IEEE Trans. Fuzzy Syst. **18**(5), 906–918 (2010)
2. Pattern Recognition with Fuzzy Objective Function Algorithms. AAPR, Springer, Boston (1981). https://doi.org/10.1007/978-1-4757-0450-1
3. Campagner, A., Ciucci, D.: Orthopartitions and soft clustering: soft mutual information measures for clustering validation. Knowl. Based Syst. **180**, 51–61 (2019)
4. Campello, R.J.: A fuzzy extension of the rand index and other related indexes for clustering and classification assessment. Pattern Recognit. Lett. **28**(7), 833–841 (2007)
5. Day, W.H.: The complexity of computing metric distances between partitions. Math. Soc. Sci. **1**(3), 269–287 (1981)
6. Denoeux, T.: Decision-making with belief functions: a review. Int. J. Approx. Reason. **109**, 87–110 (2019)
7. Huynh, V.-N., Inuiguchi, M., Le, B., Le, B.N., Denoeux, T. (eds.): IUKM 2016. LNCS (LNAI), vol. 9978. Springer, Cham (2016). https://doi.org/10.1007/978-3-319-49046-5
8. Denœux, T., Li, S., Sriboonchitta, S.: Evaluating and comparing soft partitions: an approach based on Dempster-Shafer theory. IEEE Trans. Fuzzy Syst. **26**(3), 1231–1244 (2017)
9. Denœux, T., Masson, M.H.: EVCLUS: evidential clustering of proximity data. IEEE Trans. Syst. Man Cybern. B Cybern. 34(1), 95–109 (2004)
10. Depaolini, M.R., Ciucci, D., Calegari, S., Dominoni, M.: External indices for rough clustering. In: Nguyen, H.S., Ha, Q.-T., Li, T., Przybyła-Kasperek, M. (eds.) IJCRS 2018. LNCS (LNAI), vol. 11103, pp. 378–391. Springer, Cham (2018). https://doi.org/10.1007/978-3-319-99368-3_29
11. Frigui, H., Hwang, C., Rhee, F.C.H.: Clustering and aggregation of relational data with applications to image database categorization. Pattern Recognit. **40**(11), 3053–3068 (2007)
12. Hüllermeier, E., Rifqi, M., Henzgen, S., et al.: Comparing fuzzy partitions: a generalization of the rand index and related measures. IEEE Trans. Fuzzy Syst. **20**(3), 546–556 (2011)
13. Krishnapuram, R., Keller, J.M.: A possibilistic approach to clustering. IEEE Trans. Fuzzy Syst. **1**(2), 98–110 (1993)
14. Lingras, P., West, C.: Interval set clustering of web users with rough k-means. J Intell. Inform. Syst. **23**(1), 5–16 (2004)
15. Masson, M.H., Denoeux, T.: ECM: an evidential version of the fuzzy c-means algorithm. Pattern Recognit. **41**(4), 1384–1397 (2008)
16. Peters, G.: Rough clustering utilizing the principle of indifference. Inf. Sci. **277**, 358–374 (2014)
17. Rand, W.M.: Objective criteria for the evaluation of clustering methods. J. Am. Stat. Assoc. **66**(336), 846–850 (1971)
18. Vinh, N.X., Epps, J., Bailey, J.: Information theoretic measures for clusterings comparison: variants, properties, normalization and correction for chance. J. Mach. Learn. Res. **11**, 2837–2854 (2010)
19. Xiong, H., Li, Z.: Clustering validation measures. In: Data Clustering, pp. 571–606. Chapman and Hall/CRC (2018)
20. Zhou, D., Li, J., Zha, H.: A new Mallows distance based metric for comparing clusterings. In: Proceeding of ICML 2005, pp. 1028–1035 (2005)

Causal Transfer Evidential Clustering

Kuang Zhou$^{(\boxtimes)}$ and Ming Jiang

School of Mathematics and Statistics, Northwestern Polytechnical University,
Xi'an, Shaanxi 710072, People's Republic of China
`kzhoumath@nwpu.edu.cn`, `xming@mail.nwpu.edu.cn`

Abstract. Classical prototype-based clustering algorithms usually cannot achieve satisfactory results when the data is insufficient. Transfer learning can be adopted to address this problem. For instance, in the recently proposed transfer clustering methods Transfer Evidential C-Means (TECM), the prototypes of data in the source domain are transferred to the target domain to help improve the clustering performance. However, in TECM the prototypes are calculated based on all the features of samples in the clusters in source data sets. Due to distribution shift in two domains, sometimes the prototypes obtained from all the features of samples in the source may not be a good representation for clusters in the target domain. In this paper, we propose an approach for solving this problem by exploiting causal inference, and introduce a new prototype-based causal transfer evidential clustering algorithm. The experimental results demonstrate the effectiveness of the proposed clustering approach.

Keywords: Causality · Transfer clustering · Distribution shift · Prototype-based clustering

1 Introduction

Transfer learning can utilize data or knowledge in the source domain to improve the performance of learning tasks on the target domain with only a small amount of data or even without labeled samples [6]. It relaxes the restriction of traditional machine learning methods that training and test data should follow the same probability distribution, and only requires a certain similarity between two domains. Currently, transfer learning has been successfully applied to many machine learning tasks.

Clustering is often used as a data analysis technique for discovering interesting patterns in the data sets. Sometimes, the number of samples in the clustering task is too small to construct a good cluster method. If we can get some possible help from some related domain, the clustering performance may be improved. In this sense, transfer learning can be a good try.

Evidential clustering allows ambiguity, uncertainty or doubt in the assignment of objects to clusters by using mass functions to describe the membership of objects [2]. Due to the advantages of effectively expressing uncertain cluster

S. Le Hégarat-Mascle et al. (Eds.): BELIEF 2022, LNAI 13506, pp. 13–22, 2022.
https://doi.org/10.1007/978-3-031-17801-6_2

structure in the data sets, it has been widely applied in many fields such as community detection [16], medical image segmentation [9] and so on. We have proposed an evidential transfer clustering method named TECM [15], where prototypes of clusters in the source domain are transferred to the target domain as complementary knowledge to help improve the clustering performance. However, as there is inevitable distribution shift in two domains, the prototypes calculated by all the features of the source data may not well represent the clusters in the target. This problem is related to one of the key issues in transfer learning, that is to say, what to transfer?

Feature selection, as one of the key problems in machine learning, can remove irrelevant and redundant features from the dataset [5]. Classical feature selection methods are commonly based on the correlation between features and the class attributes. But some authors claim that correlation is not robust for predictive models and propose the concept of causal features [12]. It can contribute to build interpretable and robust models by capturing the causal relationship between variables [11].

Here we will use an illustrative example to demonstrate the necessity of causal feature in the transfer clustering task. Figures 1 (a) and (b) are from the dataset in [4]. As we can see, both the source and target domains consist of pictures of cats and dogs. The task is to separate the images of cats from those of dogs in the target with the help of images in the source. In the source domain, 80% of the cats are in the grass while the remaining 20% are in the water. For the dogs, 80% are in the water, and the rest are in the grass. In the target domain, however, the opposite setup is used. In the situation, a big distribution shift occurs between the source and target.

a. Source domain   b. Target domain

Fig. 1. An example where causal feature selection is required for transfer clustering.

Admittedly, many features related to the image labels (*i.e.*, cats or dogs) can be obtained from the source domain. Here we consider three features: the length of the ear (X_1), whether to have whiskers (X_2), and the body humidity (X_3). If the correlation-based approaches are used, all the three features should be taken into account due to their strong correlation with class attributes. However, if the causality-based approaches are adopted, the body humidity will no longer be considered as it does not have a robust relationship with the animal labels. It is easy to imagine that the causal features such as X_1 and X_2 which are stably related to the animal type can be good shared knowledge to be transferred across domains. These two features are domain invariant. On the contrary, X_3 is not a good alternative as it changes significantly with the environment (domain).

To address the limitations and enable effective feature transfer across different domains in the unsupervised clustering task, we will propose a prototype-based causal transfer evidential clustering method in this paper. The remainder of this paper is organized as follows. Some related knowledge is introduced in Sect. 2. The proposed causal transfer clustering algorithm is presented in detail in Sect. 3. Some experiments are conducted in Sect. 4. Conclusions are drawn in the final section.

2   Related Work

Some related knowledge with this paper such as transfer evidential clustering and causal feature selection methods will be introduced in this section.

2.1   Transfer Evidential Clustering

Denote the n data samples in the target domain by $X = \{x_1, x_2, ..., x_n\}$, and c denotes the number of the cluster. The discernment frame of classes is $\Omega = \{\omega_1, \omega_2, ..., \omega_c\}$. The available knowledge in the source domain is represented by prototypes $V^{(s)} = \{v_1{}^{(s)}, v_2{}^{(s)}, ..., v_c{}^{(s)}\}$. The superscript (s) indicates that the prototypes are from the source domain.

TECM aims to look for the optimal credal partition $M = (m_1, m_2, ..., m_n) \in \mathbb{R}^{n \times 2^c}$ and the cluster centers $V = (v_1, v_2, ..., v_c)$ in the target domain by minimizing following objective function:

$$J_{\text{TECM}}(M, V) = \sum_{i=1}^{n} \sum_{\substack{A_j \subseteq \Omega \\ A_j \neq \emptyset}} c_j^\alpha m_{ij}^\beta d_{ij}^2 + \sum_{i=1}^{n} \delta^2 m_{i\emptyset}^\beta$$

$$+ \beta_1 \left[\sum_{i=1}^{n} \sum_{\substack{A_j \subseteq \Omega \\ A_j \neq \emptyset}} c_j^\alpha m_{ij}^\beta d_{ij}^{(s)2} + \sum_{i=1}^{n} \delta^2 m_{i\emptyset}^\beta \right] + \beta_2 \sum_{k=1}^{c} ||v_k^{(s)} - v_k||^2, \qquad (1)$$

subject to:

$$\sum_{A_j \subseteq \Omega, A_j \neq \emptyset} m_{ij} + m_{i\emptyset} = 1, \qquad (2)$$

where m_{ij} denotes $m_i(A_j)$ and $m_{i\emptyset}$ denotes $m_i(\emptyset)$. $c_j = |A_j|$ denotes the cardinal of A_j. d_{ik} denotes x_i and the barycenter $\bar{v}_k$:

$$d_{ik}^2 = ||x_i - \bar{v}_k||^2. \qquad (3)$$

Notion $\boldsymbol{v}_h$ denotes the center of samples in cluster ω_h, and parameters α, β and δ control the degree of penalization for imprecise classes with high cardinality, the fuzziness of the partition, and the amount of outliers respectively.

The Lagrange multiplier method can be used to optimize the objective function. The iterative update rule for the membership and prototypes can be shown as follows. The equations for updating the credal membership can be derived as:

$$m_{ij} = \frac{\left(1/\left(c_j^\alpha\left(d_{ij}^2 + \beta_1 d_{ij}^{2(s)}\right)\right)\right)^{\frac{1}{\beta-1}}}{\sum\limits_{\substack{A_k \subseteq \Omega \\ A_k \neq \emptyset}} \left(1/\left(c_k^\alpha\left(d_{ik}^2 + \beta_1 d_{ik}^{2(s)}\right)\right)\right)^{\frac{1}{\beta-1}} + \left(\frac{1}{\delta^2 + \beta_1 \delta^2}\right)^{\frac{1}{\beta-1}}}, \tag{4}$$

$$m_{i\emptyset} = \frac{\left(\frac{1}{\delta^2 + \beta_1 \delta^2}\right)^{\frac{1}{\beta-1}}}{\sum\limits_{\substack{A_k \subseteq \Omega \\ A_k \neq \emptyset}} \left(1/\left(c_k^\alpha\left(d_{ik}^2 + \beta_1 d_{ik}^{2(s)}\right)\right)\right)^{\frac{1}{\beta-1}} + \left(\frac{1}{\delta^2 + \beta_1 \delta^2}\right)^{\frac{1}{\beta-1}}}, \tag{5}$$

The rules for updating the prototypes in the target domain are:

$$\left(\boldsymbol{H} + \beta_2\boldsymbol{I}\right)\boldsymbol{v} = \boldsymbol{B} + \beta_2\boldsymbol{v}^{(s)}, \tag{6}$$

where $\boldsymbol{I}$ is the $(c \times c)$ identity matrix, $\boldsymbol{B}$ is a matrix of size $(c \times p)$ and $\boldsymbol{H}$ is a matrix of size $(c \times c)$:

$$\boldsymbol{B}_{lq} = \sum_{i=1}^{n} x_{iq} \sum_{\substack{A_j \subseteq \Omega \\ A_j \neq \emptyset}} c_j^{\alpha-1} m_{ij}^\beta s_{lj} = \sum_{i=1}^{n} x_{iq} \sum_{A_j \ni \omega_l} c_j^{\alpha-1} m_{ij}^\beta, \tag{7}$$

$$\boldsymbol{H}_{lk} = \sum_{i=1}^{n} \sum_{\substack{A_j \subseteq \Omega \\ A_j \neq \emptyset}} c_j^{\alpha-2} m_{ij}^\beta s_{lj} s_{kj} = \sum_{i} \sum_{A_j \supseteq \{\omega_k, \omega_l\}} c_j^{\alpha-2} m_{ij}^\beta. \tag{8}$$

2.2   Causal Feature Selection

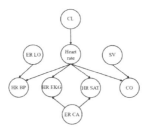

Fig. 2. The local structure of ALARM network

Over the years, many causal feature selection methods have been proposed [3,10,13]. Different from correlation-based feature selection methods designed to find strongly relevant features, causal feature selection approaches aim to learn the causality between variables. As causality is more stable in different environment, it can contribute to build more explainable and robust models [11]. Generally speaking, the goal of causal feature selection is to identify the Markov blanket (MB) of the class attributes or a subset of the MB which can be used as the optimal subset for feature selection. When a Bayesian network is used to represent the relationship between feature variables, the MB of a variable includes its parents (direct causes), children (direct effects), and spouses (the other parents of its children) under the faithfulness assumption [7]. Besides, it has been shown that the set of direct causes (parents) of the class attributes can be used as a stable or invariant feature subset when the data of the source domain and the target domain are from different distributions [13]. Figure 2 shows the MB of variable "Heart rate" which is from A Logical Alarm Reduction Mechanism (ALARM) network [1], including four children (HR BP, HR EKG, HR SAT and CO), one parent (CL) and three spouses (ER LO, ER CA and SV).

3   Causal Transfer Evidential Clustering

In order to show the proposed causal transfer evidential clustering method clearly, we continue with the illustrative example with the animal image data set. In causal inference, the Directed Acyclic Graph (DAG) is often used to represent causal relationship between variables. Generally speaking, a DAG consists of node set V and edge set E, where edge $V_i(\in V) \rightarrow V_j(\in V)$ denotes that V_i is the parent of V_j [11].

Recall that we have three variables to identify the animal labels in the images. Let X_4 denote the animal labels ($X_4 = 0$ for dogs, while $X_4 = 1$ for cats). X_1 and X_2 denote two physical features, i.e., "the length of the ear" and "whether to have whiskers" respectively. It is easy to know that they are the parents of X_4. We introduce a context variable C, representing the environment. When $C = 1$, we say it is the source domain with most of the cats are in the grass. On the contrary, when $C = 0$, we can say this is the case of target domain with most of the dogs are in the grass. Let X_3 denote the body humidity. We know the value of X_3 is up to C and X_4. Thus, we can get a DAG as shown in Fig. 3.

As there is strong correlation between the body humidity and the class attributes, it is clear that the correlation-based feature selection method will not remove the humidity from the feature subset. However, if we eventually select the feature subset which contains the humidity, the clustering performance in the target data will be greatly degraded. On the one hand, when clustering the samples with the humidity, the samples are largely grouped by the context (grass or water). On the other hand, the humidity leads to larger distribution shift between domains, i.e., inconsistent conditional distributions in the source and target domains as shown in Eq.(9), which further affects the performance of transfer clustering.

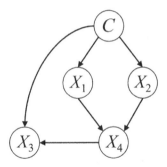

Fig. 3. A simplified directed acyclic graph with the animal image data

$$P_s\{X_4|X_3\} \neq P_t\{X_4|X_3\}. \tag{9}$$

where P_s and P_t represent the distribution of the source domain and the target domain, respectively.

Causality-based feature selection methods can avoid both of the above problems. According to the causal graph model shown in Fig. 3, causal feature selection will select $X1, X2$ as the feature subsets. On the one hand, X_1, X_2 represent the physical features which are completely correct for the partition of the animals. On the other hand, from the concept of d-separation, we know that the conditional distribution of the class attributes X_4 with respect to X_1 and X_2 is consistent no matter in the source or target domain as shown in Eq. (10)

$$P_s\{X_4|X_1, X_2\} = P_t\{X_4|X_1, X_2\}. \tag{10}$$

Following the discussion above, we propose the Causal Transfer Evidential Clustering (CTEC) approach, which can be divided into two steps:

(1) Identify causal features of class attributes based on the source domain data. Even though learning the DAG from data is an NP-hard problem because of the combinatorial acyclicity constraint [14], some causal structure learning and DAG learning methods have been proposed to select causal features [8, 14]. In this paper, we consider the adjacency matrix based numerical method to estimate the DAG and identify the causal invariant features [14]. The objective function of this causal feature selection algorithm can be defined as follows:

$$\min_{W \in R^{d \times d}} f(W) = \frac{1}{2n}||\boldsymbol{X} - \boldsymbol{X}W||_F^2 + \lambda||W||_1, \tag{11}$$

subject to:

$$h(W) = tr(e^{W \circ W}) - d = 0, \tag{12}$$

where $\boldsymbol{X} \in R^{n \times d}$ denotes the sample data, $W \in R^{d \times d}$ denotes the adjacency matrix of the DAG, $||\cdot||_F$ is Frobenius norm and $\circ$ is the Hadamard product. The acyclicity of DAG can be guaranteed by the constraint $h(W) = 0$.

The above optimization problem can be solved by the use of the augmented Lagrange multiplier method given the source domain data. After that we can obtain the adjacency matrix W of the corresponding DAG between the features and class attributes. Then the causal features can be selected based on the adjacency matrix from the source and target domain data respectively.

(2) Perform the prototype-based evidential transfer clustering TECM.

With the selected features in the last step, we can use the prototype-based transfer clustering method. It is remarked here that with the causal feature selection step, the transferred knowledge from the source domain is not based on the whole set of features. We just find some transferable knowledge from the perspective of causal learning.

4 Experiments

Some experiments are provided in this section to demonstrate the effectiveness of the proposed causal transfer clustering method. The Adjusted Rand Index (ARI) and Normalized Mutual Information (NMI) are used as the metrics to evaluate the performance of the clustering methods. Denote the number of data instances in the source and target domain by N_s and N_t respectively. The causal feature selection method in proposed CTEC method and correlation test method are applied respectively to obtain the transferred feature subset across domains based on the source domain data. Denote the obtained correlation-based and causality-based feature subsets by A and B respectively. In the following experiments, the number of features in source and target domains is the same.

4.1 Synthetic Datasets

We continue with the example with the DAG in Fig. 3. Based on this DAG, the simulated datasets are generated by the Structure Equation Model (SEM) defined in the following equation:

$$X_j = f_j(X_j(pa)) + \epsilon, \tag{13}$$

where $X_j(pa)$ denotes the parents of X_j and the noise $\epsilon \sim N(0, 1)$. Both the source and target domain have three feature variables (X_1, X_2, X_3) and one class variable (X_4).

In this experiment, let $N_s = 1000$, $N_t = 100$ and the number of the cluster be $c = 2$. The functions $f_j, j = 1, 2, 3, 4$ are defined as follows:

$$f_1(C) = C + p, f_2(C) = C + p,$$
$$f_4(X_1, X_2) = X_1 + X_2, f_3(C, X_4) = X_4 + 2 \cdot C.$$

In the above equations, C is the context variable describing the two domains and p denotes a binary vector used to divide the data into two clusters. Let $C \sim b(N_s, 0.8)$ in the source domain while $C \sim b(N_t, 0.2)$ in the target domain.

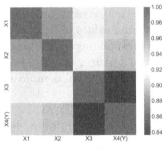 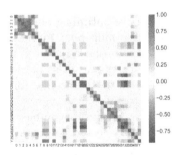

a. Synthetic datasets   b. ALARM network datasets

Fig. 4. Correlation matrices of the variables from the synthetic datasets snd ALARM network datasets

The correlation feature set A can be obtained based on the correlation coefficient between feature variables and the class variable as shown in Fig. 4(a), while the causality feature set B can be got by the first step of CTEC. The obtained feature sets are as follows:

$$A = \{X_1, X_2, X_3\}, B = \{X_1, X_2\}. \tag{14}$$

Then we examine the clustering results of the target domain with different feature subsets by TECM. The transfer evidential clustering approach with the two feature sets are denoted by TECM-A and TECM-B respectively.

Table 1. The clustering results on the synthetic dataset.

Method	TECM-A	TECM-B	Method	TECM-A	TECM-B
ARI	0.5433	0.9208	NMI	0.5373	0.8586

The ARI and NMI values of the clustering results provided by TECM with feature subset A or B are listed in Table. 1. The clustering results with the feature subset B are significantly better than those with the feature subset A.

4.2 ALARM Network Dataset

In the experiment, we use the datasets from the ALARM network which includes 37 variables and choose the "Heart rate" as the class variable which have three class labels. Set $N_s = 5000$, $N_t = 500$ and $c = 3$. Five features are selected to form the correlation feature set A by comparing the correlation coefficients between features and the class variable as shown in Fig. 4(b). Based on the DAG and MB of the class variable learned by the first step of CTEC, seven features are selected as the causality feature set B.

Table 2. The clustering results on ALARM network dataset.

Method	TECM-A	TECM-B	Method	TECM-A	TECM-B
ARI	0.5519	0.6080	NMI	0.4802	0.5321

The ARI and NMI values of the clustering results based on ALARM network datasets are listed in Table 2. The clustering results of TECM-B are also better than those of TECM-A.

From these two experiments we can see that the causal features are able to transfer stably in the cross-domain tasks, and consequently can improve the performance of transfer clustering methods.

5 Conclusion

In this study, the causal features have been used to develop causal prototype-based transfer evidential clustering methods named CTEC for the application of clustering task when the target data are uncertain or insufficient. The proposed CTEC algorithm can effectively learn from not only the data of the target domain but also from the knowledge of the source domain in the form of prototypes as well. Compared with the existing transfer clustering methods, with the help of an additional causal feature selection process, the transferred knowledge in the form of prototypes in CTEC is more effective, consequently the clustering performance has been greatly improved. The experiments on the synthetic datasets and ALARM network datasets have demonstrated the effectiveness of the proposed causal transfer evidential clustering method.

The success of the proposed causal prototype-based transfer clustering algorithm indicates that causal features are robust across different domains and that learning them is beneficial for transfer clustering task. However, this work is only a simple work to demonstrate the role of causal features in transfer clustering. More experiments will be conducted in the future. We hope that this work will inspire further research on transfer clustering from a causal perspective. Moreover, in this work we divide the causal feature learning and transfer clustering into two separate steps. In the future, we will study how to combine the two steps together to further improve the robustness of the transfer clustering task.

Acknowledgements. This work was supported by the National Natural Science Foundation of China (No. 61701409), the Aero Science Foundation of China (No. 20182053023), the Science Research Plan of China (Xi'an) Institute for Silk Road Research (2019ZD02).

References

1. Beinlich, I.A., Suermondt, H.J., Chavez, R.M., Cooper, G.F.: The ALARM monitoring system: A case study with two probabilistic inference techniques for belief networks. In: Hunter, J., Cookson, J., Wyatt, J. (eds.) AIME 89. Lecture Notes in Medical Informatics, vol. 38, pp. 247–256. Springer (1989). https://doi.org/10.1007/978-3-642-93437-7_28

2. Denœux, T., Kanjanatarakul, O.: Evidential clustering: a review. In: Huynh, V.-N., Inuiguchi, M., Le, B., Le, B.N., Denoeux, T. (eds.) IUKM 2016. LNCS (LNAI), vol. 9978, pp. 24–35. Springer, Cham (2016). https://doi.org/10.1007/978-3-319-49046-5_3

3. Gao, T., Ji, Q.: Efficient Markov blanket discovery and its application. IEEE Trans. Cybern. **47**(5), 1169–1179 (2016)

4. He, Y., Shen, Z., Cui, P.: Towards non-iid image classification: a dataset and baselines. Pattern Recogn. **110**, 107383 (2021)

5. Li, Y., Li, T., Liu, H.: Recent advances in feature selection and its applications. Knowl. Inf. Syst. **53**(3), 551–577 (2017)

6. Pan, S.J., Yang, Q.: A survey on transfer learning. IEEE Trans. Knowl. Data Eng. **22**(10), 1345–1359 (2009)

7. Pearl, J.: Probabilistic Reasoning in Intelligent Systems: Networks of Plausible Inference. Elsevier, Amsterdam (2014)

8. Ramsey, J., Glymour, M., Sanchez-Romero, R., Glymour, C.: A million variables and more: the fast greedy equivalence search algorithm for learning high-dimensional graphical causal models, with an application to functional magnetic resonance images. Int. J. Data Sci. Anal. **3**(2), 121–129 (2017)

9. Wang, F., Lian, C., Vera, P., Ruan, S.: Adaptive kernelized evidential clustering for automatic 3d tumor segmentation in FDG-PET images. Multimedia Syst. **25**(2), 127–133 (2019). https://doi.org/10.1007/s00530-017-0579-0

10. Wu, X., Jiang, B., Yu, K., Miao, C., Chen, H.: Accurate Markov boundary discovery for causal feature selection. IEEE Trans. Cybern. **50**(12), 4983–4996 (2019)

11. Yu, K., Guo, X., Liu, L., Li, J., Wang, H., Ling, Z., Wu, X.: Causality-based feature selection: methods and evaluations. ACM Comput. Surv. (CSUR) **53**(5), 1–36 (2020)

12. Yu, K., Liu, L., Li, J.: A unified view of causal and non-causal feature selection. ACM Trans. Knowl. Disc. Data (TKDD) **15**(4), 1–46 (2021)

13. Yu, K., Liu, L., Li, J., Ding, W., Le, T.D.: Multi-source causal feature selection. IEEE Trans. Pattern Anal. Mach. Intell. **42**(9), 2240–2256 (2019)

14. Zheng, X., Aragam, B., Ravikumar, P.K., Xing, E.P.: DAGs with no tears: Continuous optimization for structure learning. Adv. Neural Inf. Process. Syst. **31**, 9492–9503 (2018)

15. Zhou, K., Guo, M., Martin, A.: Evidential clustering based on transfer learning. In: Denœux, T., Lefèvre, E., Liu, Z., Pichon, F. (eds.) Belief Functions: Theory and Applications, pp. 56–65. Springer, Cham (2021). https://doi.org/10.1007/978-3-030-88601-1_6

16. Zhou, K., Martin, A., Pan, Q., Liu, Z.G.: Median evidential c-means algorithm and its application to community detection. Knowl.-Based Syst. **74**, 69–88 (2015)

A Variational Bayesian Clustering Approach to Acoustic Emission Interpretation Including Soft Labels

Martin Mbarga Nkogo, Emmanuel Ramasso$^{(\boxtimes)}$, Patrice Le Moal,
and Gilles Bourbon

Department of Applied Mechanics, UFC/CNRS/ENSMM/UTBM,
Univ. Bourgogne Franche-Comté, FEMTO-ST Institute, 25000 Besançon, France
`emmanuel.ramasso@femto-st.fr`

Abstract. We investigate Gaussian Mixture Models (GMM) with uncertain parameters to evaluate whether this model can help in interpreting acoustic emission data used in non-destructive testing. This model, called VBGMM (variational Bayesian GMM) allows the end-user to automatically determine the number of clusters which makes it relevant for this type of application where clusters are related to damages. In this work, we modify the training procedure to include prior knowledge about clusters. Experiments are made on a recently published benchmark, ORION-AE, that aims at estimating the tightening levels in a bolted structure under vibrations. Preliminary results of the VBGMM with soft priors (VBGMM-SOFT) show good improvement over the standard VBGMM.

Keywords: Clustering · Soft labels · Acoustic emission · Approximate inference · Structural health monitoring · Loosening of bolted joints · ORION-AE benchmark

1 Introduction

The idea of using soft labels in clustering introduced in [4,20,21] for the Gaussian Mixture Model (GMM) and more general models [7] was motivated by the possibility to introduce the available knowledge on the components of a mixture model used to generate each observation. By using belief functions, the end-user can encode imprecision and uncertainty on the labels used in training and inference. Several studies demonstrated that the use of soft labels (using belief functions and probability theories) in various clustering and classification methods improves not only the global performance [6], but also the interpretation of clusters by providing insights about the decision frontiers [12], and the robustness against mislabelling [4,15].

The present study aims at investigating how the Variational Bayesian GMM [2] (VBGMM) behaves when soft labels are introduced. This work is motivated by an application related to Structural Health Monitoring (SHM) based on the Acoustic Emission (AE) non-destructive technique.

S. Le Hégarat-Mascle et al. (Eds.): BELIEF 2022, LNAI 13506, pp. 23–32, 2022.
https://doi.org/10.1007/978-3-031-17801-6_3

The AE technique relies on permanently attached piezoelectric sensors glued on the surface of a material. Under loading, damage occurring within the material releases energy, and a part of it takes the form a high frequency (possibly between 20 kHz and 2 MHz) transient elastic wave propagating on the surface, converted into a voltage signal by the sensors. The technique is widely used in industry to detect anomalies, such as in civil infrastructures and aeronautics [9,10,19]. The advantage of this technique is the high sensivity of the available sensors which allows to get a lot of details about the damages. With a sampling frequency generally around 5 MHz on multiple sensors, the AE technique provides from thousands to millions of signals during mechanical tests.

There is no physics-based model able to interpret all the collected transient signals due to the several difficulties, in particular the unknowns about the influence of damages on the content of transient elastic waves. Therefore, the main methodologies to interpret AE data are mainly based on clustering, where the clusters are analyzed *a posteriori* to assess their relevance for a given application. The most widely used algorithms are the K-means [3], the fuzzy C-means (FCM) [11], the Gustafson-Kessel (GK) algorithm [13] and Gaussian Mixture Models (GMM) [18].

In previous studies, authors made use of clustering validity indices to estimate the number of clusters [22,23], which is of paramount interest in Material Science and for AE users because it indicates the number of AE sources which are related to damages. One of the advantage of the VBGMM is its ability to automatically estimate the number of clusters and can thus be of interest for interpreting AE data. By allowing the introduction of soft labels in this model, we expect to improve the results even with small amount of prior.

Section 2 presents how to introduce soft priors in a VBGMM. Our first results are illustrated in Sect. 3 on a benchmark recently proposed for AE data clustering and classification.

2   Use of Soft Labels in a Variational Bayesian GMM

2.1   Directed Acyclic Graph

A Bayesian Gaussian Mixture Model (GMM) is represented by the directed graph in Fig. 1 where $\mathbf{y}_n$ is the value taken by an observed variable $\mathbf{Y}$, made of D-dimensional features in $\Re^D$ and $n = 1 \ldots N$ the number of feature vectors, $\mathbf{x}_n$ is the value taken by a latent variable $\mathbf{X}$. Like in a standard GMM, $\boldsymbol{\pi}$, $\boldsymbol{\mu}$ and $\boldsymbol{\Lambda}$ are the mixing proportions, means and precision (inverse of covariance).

In a Bayesian GMM, the three last parameters are uncertain and are considered as random variables to which the end-user assigns a prior distribution: A Dirichlet prior over $\boldsymbol{\pi}$, and an independent Gaussian-Wishart prior [2, Chap. 2] on $(\boldsymbol{\mu}, \boldsymbol{\Lambda})$ (mean and precision) of each Gaussian component [2, Chap. 10]. These particular priors are said conjugate because the posterior distributions have the same functional form as the priors through Bayes rule. The learning process consists in estimating the distribution over the uncertain parameters using a Bayesian Expectation-Maximization algorithm [1].

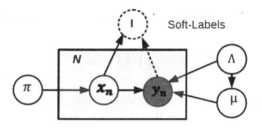

Fig. 1. Directed graph of a VBGMM-SOFT with prior on the latent variables **x**.

In this graphical model, the prior (expert judgments) is represented by variable **I**. In this "data-driven Bayesian network with expert judgments" [5], the values I_n taken by this variable are dependent on the values of $\mathbf{x}_n$ and may be dependent on $\mathbf{y}_n$, for example when the end-user tunes I_n according to the observed values and what he expects on $\mathbf{x}_n$.

2.2 Learning Problem Under *pl*

Solving the learning problem relies on a process detailed in [1] and [2, Chap. 10] and consists in maximizing the lower bound of the log-marginal probability of the data $p(\mathbf{X})$ subject to a factorization constraint. Indeed, a solution to this learning problem, called Variational Bayesian Expectation-Maximization, assumes a factorization between, on one side, a first factor that is the distribution over the parameters ($\boldsymbol{\pi}$, $\boldsymbol{\mu}$ and $\boldsymbol{\Lambda}$) and, on the other side, a second factor which is the distribution over the latent variables $\mathbf{x}_n, n = 1 \ldots N$. Then, for one of those factors, we compute the *expectation of the logarithm of the joint distribution over all hidden and visible variables and then take the expectation with respect to the other factor*. The process is general can be applied to any mixture model.

The expectations are not trivial, the reader can refer to [2, Chap. 10] for more details. We here remind some of the results helpful to understand how to introduce the prior on $\mathbf{x}_n$. The other equations remain the same.

The first step is to compute the joint distribution on all variables in this model. Following Fig. 1, it is given by:

$$p(\mathbf{X}, \mathbf{Y}, \boldsymbol{\pi}, \boldsymbol{\mu}, \boldsymbol{\Lambda}, \mathbf{I}) = p(\mathbf{X} \mid \boldsymbol{\pi})p(\mathbf{Y} \mid \mathbf{X}, \boldsymbol{\mu}, \boldsymbol{\Lambda})p(\boldsymbol{\pi})p(\boldsymbol{\mu} \mid \boldsymbol{\Lambda})p(\boldsymbol{\Lambda})p(\mathbf{l}|\mathbf{X}) \quad (1)$$

where

$$p(l_n|x_{nk} = 1) = pl_{nk} \quad (2)$$

represents the prior on the latent variable for the n-th feature vector and the k-th component in the mixture. This prior is represented by a plausibility contour function for each feature vector, generated from a belief mass over the set of K components. The way of introducing the prior using an auxiliary variable (here **I**) was proposed in [8].

Using the cognitive independence assumption [7], we can write

$$p(\mathbf{l}|\mathbf{X}) = \prod_{n=1}^{N} \prod_{k=1}^{K} [pl_{nk}]^{x_{nk}} \tag{3}$$

Introducing the prior on the latent variables modifies the expression of the expectation of x_{nk}, and does not change the Maximization step. Only the expectation on the latent variable (the values taken by x_{nk}) are modified by the plausibilities as follows:

$$\mathbb{E}[x_{nk}] \propto pl_{nk}\tilde{\pi}_k |\tilde{\boldsymbol{\Lambda}}_k|^{1/2} \exp\left[-\frac{D}{2\beta_k} - \frac{\nu_k}{2}(\mathbf{y}_n - \mathbf{m}_k)^{\mathsf{T}} \mathbf{W}_k (\mathbf{y}_n - \mathbf{m}_k)\right] \tag{4}$$

with

$$\log \tilde{\pi}_k = \psi(\alpha_k) - \psi\left(\sum_{k=1}^{K} \alpha_k\right) \tag{5a}$$

$$\log |\tilde{\boldsymbol{\Lambda}}_k| = \sum_{i=1}^{D} \psi\left(\frac{\nu_k + 1 - i}{2}\right) + D \log 2 + \log |\mathbf{W}_k| \tag{5b}$$

and $\psi(a) = \frac{d}{da}\Gamma(a)$ is the digamma function, $(\alpha_k, \beta_k, \nu_k \mathbf{W}_k, \mathbf{m}_k)$ are the parameters of the Dirichlet and Gaussian-Wishart distributions [2, Chap. 2,10]. The interesting point with this result is that Eq. 4 boils down to the expression found for non Bayesian GMM in [4] when uncertainty on parameters tends to 0.

2.3   Algorithm and Automatic Relevance Determination

The algorithm starts with random initial values and updates the parameters iteratively until the maximum number of iterations is reached (2000) or the evolution of the likelihood becomes less than 10^{-8}. The general algorithm is provided in Algorithm 1.

One of the interests of this algorithm lies in the way some of the components vanish during learning. And this is possible without numerical instabilities as observed in standard GMM when K, for example, is too large. This phenomenon qualified as "Automatic Relevance Determination" allows the end-user to actually set the number of clusters to a large value, and, after convergence, some of the clusters can be removed due to the fact that several parameters tend to their prior. In particular, the *expected values of the mixing coefficients* $\mathbb{E}[\boldsymbol{\pi}]$ *in the posterior distribution* tend to $\alpha_0/(K\alpha_0 + N)$ for very small cluster and N data points. In this expression, α_0 (the same for all components, set to 1 in experiments) is the parameter of the Dirichlet distribution over the mixing coefficients $\boldsymbol{\pi}$, which is the effective prior number of observations associated with each component of the mixture. Therefore, a cluster made of $\mathbb{E}[\pi_k]$ elements can be removed.

Algorithm 1. General algorithm of VBGMM-SOFT.

1: Generate plausibilities with ρ provided by end-user
2: **while** max_iterations not reached and evolution of likelihood above threshold **do**
3:    E-Step: Compute $\log q^\star(\mathbf{X}) = \mathbb{E}_{\pi,\mu,\Lambda}[\log p(\mathbf{X}, \mathbf{Y}, \pi, \mu, \Lambda, \mathbf{I})] + c$ (c is a constant)
4:    For this step, use the same equations as in [2] except for $\mathbb{E}[x_{nk}]$ where Eq. 4 should be used instead.
5:    M-Step: Compute $\log q^\star(\pi)$, $\log q^\star(\mu)$ and $\log q^\star(\Lambda)$ by taking the expectation of Eq. 1 on, respectively, π, μ and Λ, with respect to $\mathbf{X}$ (using the results of the E-Step).
6:    Compute the likelihood (expression given in [2, p. 481] and using Eq. 4).
7: **end while**

3   First Results and First Conclusion

This section presents preliminary results on the capacity of the VBGMM with soft labels to provide relevant clusters. For that we used a benchmark, called ORION-AE, obtained from a real system, and described in [17].

3.1   Data Set Description

The system is composed of a two metallic plates jointed by three bolts and was designed to reproduce the loosening phenomenon observed on structures made of assemblies, in particular when submitted to vibrations. One of the bolts was untightened manually to simulate the loosening. The lower plate was submitted to 120 Hz harmonic force by means of a shaker. Seven levels of tightening were considered, and for each level, an acoustic emission sensor recorded the transients liberated during the test. Each tightening level was maintained during 10 s. The test was repeated 5 times, leading to 5 datasets, each made of 7 seven classes with 70 s of data for different sensors (in this paper only the second sensor was used).

The seven tightening levels can be used as a ground truth when designing learning methods. This makes this dataset useful for developing and testing clustering and classification methods for interpreting acoustic emission data.

The ORION-AE data are raw time-series. To be used in a VBGMM, we need a preprocessing stage. We used a similar preprocessing to [16] with a first step consisting in detecting the transients in the data stream and followed by a step of feature extraction. Conversely to [16], the Principal Components Analysis (PCA) was not used. Thirteen features were kept, and all combinations of four features were considered (four were used to decrease the amount of time of tests since all combinations were considered). The VBGMM was applied for each combination while also varying the amount of prior.

3.2   The Priors

The priors (pl) were generated as proposed by Côme et al. [4, Section 5.2.1] using the true labels. For each training sample i, a number p_i was drawn from a specific Beta distribution with expected value equal to $\rho \in \{0, 0.3, 0.6, 0.9, 1\}$ and variance 0.1, *used to define the doubt expressed by a hypothetical expert on the class of that sample*. With probability p_i, the label of sample i was changed (to any other class with equal probabilities). Therefore ρ controls the amount of prior introduced: When $\rho = 0$, all the labels are used as priors corresponding to supervised learning, whereas $\rho = 1$ corresponds to the unsupervised learning situation[1].

3.3   Sorting the Partitions

The partitions obtained for all combinations for a given ρ were then sorted according to a criterion proposed in [14]. The criterion works as follows. For each partition, the onset time (first occurrence) of each cluster was determined. Then, each cluster was re-labelled according to their order of occurrence: the first cluster to occur was labelled "1", the second cluster labelled "2", and so on. This co-association allows the fusion of partitions since all clusters with the same label are assumed to correspond to the same source [13].

After re-labelling, each partition was ranked by :

$$C(\mathcal{S}) = \sum_{k=1}^{\kappa-1} \Delta_{\text{onset}}(k, k+1) \log \Delta_{\text{onset}}(k, k+1) \tag{6a}$$

$$\Delta_{\text{onset}}(k, k+1) = t_{\text{onset},k+1} - t_{\text{onset,k}} \tag{6b}$$

where κ is the number of cluster, $\mathcal{S}$ is a subset of features used to compute the partition, $t_{\text{onset},\kappa+1}$ is equal to the timestamp of the last AE signal. This criterion assumes that the onsets of all clusters in a given partition should be spread onto the time (or load) axis as uniformly as possible.

Once the partitions have been sorted according to this criterion, the best partition is taken for evaluating the quality of the clusters. In order to evaluate the performances, we used the Adjusted Rand Index [24], which provides a value between 0 and 1, where a "1" is obtained for perfect correspondence between the clusters estimated and the true labels.

3.4   Results

Several tests were performed, using various ρ, considering uncertain and noisy priors, on the five datasets available in ORION-AE, with a comparison to clustering algorithms used in the literature. In this communication, results are only shown for the first dataset.

Figure 2 illustrates the results for the first dataset and for the second sensor. The priors were considered uncertain. Each curve represents the *decadic logarithm* of the cumulated number of acoustic emission transients in each clusters. For example, consider the left-hand side of the top-left figure (Fig. 2a, corresponding to the unsupervised case so with $\rho = 1$). The blue curve, for example,

---

[1] The code can be found on T. Denœux's homepage in software/E2M/add_noise.m.

is the cumulated number of the acoustic emission transients assigned to the first cluster. We can see that the curve reaches a plateau around 10 s, knowing that interval $[0, 10]$ seconds corresponds to the first level of tightening (the second interval, $[10, 20]$ is the second level, and so on). Therefore, this first cluster is relevant and can be assigned to the first tightening level because the blue curve evolves only in the first period. The yellow, purple, green and light blue clusters correspond to tightening levels 2, 3, 5 and 6 respectively. We can see that cluster 6 starting at 60 s does not stop increasing in (level 7), which means that cluster 6 gathers data from both intervals. We can also see that the red cluster starts within $[0, 10]$ which means that the first level of tightening is split in two clusters. This red cluster stops increasing until 30 s where it grows again. This cluster is certainly related to the fourth level (the vertical axis is in logarithmic scale) and shares common features with the first one (according to the clustering method).

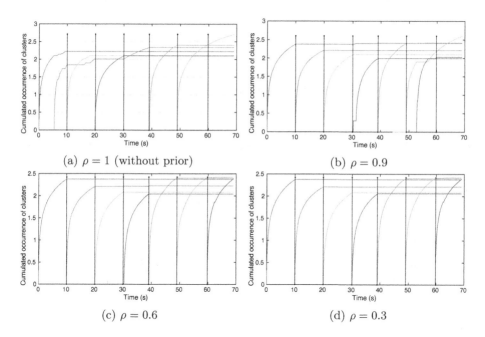

Fig. 2. Cumulated number of transient signals assigned to each cluster for $\rho = \{1, \ 0.9, \ 0.6, \ 0.3\}$.

The second figure (top right, Fig. 2b) corresponds to a small amount of prior, $\rho = 0.9$ (all labels are true but with large uncertainty). We can see that the red cluster is now at the good location, while a new cluster was positioned between 50 and 60 s. Therefore, there is a mixing between the two last levels. Then from $\rho = 0.6$ (Fig. 2c) downwards, the clusters correspond quite precisely to the tightening levels, with no difference between $\rho = 0.6$ and $\rho = 0.3$.

These results show that the cumulated plots of clusters bring two main information: The starting points (called onsets) of the accumulation, and the steady phase. When an onset is well located, it means that the clustering is able to assign the first transients of a given tightening level to the correct cluster. Concerning the steady phase, when it starts at the right location, it means that we are able to locate when a cluster stops increasing, therefore when a possible damage or functioning condition stops occurring. The height of the steady phase depends on how active was a damage and this can be useful for monitoring.

Figure 3 illustrates the evolution of the ARI for different values of ρ and considering all combinations of features (715). The ARI values were sorted by descending order. Circled-markers represent the 10 best ARI values corresponding to the 10 best partitions (estimated by the VBGMM-SOFT method and selected by the criterion proposed in [14]). We can see that the markers are located on the left-hand side of these curves corresponding to quite high ARI values.

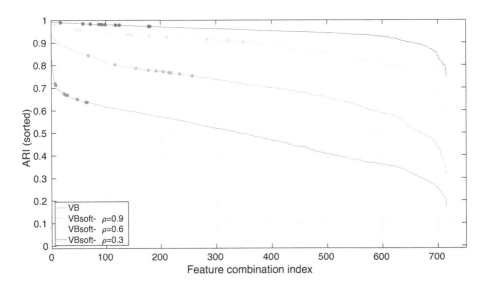

Fig. 3. Adjusted Rand Index (ARI) sorted for all partitions. The ten best partitions obtained by a VBGMM with different amounts of priors are superimposed with markers.

4   Conclusion

The first results obtained with a VBGMM on acoustic emission data are encouraging. We propose to add some prior which results in a small modification of the original algorithm.

With a small amount of prior the performance are greatly improved. The clusters obtained are interpreted by means of cumulated plots. For the application targeted, these plots underline two important pieces of information: the onsets of clusters and their steady phase. As a future work, we plan to exploit both of them for Structural Health Monitoring since it informs about the damage process taking place within the material.

The way to generate the prior remains a key problem. One idea followed by the acoustic emission community consists in getting prior knowledge from numerical simulations. However, in addition to the computational burden involved in such simulations, another difficulty holds in the fact that these simulations also require knowledge about the material properties which evolve during a test.

Acknowledgments. This work was supported by the EIPHI Graduate school (contract "ANR-17-EURE-0002").

References

1. Beal, M.J.: variational algorithms for approximate Bayesian inference. Ph.D. thesis, Gatsby Computational Neuroscience Unit, University College London (2003)
2. Bishop, C.: Pattern Recognition and Machine Learning. Springer, Cham (2006)
3. Chai, M., Zhang, J., Zhang, Z., Duan, Q., Cheng, G.: Acoustic emission studies for characterization of fatigue crack growth in 316LN stainless steel and welds. Appl. Acoust. **126**, 101–113 (2017)
4. Côme, E., Oukhellou, L., Denoeux, T., Aknin, P.: Learning from partially supervised data using mixture models and belief functions. Pattern Recogn. **42**(3), 334–348 (2009)
5. Constantinou, A.C., Fenton, N., Neil, M.: Integrating expert knowledge with data in Bayesian networks: preserving data-driven expectations when the expert variables remain unobserved. Expert Syst. Appl. **56**, 197–208 (2016)
6. Cour, T., Sapp, B., Taskar, B.: Learning from partial labels. J. Mach. Learn. Res. **12**, 1501–1536 (2011)
7. Denoeux, T.: Maximum likelihood estimation from uncertain data in the belief function framework. Knowl. Data Eng. IEEE Trans. **25**(1), 119–130 (2013)
8. Denœux, T.: Maximum likelihood estimation from fuzzy data using the EM algorithm. Fuzzy Sets Syst. **183**(1), 72–91 (2011)
9. Giannì, C., Balsi, M., Esposito, S., Ciampa, F.: Low-power global navigation satellite system-enabled wireless sensor network for acoustic emission localisation in aerospace components. Struct. Control. Health Monit. **27**(6), e2525 (2020)
10. Grigg, S., et al.: Development of a low-power wireless acoustic emission sensor node for aerospace applications. Struct. Control. Health Monit. **28**(4), e2701 (2021)
11. Omkar, S.N., Suresh, S., Raghavendra, T.R., Mani, V.: Acoustic emission signal classification using fuzzy c-means clustering. In: Proceedings of the 9th International Conference on Neural Information Processing, 2002. ICONIP 2002, vol. 4, pp. 1827–1831 (2002). https://doi.org/10.1109/ICONIP.2002.1198989
12. Quost, B., Denœux, T.: Learning from data with uncertain labels by boosting credal classifiers. In: Proceedings of the 1st ACM SIGKDD Workshop on Knowledge Discovery from Uncertain Data, pp. 38–47 (2009)
13. Ramasso, E., Placet, V., Boubakar, M.: Unsupervised consensus clustering of acoustic emission time-series for robust damage sequence estimation in composites. IEEE Trans. Instr. Meas. **64**(12), 3297–3307 (2015)

14. Ramasso, E., et al.: Learning the representation of raw acoustic emission signals by direct generative modelling and its use in chronology-based clusters identification. Eng. Appl. Artif. Intell. **90**, 103478 (2020)
15. Ramasso, E., Denoeux, T.: Making use of partial knowledge about hidden states in HMMs: an approach based on belief functions. IEEE Trans. Fuzzy Syst. **22**(2), 395–405 (2014)
16. Ramasso, E., Denoeux, T., Chevallier, G.: Clustering acoustic emission data streams with sequentially appearing clusters using mixture models. arXiv preprint arXiv:2108.11211 (2021)
17. Ramasso, E., Verdin, B., Chevallier, G.: Monitoring a bolted vibrating structure using multiple acoustic emission sensors: a benchmark. Data **7**(3), 31 (2022)
18. Sawan, H.A., Walter, M.E., Marquette, B.: Unsupervised learning for classification of acoustic emission events from tensile and bending experiments with open-hole carbon fiber composite samples. Compos. Sci. Technol. **107**, 89–97 (2015)
19. Tonelli, D., et al.: Effectiveness of acoustic emission monitoring for in-service pre-stressed concrete bridges. In: Huang, H., Zonta, D., Su, Z. (eds.) Sensors and Smart Structures Technologies for Civil, Mechanical, and Aerospace Systems 2021, vol. 11591, pp. 178–192. International Society for Optics and Photonics, SPIE (2021). https://doi.org/10.1117/12.2585527
20. Vannoorenberghe, P.: Finite mixture models estimation with a credal EM algorithm. Traitement du Signal **24**(2), 103–113 (2007)
21. Vannoorenberghe, P., Smets, P.: Partially supervised learning by a credal EM approach. In: Godo, L. (ed.) Symbolic and Quantitative Approaches to Reasoning with Uncertainty. Lecture Notes in Computer Science, vol. 3571, pp. 956–967. Springer, Berlin Heidelberg (2005)
22. Vendramin, L., Campello, R., Hruschka, E.: Relative clustering validity criteria: a comparative overview. Stat. Anal. Data Min. **3**(4), 209–235 (2010)
23. Vendramin, L., Jaskowiak, P.A., Campello, R.J.: On the combination of relative clustering validity criteria. In: Proceedings of the 25th International Conference on Scientific and Statistical Database Management, p. 4. ACM (2013)
24. Vinh, N., Epps, J., Bailey, J.: Information theoretic measures for clusterings comparison: is a correction for chance necessary? In: Proceedings of the 26th Annual International Conference on Machine Learning, pp. 1073–1080. ACM, New York (2009)

Evidential Clustering by Competitive Agglomeration

Lulu Xu[1], Qian Wang[2], Pei-hong Wang[1], and Zhi-gang Su[1(✉)]

[1] National Engineering Research Center of Power Generation Control and Safety, School of Energy and Environment, Southeast University, Nanjing 210096, China
zhigangsu@seu.edu.cn

[2] School of Energy and Power, Jiangsu University of Science and Technology, Zhenjiang 212003, China

Abstract. A new clustering method, named Evidential clustering by Competitive Agglomeration (ECA), is introduced by applying the framework of belief functions to a competitive strategy. It has two-fold advantages: Firstly, with the help of the credal partition, it has a good ability to deal with noise objects since it can mine the ambiguity and uncertainty of the data structure; secondly, through a competitive strategy, it can automatically gain the number of clusters under the rule of intra-class compactness and inter-class dispersion. Results demonstrate the effectiveness of the proposed method on synthetic and real-world datasets.

Keywords: Belief functions · Credal partition · Competitive agglomeration

1 Introduction

Clustering analysis, as an important data mining technology [1–3], is widely used in pattern recognition [4], image processing, machine learning[5], information retrieval and other fields [6]. The most classical fuzzy partition method is the Fuzzy C-Means (FCM) algorithm [7]. But for FCM, the number of clustering centers is difficult to determine. Concerning this problem, Frigui and Krishnapuram [8] propose the Competitive Agglomeration (CA) algorithm, which combines the advantages of hierarchical clustering and partition clustering technology. In order to improve performance of the CA algorithm, Grira et al. [9,10] effectively propose an active fuzzy constrained clustering (AFCC) algorithm. Gao et al. [11] further improve the objective function of AFCC algorithm and propose a semi-supervised fuzzy clustering algorithm with pairwise constraints (SCAPC). However, all these CA and CA-based algorithms confront a common problem of poor robustness to noise and outliers [12].

Belief function theory [13] adds the concept of credal partition to the existing hard and fuzzy, which allows for a deeper understanding of the data and improves the robustness of outliers. Therefore, the credal partition algorithm has attracted more and more scholars' attention recently. Masson and Denoeux

S. Le Hégarat-Mascle et al. (Eds.): BELIEF 2022, LNAI 13506, pp. 33–43, 2022.
https://doi.org/10.1007/978-3-031-17801-6_4

[14] solve the problem of calculating credal partition from object data and propose a new algorithm called Evidential C-Means (ECM). Su [15] proposes a new ECM evolutionary algorithm (E2CM), in which the objective function J and standardized specificity index N^* are simultaneously optimized in ECM. Inspired by a novel clustering method DPC [16], Su [17] introduces Belief-Peaks Evidential Clustering (BPEC) in the framework of belief functions. To improve the clustering performance, Gong [18] proposes the cumulative belief peaks evidential K-nearest neighbor clustering (CBP-EKNN).

Based on the above discussion, in order to better mine the ambiguity and uncertainty of the data structure, this paper proposes Evidential clustering by Competitive Agglomeration (ECA) by applying the framework of belief functions to a competitive strategy. Interestingly, the proposed ECA not only retains the unique ability of CA to automatically gain the number of clusters, but also has the powerful ability to reveal the data structure. It can deal with noise objects, which improves the robustness of CA.

The rest of this article is organized as follows. The basic concepts of CA and belief functions are briefly introduced in Sect. 2. The ECA method is presented in Sect. 3. The performance of ECA is verified by an numerical example and compared with other clustering algorithms in Sect. 4. The last part gives the conclusion.

2   Background

We recommend that readers become familiar with some basic concepts of Competitive Agglomeration (CA) and belief functions through a brief overview of the paper.

2.1   Competitive Agglomeration (CA)

Let $X = \{x_j | j = 1, ..., n\} \in R^{n \times p}$. Let $V = \{v_1, ..., v_c\}' \in R^{c \times p}$ represents a c-tuple of prototypes each of which characterizes one of the c clusters. Each v_i consists of a set of parameters.

The CA algorithm first divides a dataset into a large number of small clusters. With the development of the algorithm, neighboring clusters compete for data points, and some clusters that lose competition gradually dry up and disappear. The final partition has the "optimal" cluster number. It should be noted that the number of clusters c is dynamically updated in each iteration. The CA algorithm minimizes the following objective function, which is shown in Eq. (1), and Eq. (2) are constraints.

$$J_{CA}(U, V) = \sum_{i=1}^{c} \sum_{j=1}^{n} u_{ij}^m d_{ij}^2 - \alpha \sum_{i=1}^{c} [\sum_{j=1}^{n} u_{ij}]^2, \tag{1}$$

$$s.t. \sum_{i=1}^{c} u_{ij} = 1, \forall j = 1, 2, \cdots, n, \tag{2}$$

where u_{ij} is the membership degree of the sample x_j to the ith cluster, and $U = [u_{ij}]$ is a $c \times n$ matrix called a constrained fuzzy c-partition matrix. m is the degree of ambiguity of the algorithm, which is generally set to 2, d_{ij} is the Euclidean distance between the center of ith cluster and the sample x_j, α is to balance the first and second terms.

2.2 Basic Concepts of Belief Functions

Given a *frame of discernment* $\Omega = \{\omega_1, \omega_2, \cdots, \omega_c\}$, a mass function is defined as a mapping from 2^Ω to $[0, 1]$, such that

$$\sum_{A \subseteq \Omega} m^\Omega(A) = 1, \tag{3}$$

where the subsets A of Ω with $m^\Omega(A) > 0$ are called the focal sets of m. A mass function is said to be Bayesian if it only has singletons (i.e., $|A| = 1$) as focal sets, where $|\cdot|$ denotes the cardinality of a focal set or size of a dataset. It is said to be Nondogmatic belief function if it has Ω as one focal set. In particular, the vacuous mass function, verifying $m^\Omega(\Omega) = 1$, corresponds to total ignorance.

A mass function has other equivalent representations such as *belief function* and *plausibility function*, defined as:

$$bel^\Omega(A) = \sum_{\emptyset \neq B \subseteq A} m^\Omega(B), \tag{4}$$

$$pl^\Omega(A) = \sum_{A \cap B \neq \emptyset} m^\Omega(B), \tag{5}$$

for all $A \subseteq \Omega$.

The combination of mass functions plays an important role in theory of belief functions. Given two mass functions m_1 and m_2, the combination of mass functions is defined as follows:

$$m_{1 \cap 2}^\Omega(A) = \sum_{B \cap C = A} m_1^\Omega(B) m_2^\Omega(C), A \subseteq \Omega. \tag{6}$$

The normality condition $m^\Omega(\emptyset) = 0$ is recovered by dividing each mass $m_{1 \cap 2}^\Omega(A)$ by $1 - m_{1 \cap 2}^\Omega(\emptyset)$. This operation is noted $\oplus$ and called Dempster's rule of combination:

$$m_{1 \oplus 2}^\Omega(A) = \frac{m_{1 \cap 2}^\Omega(A)}{1 - m_{1 \cap 2}^\Omega(\emptyset)}, \emptyset \neq A \subseteq \Omega. \tag{7}$$

3 Main Results

3.1 Basic Idea and Motivations

The basic idea of ECA can be expressed as follows: the belief functions are applied to a competitive strategy to obtain an objective function similar to CA and the required parameters are solved by minimizing the objective function. Here we define the sum of pl values of each sample corresponding to the tth cluster in clusters c as the cardinality N'_t of this cluster. In general, there are few samples near the spurious clusters and they are influenced by other clusters, so the support degree of samples to this cluster is not high, that is, the sum of the sample pl values corresponding cluster is low. The competitive strategy proposed in this paper can further devalue their pl value to such clusters during iteration, which leads to a gradual reduction of the cardinality of these spurious clusters. When the cardinality of a cluster drops below a threshold, we discard the cluster and update the number of clusters. In the process, neighboring clusters compete with each other to obtain the "optimal" cluster number.

The motivation of ECA is to improve the performance of CA by introducing belief functions to mine the ambiguity and uncertainty of data structure. It can reveal data structures in the form of credal partition, from which hard, fuzzy, possibilistic and rough partitions can be derived, thus improving the robustness of CA.

3.2   The Proposed Method

For a given dataset $X = \{x_i | i = 1, ..., n\} \in R^{n \times p}$, the recognition framework is $\Omega = \{\omega_1, \omega_2, ..., \omega_c\}$. For each sample x_i, we can construct a piece of evidence m_i to represent the cluster to which x_i belongs. Each mass $m_i(A_j)$ indicates the degree that x_i belongs to cluster A_j. The set of n pieces of evidence constitutes a credal partition $M = \{m_1, m_2, \cdots, m_n\}' \in R^{n \times 2^c}$. $V = \{v_1, ..., v_c\}' \in R^{c \times p}$ is the set of cluster centers.

The objective function of ECA is composed of two components, i.e., $J_{ECA} = J_1 + J_2$. Its objective function and the constraints are shown in Eq. (8)–(11). The first component J_1 is similar to the objective function of ECM, which allows us to control the shape and size of clusters and obtain compact clusters. When the number of clusters c equals the number of samples n, that is, each cluster contains only one data point, the global minimum value of this component can be obtained. Here we define the sum of m_{ij} corresponding to the jth subset (i.e., focal set A_j) as the cardinality N_j of this subset, so the second component J_2 is the sum of the β power of the cardinality of each subset plus a minus sign, which allows us to control the number of clusters. The global minimum for the item (including the minus sign) can be achieved when all points are in one cluster and all other clusters are empty. When the two components are combined and γ is correctly selected, the final partition minimizes the sum of distances within the cluster while dividing the dataset into the smallest possible number of clusters. It should be noted that the number of clusters c is dynamically updated in each iteration.

$$J_{ECA}(M, V) = J_1 + J_2, \tag{8}$$

$$J_1 = \sum_{i=1}^{n} \sum_{j | \emptyset \neq A_j \subseteq \Omega} |A_j|^\alpha m_{ij}^\beta d_{ij}^2 + \sum_{i=1}^{n} \delta^2 m_{i\emptyset}^\beta \tag{9}$$

$$J_2 = -\gamma[\sum_{j|\emptyset \neq A_j \subseteq \Omega} (\sum_{i=1}^{n} m_{ij})^\beta + (\sum_{i=1}^{n} m_{i\emptyset})^\beta], \tag{10}$$

$$s.t. \sum_{j|\emptyset \neq A_j \subseteq \Omega} m_{ij} + m_{i\emptyset} = 1, \forall i = 1, 2, \cdots, n, \tag{11}$$

where d_{ij} represents the Euclidean distance between the sample x_i and the center of gravity $\bar{v}_j$ (the calculation of $\bar{v}_j$ is shown in Eq. (12)), m_{ij} represents the mass of the jth subset assigned to the sample x_i, $m_{i\emptyset}$ represents the quality of the empty set assigned to the sample x_i, α is the weight coefficient used to compensate for the focal elements of the high primitives in the subset, the coefficient β controls the ambiguity of the belief partition degree, usually 2, δ coefficient mainly controls the number of noise points considered, and usually the larger the δ, the fewer points will be classified as noise points.

Based on the basic concept of belief functions, the centers of gravity $\bar{v}_j$ are defined as follows:

$$\bar{v}_j = |A_j|^{-1} \sum_{t=1}^{c} s_{tj} v_t, \tag{12}$$

where $w_t \in A_j$, $s_{tj} = 1$, otherwise, $s_{tj}=0$, v_t is the center of the single cluster (primitive number $|A_j| = 1$).

To minimize the objective function in Eq. (8) with respect to M, we apply Lagrange multipliers and obtain

$$L(M, \lambda_1, \lambda_2, \cdots, \lambda_n) = J_{ECA} - \sum_{i=1}^{n} \lambda_i (\sum_{j|\emptyset \neq A_j \subseteq \Omega} m_{ij} + m_{i\emptyset} - 1). \tag{13}$$

Making the first-order partial derivative of $L(\cdot)$ about m_{ij}, $m_{i\emptyset}$ equal to 0, we get:

$$\begin{cases} \frac{\partial L}{\partial m_{ij}} = \beta |A_j|^\alpha m_{ij}^{\beta-1} d_{ij}^2 - \beta\gamma(\sum_{s=1}^{n} m_{sj})^{\beta-1} - \lambda_i = 0, \\ \frac{\partial L}{\partial m_{i\emptyset}} = \beta\delta^2 m_{i\emptyset}^{\beta-1} - \beta\gamma(\sum_{s=1}^{n} m_{s\emptyset})^{\beta-1} - \lambda_i = 0. \end{cases} \tag{14}$$

The solution can be greatly simplified by assuming that the credal partition M does not change significantly from one iteration to the next and by computing the term $(\sum_{s=1}^{n} m_{sj})^{\beta-1}$ and $(\sum_{s=1}^{n} m_{s\emptyset})^{\beta-1}$ in Eq. (14) using the credal partition M from the previous iteration. Under this assumption, we get:

$$\begin{cases} m_{ij} = [\frac{\lambda_i + \beta\gamma N_j}{\beta |A_j|^\alpha d_{ij}^2}]^{\frac{1}{\beta-1}}, \\ m_{i\emptyset} = [\frac{\lambda_i + \beta\gamma N_\emptyset}{\beta\delta^2}]^{\frac{1}{\beta-1}}, \end{cases} \tag{15}$$

where

$$\begin{cases} N_{j|\emptyset \neq A_j \subseteq \Omega} = (\sum_{s=1}^{n} m_{sj})^{\beta-1}, \\ N_\emptyset = (\sum_{s=1}^{n} m_{s\emptyset})^{\beta-1}. \end{cases} \tag{16}$$

According to the constraints, we get:

$$\sum_{k|\emptyset \neq A_k \subseteq \Omega} [\frac{\lambda_i + \beta\gamma N_k}{\beta|A_k|^\alpha d_{ik}^2}]^{\frac{1}{\beta-1}} + [\frac{\lambda_i + \beta\gamma N_\emptyset}{\beta\delta^2}]^{\frac{1}{\beta-1}} = 1. \tag{17}$$

In this paper, in order to simplify the calculation, we set the coefficient β to be 2, then the equation is simplified to :

$$\lambda_i \sum_{k|\emptyset \neq A_k \subseteq \Omega} \frac{1}{2|A_k|^\alpha d_{ik}^2} + \frac{\lambda_i}{2\delta^2} + \gamma \sum_{k|\emptyset \neq A_k \subseteq \Omega} \frac{N_k}{|A_k|^\alpha d_{ik}^2} + \frac{\gamma N_\emptyset}{\delta^2} = 1. \tag{18}$$

Thus the expression λ_i is obtained:

$$\lambda_i = \frac{1 - \gamma \sum_{k|\emptyset \neq A_k \subseteq \Omega} \frac{N_k}{|A_k|^\alpha d_{ik}^2} - \frac{\gamma N_\emptyset}{\delta^2}}{\sum_{k|\emptyset \neq A_k \subseteq \Omega} \frac{1}{2|A_k|^\alpha d_{ik}^2} + \frac{1}{2\delta^2}}. \tag{19}$$

After substituting λ_i into the expression of m_{ij}, we can get $m_{ij} = m_{ij}^{ECM} + m_{ij}^{Bias}$ and $m_{i\emptyset} = m_{i\emptyset}^{ECM} + m_{i\emptyset}^{Bias}$:

$$\begin{cases} m_{ij}^{ECM} = \frac{1}{\frac{|A_j|^\alpha d_{ij}^2}{\sum_{k|\emptyset \neq A_k \subseteq \Omega} |A_k|^\alpha d_{ik}^2} + \frac{|A_j|^\alpha d_{ij}^2}{\delta^2}}; & m_{ij}^{Bias} = \frac{\gamma}{|A_j|^\alpha d_{ij}^2}(N_j - \bar{N}_i), \\ m_{i\emptyset}^{ECM} = \frac{1}{\frac{\delta^2}{\sum_{k|\emptyset \neq A_k \subseteq \Omega} |A_k|^\alpha d_{ik}^2} + 1}; & m_{i\emptyset}^{Bias} = \frac{\gamma}{\delta^2}(N_\emptyset - \bar{N}_i), \end{cases} \tag{20}$$

where $\bar{N}_i = \dfrac{\sum_{k|\emptyset \neq A_k \subseteq \Omega} \frac{N_k}{|A_k|^\alpha d_{ik}^2} + \frac{N_\emptyset}{\delta^2}}{\sum_{k|\emptyset \neq A_k \subseteq \Omega} \frac{1}{|A_k|^\alpha d_{ik}^2} + \frac{1}{\delta^2}}$ is the weighted average of the cardinality of each subset from the point of view of sample x_i, the m_{ij}^{Bias} and $m_{i\emptyset}^{Bias}$ are the signed bias terms represented as the difference between the cardinality of the jth subset and the weighted average of the cardinality. For the cardinality of jth subset in the higher than average, the bias term is positive, thus appreciating the belief mass value. On the other hand, for low cardinality subset, the bias term is negative, thus depreciating the belief mass value.

Then we minimize the objective function in Eq. (8) with respect to V. As the second term J_2 of objective function in Eq. (8) is not affected by distance, the update equation of centers obtained is the same as those of ECM, i.e., Eq. (21)

$$HV = B, \tag{21}$$

where

$$\begin{cases} B_{lq} = \sum_{i=1}^{n} x_{iq} \sum_{\omega_l \in A_j} |A_j|^{\alpha-1} m_{ij}^\beta, \\ H_{lk} = \sum_{i=1}^{n} \sum_{\omega_l, \omega_k \subseteq A_j} |A_j|^{\alpha-2} m_{ij}^\beta, l, k = 1, 2, \cdots, c, q = 1, 2, \cdots, p. \end{cases} \tag{22}$$

In ECA algorithm, it is important to choose γ in Eq. (8) so that the two terms (i.e., J_1 and J_2) are of the same order of magnitude. If γ is too large, the first item will be ignored and all samples will be grouped into a cluster. If γ is too small, the second term will be ignored and the number of clusters will not decrease. In this paper, we choose γ to be

$$\gamma(k) = \eta(k) \frac{\sum_{i=1}^{n} \sum_{j | \emptyset \neq A_j \subseteq \Omega} |A_j|^\alpha m_{ij}^2 d_{ij}^2 + \sum_{i=1}^{n} \delta^2 m_{i\emptyset}^2}{\sum_{j | \emptyset \neq A_j \subseteq \Omega} (\sum_{i=1}^{n} m_{ij})^2 + (\sum_{i=1}^{n} m_{i\emptyset})^2}, \tag{23}$$

where $\eta(k) = \eta_0 \exp(-k/\tau)$, k is the number of iterations, η_0 is the initial value constant, and τ is the number of iterations constant.

Different from CA, u_{ij} is replaced by m_{ij} in ECA, so the expression form of the cardinality N_t' evolves as follows:

$$N_t' = \sum_{i=1}^{n} pl_{it}, \tag{24}$$

where pl can be obtained by Eq. (5), N_t' is the cardinality of tth cluster in clusters c and ε_1 (i.e., the threshold of N_t') is an evaluation index used to retain the correct clusters and eliminate spurious clusters.

Interestingly, it can be found from the above discussion that ECA and CA or ECM have a conversion relationship: when $m_{ij} = u_{ij}$, $\delta = 0$, ECA degenerates into CA, and the algorithm loses the function of revealing data structure deeply; when $\gamma = 0$, ECA degenerates into ECM, and the algorithm loses the ability to gain the number of clusters automatically.

Based on the above explanations, the ECA is summarized in Algorithm 1:

4   Experimental Evaluation

In this section, we conduct experiments on synthetic and real-world datasets. In Sect. 4.1, an numerical example is used to show the effectiveness of the ECA algorithm. In Sect. 4.2, the performance of ECA is compared with that of CA and ECM.

4.1   An Numerical Example: Four-Class Dataset

The Four-class dataset is divided into 4 clusters, each containing 100 2-dimensional data points. The input parameters of the algorithm simulation are set to $k = 0$, $k_{max} = 30$, $c_{max} = 8$, $\alpha = 2$, $\beta = 2$, $\delta^2 = 20$, $\eta_0 = 3$, $\tau = 10$, $\varepsilon_1 = 30$.

Figure 1(a)-(f) are the implementation process of ECA in the Four-class dataset. The position of centers are shown as "+" signs superimposed on the dataset. Figure 1(a) is the dataset verified in this paper. Figure 1(b) shows the

---

Algorithm 1: ECA Clustering Algorithm

Input: Determine the maximum number of clusters c_{max}, α, β, δ, dataset X, maximum number of iterations k_{max}, ε_1(i.e., the threshold of N'_t), the initial value constant η_0 and the number of iterations constant τ.

Output: the cluster centers V and the belief partition matrix M.

1 the initial $V^{(0)}$ and $M^{(0)}$ are obtained by ECM algorithm;

2 $k = 0$

3 Repeat

4 calculate $d^2(x_i, v_j)$;

5 use Eq. (23) to update $\gamma(k)$;

6 use Eq. (20) to update M to obtain the likelihood function pl;

7 use Eq. (24) to calculate the cardinal N'_t of each cluster $(1 < t < C)$, if $N'_t <$ threshold ε_1, discard the v_t;

8 update the number of cluster C and get V after discarding v_t;

9 use Eq. (20) $M^{(k)}$ again;

10 use Eq. (21) to update $V^{(k)}$;

11 $k = k + 1$;

12 Until $(k = k_{max}$, M and V are stable).

---

initial parameters of the prototype partition. The dataset is broken up into many small clusters. Figure 1(c) shows that the ECA discards three spurious clusters after one iteration. Figure 1(d) shows that after the second iteration, one spurious cluster is discarded again and the number of clusters is reduced to four. Figure 1(e)-(f) show that in the subsequent iteration, the number of clusters is no longer reduced and optimization is carried out on the basis of four clusters. After reaching the set maximum number of iterations, the algorithm converges. The result verifies that ECA can find the "optimal" number of clusters without prior knowledge.

The credal partition for the Four-class dataset is shown in Fig. 2, where ω_{jk} means $\{\omega_j, \omega_k\}$. As can be seen, in addition to the four obvious clusters (the cluster center V is represented by the red '+'), credal partition also displays its powerful data analysis structure, including the description of the transition data points between adjacent clusters, represented by $\{\omega_{12}, \omega_{23}, \omega_{34}, \omega_{14}\}$. Even more complex cases involving $\omega_{ijk} = \{\omega_i, \omega_j, \omega_k\}$ (primitive number $|A_j| = 3$) or complete ignorance of $\Omega = \{\omega_1, \omega_2, \omega_3, \omega_4\}$ can also be expressed. Finally, ECA has the same function as the ECM and can identify the existence of noise point, namely $X(\emptyset) := \{i | m_i(\emptyset) = \max_k m(A_k)\}$, which is framed by a red box in Fig. 2.

4.2 Compared with Other Clustering Methods

In this section, the used datasets are summarized in Table 1, among which the first three real world datasets are from UCI database [19], and the last simulation dataset is from [20]. To ensure the effective execution of the algorithm, the parameters are set to $\alpha = 2$, $\beta = 2$, $\delta^2 = 20$.

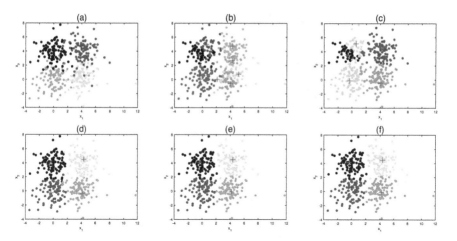

Fig. 1. Original image of spherical cluster on a dataset of intermediate results of ECA algorithm (a) original dataset (b) prototype used for initialization (c) 1 iteration (d) 2 iterations (e) 20 iterations (f) 30 iterations (convergence).

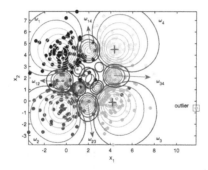

Fig. 2. Contours of credal partition for the Four-class dataset via the ECA.

The CA and ECM are chosen as the comparing algorithms, where the ECM needs to set the number of clusters in advance while the CA does not. In order to be fair, we take the number of clusters obtained by ECA as the prerequisite of ECM. For CA and ECA, the parameters are set to $c_{max} = 30$, $\eta_0 = 3$, $\tau = 10$, $\varepsilon_1 = 15$ for the first two data datasets, $\varepsilon_1 = 30$ for the last two datasets.

We use ARI (Adjusted Rand Index) to measure the performance of these clustering algorithms. Table 2 shows the ARI values (mean $\pm$ variance) simulated by ECA, CA and ECM after 30 runs respectively. The bold and underlined value(s) in each row indicates the best performance, and the number in parentheses represents the number of clusters that the algorithm results in when applied to this dataset. As can be seen that ECA performs better than CA in most cases according to the ARI values. Compared with ECM, ECA is slightly better or equivalent. Considering that ECA can automatically gain the number of clusters

Table 1. Dataset description

Dataset	#Samples	#Dimensions	#Clusters
Ecoli	272	7	3
Seeds	210	7	3
WDBC	569	30	2
Flame	240	2	2

Table 2. Comparison of ARI values of ECA, CA and ECM

Dataset	#ECA	#CA	#ECM
Ecoli	**0.7696 ± 0 (3)**	0.7163 ± 0 (3)	0.7658 ± 0 (3)
Seeds	0.7415 ± 0 (3)	0.7149 ± 0 (3)	**0.7459 ± 0.0061(3)**
WDBC	**0.7366 ± 0 (2)**	0.7361 ± 0 (2)	**0.7366 ± 0 (2)**
Flame	**0.4880 ± 0 (2)**	**0.4880 ± 0 (2)**	0.4649 ± 0 (2)

while ECM fails, ECA is superior to ECM. As mentioned above, ECA yields the best comprehensive performance here.

5   Conclusion

This paper proposes a new clustering method named Evidential clustering by Competitive Agglomeration (ECA) by applying the framework of belief functions to a competitive strategy. The ECA algorithm can find the "optimal" number of clusters without prior knowledge. In addition, compared to CA, ECA can have a credal partition that displays data structures in a way that makes more sense than classical partition clustering. An numerical example is given to illustrate the effectiveness of the proposed method, and it is compared with other clustering algorithms on four other datasets.

Some practical issues as well as more simulations have not been raised and will be discussed in the near future. In addition, there is some further work. The first method is that the combination of CA and belief functions can be considered from another perspective to achieve more appropriate performance. The second method extends ECA to deal with instance constraints to improve the accuracy of the algorithm.

References

1. Gong, C., Su, Z.G., Wang, P.H., Wang, Q.: An evidential clustering algorithm by finding belief-peaks and disjoint neighborhoods. Pattern Recogn. **113**, 2 (2020)
2. Gong, C.Y., Su, Z.G., Wang, P.H., You, Y.: Distributed evidential clustering toward time series with big data issue. Expert Syst. Appl. **191**, 116279 (2022)

3. Gong, C.Y., Wang, P.H., Su, Z.G.: An interactive nonparametric evidential regression algorithm with instance selection. Soft. Comput. **24**(5), 3125–3140 (2020)
4. Gong, C.Y., Su, Z.G., Wang, P.H., Wang, Q., You, Y.: Evidential instance selection for k-nearest neighbor classification of big data. Int. J. Approximate Reasoning **138**, 123–144 (2021)
5. Gong, C., Su, Z.G., Wang, P.H., Wang, Q., You, Y.: A sparse reconstructive evidential-nearest neighbor classifier for high-dimensional data. IEEE Trans. Knowl. Data Eng. (2022)
6. Han, J., Kamber, M., Pei, J.: EnglishData mining: concepts and techniques (2012)
7. Bezdek, J.C.: Pattern recognition with fuzzy objective function algorithms. Kluwer academic (1981)
8. Frigui, H., Krishnapuram, R.: Clustering by competitive agglomeration. Pattern Recogn. **30**(7), 1109–1119 (1997)
9. Grira, N., Crucianu, M., Boujemaa, N.: Semi-supervised fuzzy clustering with pairwise-constrained competitive agglomeration. In: Fuzzy Systems, 2005. FUZZ 2005. The 14th IEEE International Conference on (2005)
10. Grira, N., Crucianu, M., Boujemaa, N.: Active semi-supervised fuzzy clustering. Pattern Recogn. **41**(5), 1834–1844 (2008)
11. Gao, C.F., Wu, X.J.: A new semi-supervised clustering algorithm with pairwise constraints by competitive agglomeration. Appl. Soft Comput. J. **11**(8), 5281–5291 (2011)
12. Krishnapuram, R., Keller, J.: A possibilistic approach to clustering. IEEE Trans. Fuzzy Syst. **1**(2), 98–110 (1993)
13. Shafer, G.: A Mathematical Theory of Evidence, vol. 1 (1976)
14. Masson, M.H., Denoeux, T.: ECM: an evidential version of the fuzzy c -means algorithm. Pattern Recogn. **41**(4), 1384–1397 (2008)
15. Su, Z.G., Zhou, H.Y., Wang, P.H., Zhao, G., Zhao, M.: E2cm: an evolutionary version of evidential c-means clustering algorithm. In: Destercke, S., Denoeux, T., Cuzzolin, F., Martin, A. (eds.) Belief Functions: Theory and Applications, pp. 234–242. Springer International Publishing, Cham (2018). https://doi.org/10.1007/978-3-642-29461-7
16. Rodriguez, A., Laio, A.: Clustering by fast search and find of density peaks. Science **344**(6191), 1492–1496 (2014)
17. Su, Z.G., Denoeux, T.: BPEC: Belief-peaks evidential clustering. IEEE Trans. Fuzzy Syst. **27**(1), 111–123 (2018)
18. Gong, C.Y., Su, Z.G., Wang, P.H., Wang, Q.: Cumulative belief peaks evidential k-nearest neighbor clustering. Knowl.-Based Syst. **200**, 105982 (2020)
19. Bache, K., Lichman, M.: UCI machine learning repository (2013)
20. Chang, H., Yeung, D.-Y.: Robust path-based spectral clustering. Pattern Recogn. **41**(1), 191–203 (2008)

Imperfect Labels with Belief Functions for Active Learning

Arthur Hoarau[✉], Arnaud Martin, Jean-Christophe Dubois,
and Yolande Le Gall

Univ Rennes, CNRS, IRISA, DRUID, Rennes, France
`arthur.hoarau@outlook.fr`

Abstract. Classification is used to predict classes by extracting information from labeled data. But sometimes the collected data is imperfect, as in crowdsourcing where users have partial knowledge and may answer with uncertainty or imprecision. This paper offers a way to deal with uncertain and imprecise labeled data using Dempster-Shafer theory and active learning. An evidential version of K-NN that classifies a new example by observing its neighbors was earlier introduced. We propose to couple this approach with active learning, where the model uses only a fraction of the labeled data, and to compare it with non-evidential models. A new computable parameter for EK-NN is introduced, allowing the model to be both compatible with imperfectly labeled data and equivalent to its first version in the case of perfectly labeled data. This method increases the complexity but provides a way to work with imperfectly labeled data with efficient results and reduced labeling costs when coupled with active learning. We have conducted tests on real data imperfectly labeled during crowdsourcing campaigns.

Keywords: Belief functions · Imperfect labels · Active learning

1 Introduction

In supervised classification, where the aim is to find the class of an observation, one still works largely with hard labels *i.e.* if a label exists for an observation, this label is defined in a categorical way. The labeling process often is carried out by humans [7,10]; without making any difference between a label given by someone who has hesitated for a long time and someone who has no doubt.

Using hard labels might be convenient for many machine learning and deep learning problems but is never completely representative of the reality. Imperfection, on the other hand, can help us fill in this lack of information. It can be represented by many criteria but only uncertainty and imprecision will be discussed in this paper. Ignorance is then derived from imprecision. Such information can be modeled with the theory of belief functions, introduced in [1,12]. This paper proposes to compare a non-evidential model with its evidential version and to observe the impact of imperfect labeling on classification. The widely used non-parametric model K-NN [5] will then be compared with EK-NN, an evidential version presented in [2–4]. A new parameter will be proposed for EK-NN to work with imperfectly labeled data and to maintain equivalence with the original model. In a context where data labeling is not only imperfect but also expensive, active learning [11] is particularly interesting. Indeed, a small volume of labeled data is sufficient to obtain good performance. Very little research has been done to couple belief functions with active learning. The main difference with [14] is the use of an imperfectly labeled data set instead of using only noise. The plan of the paper is as follows. Section 2 reviews the theory of belief functions, K-NN and EK-NN algorithms and then ends with an overview of active learning. Section 3 describes the proposed method, and the contribution concerning the parameters of EK-NN. A new credibilist dataset is also presented. Experiments on datasets composed of noisy real data and imperfectly labeled data are discussed in Sect. 4. Finally, Sect. 5 concludes the article.

2   Background

2.1   Reminder on Belief Functions

The theory of belief functions, also called Dempster-Shafer theory [1,12], is used in this study in order to model both imprecision and uncertainty.

One considers $\Omega = \{\omega_1, \ldots, \omega_M\}$ the frame of discernment for M exclusive and exhaustive hypotheses. The power set 2^Ω is the set of all subsets of Ω. A Basic Belief Assignment (BBA) is the belief that a source may have about the elements of the power set of Ω, this function assigns a mass to each element of this power set such that the sum of all masses is equal to 1.

$$m : 2^\Omega \rightarrow [0,1],$$
$$\sum_{A \in 2^\Omega} m(A) = 1. \tag{1}$$

Each subset $A \in 2^\Omega$ such as $m(A) > 0$ is called a *focal element* of m. If $m(A) = 1 - \delta$ and $m(\Omega) = \delta$ with $A \in 2^\Omega \backslash \emptyset$ and $\delta \in [0,1]$, m is called a *simple support mass function*.

A source might not be trustworthy, a discounting coefficient α is then introduced to transfer some belief into Ω, also called the ignorance, such that:

$$\begin{cases} m_\alpha(A) = \alpha m(A) , & \forall A \in 2^\Omega, A \neq \Omega, \\ m_\alpha(\Omega) = 1 - \alpha(1 - m(\Omega)), \end{cases} \tag{2}$$

where m_α is the new discounted mass.

The normalized conjunctive combination of Basic Belief Assignments (BBAs) m^j derived from $\mathcal{N}$ sources is given by:

$$\begin{cases} m(A) = \dfrac{1}{1-\kappa} \displaystyle\sum_{B_1 \cap \ldots \cap B_\mathcal{N} = A} \prod_{j=1}^{\mathcal{N}} m^j(B_j) & \text{if } A \neq \emptyset, \\ m(\emptyset) = 0, \end{cases} \tag{3}$$

with $\emptyset$ the empty set and:

$$\kappa = \sum_{B_1 \cap \ldots \cap B_{\mathcal{N}} = \emptyset} \prod_{j=1}^{\mathcal{N}} m^j(B_j). \tag{4}$$

On decision level, the pignistic probability $BetP$ helps decision making on singletons:

$$BetP(\omega) = \sum_{A \in 2^{\Omega}, \, \omega \in A} \frac{m(A)}{|A|}. \tag{5}$$

2.2 K-Nearest Neighbors

When dealing with perfectly labeled data, a non-parametric discrimination model known as the K-Nearest Neighbors (K-NN) is introduced in [6]. This is a popular classification model in which the label of an incoming sample is predicted according to its K nearest neighbors. The main drawback of this algorithm is that it assumes that there are close neighbors of the incoming sample. It is then proposed in [5] a distance-weighted K-NN where each neighbor is weighted according to its closeness to the incoming sample.

2.3   EK-NN

An evidential version of K-NN is introduced in [3], this Evidential K-Nearest Neighbors (EK-NN) uses belief functions to assign a label to a new sample. It is presented in the original paper as working with perfectly labeled data, but some work has subsequently been done to make this algorithm work with imperfectly labeled data. A version of EK-NN [4] is proposed with data labeled with possibility theory and [2] allows to calculate the parameters when dealing with imperfectly labeled data coupled with the theory of belief functions. However, it then loses the equivalence with the previous model in the particular case of perfectly labeled data.

2.4   Active Learning

Imperfect labels can be modeled by belief functions, and EK-NN can be a tool for learning from imperfectly labeled data, but we are also interested in reducing the number of labeled instances. Active learning [11] is a part of machine learning where the learner can choose which observation to label in order to work with only a fraction of the labeled data to reduce the labeling cost. Observations are called *Instances*, the act of requesting for the label of an instance is a *Query* and the entity giving its label to an instance is called the *Oracle*.

The difficulty is therefore to determine which instances should be labeled first. This process is called *Sampling*, the best known being Random Sampling where queries to the Oracle are made on random instances. Uncertainty Sampling, on the other hand, aims to perform a query on the sample for which the model is the least certain.

3   Classification of Imperfectly Labeled Data with EK-NN and Active Learning

Let $\mathcal{X}$ be a P features collection of N samples such as $\mathcal{X} = \{x^n = (x_1^n, \ldots, x_P^n) \mid n = 1, \ldots, N\}$, and Ω a set of M classes as $\Omega = \{\omega_1, \ldots, \omega_M\}$. Let $d^{s,i}$ be the distance between x^s and x^i with x^s an incoming sample to be classified using the information contained in the training set and x^i one of its K nearest neighbors. Classifying x^s means assigning it one class in Ω. Let Φ^s be the set of the K-nearest neighbors of x^s in $\mathcal{X}$ and m^i the BBA associated to x^i.

3.1   EK-NN for Imperfectly Labeled Data

In [3], the author introduces an equation of the BBA between an unclassified sample x^s and a neighbor x^i when it comes to imperfectly labeled data. This section results from the following proposition. If x^s is a sample to be classified, one's belief about the class of x^s induced by knowing that $x^i \in \Phi^s$ can be represented by a basic belief assignment $m^{s,i}$ deduced from m^i and $d^{s,i}$:

$$m^{s,i}(A) = \alpha_0 \phi(d^{s,i}) m^i(A),$$
$$m^{s,i}(\Omega) = 1 - \sum_{A \in 2^{\Omega} \setminus \Omega} m^{s,i}(A), \tag{6}$$

with ϕ a monotonically decreasing function and:

$$0 < \alpha_0 < 1,$$
$$\phi(0) = 1, \tag{7}$$
$$\lim_{d \to \infty} \phi(d) = 0.$$

As a decreasing function ϕ, [3] suggests to choose:

$$\phi(d) = e^{-\gamma d^{\beta}}, \tag{8}$$

with $\gamma > 0$ and $\beta \in \{1, 2, \ldots\}$ possibly fixed to a small value. When ϕ is first introduced, it depends on γ_q with ω_q the class of x^i and there are as many γ_q as different classes. As each point x^i no longer has a unique label since we are using imperfectly labeled data, γ_q cannot be calculated. This specificity forces the model to differ from a model using hard labeled data. It is discussed in Sect. 3.2.

Each BBA is now combined using (3):

$$\bar{m}^s(A) = \sum_{B_1 \cap \ldots \cap B_K = A} \prod_{x^i \in \Phi^s} \alpha_0 \phi(d^{s,i}) m^i(B_i), \forall A \in 2^{\Omega}. \tag{9}$$

Considering the closed world, the mass of the empty set must be forced to be null. The new normalized combined BBA, denoted m^s is obtained as:

$$\begin{cases} m^s(A) = \dfrac{1}{1 - \kappa} \bar{m}^s(A), & A \neq \emptyset, \\ m^s(\emptyset) = 0. \end{cases} \tag{10}$$

with κ the fusion inconsistency given at Eq. (4). Each new sample is then classified by maximizing the pignistic probability.

3.2   Parameters Optimization and γ_i-EKNN

This part deals with the calculation of the parameters of the model. They are: K, α_0, γ and β. The number K of nearest neighbors can be optimized identically to K-NN, using for example cross-validation. Furthermore, the use of variable size datasets within active learning has an impact on the optimal K. From preliminary experimental results (not given here), the parameter α_0 is set to 0.8, but might depend on the knowledge of the sources, which modifies the results very slightly; $\beta = 2$ gave satisfying results with little impact when changed. When dealing with imperfectly labeled data, the use of one γ parameter per class becomes meaningless, as there are no longer any classes but only samples with BBAs that more or less belong to a class. Several options have been proposed in [2–4] and compared in [2].

- In its first version [3], here renamed γ_q-EKNN, the model is presented with a γ_q parameter depending on the class ω_q of the neighbor x^i. The computable formula given for γ_q is $1/d_q^\beta$ with d_q the mean distance between two training vectors of the same class.
- A one γ version of the model [4], γ-EKNN, is later presented in a possibilistic environment and suitable for imperfectly labeled data. The use of a single γ parameter leads to the loss of equivalence with the initial model.
- Finally, a contextual-discounting based model [2] with M learnable γ_q is introduced, and will be referred to as CD-EKNN.

In this paper, we propose γ_i-EKNN, a version with K computable γ_i parameters, allowing to recover the equivalence, both theoretical and practical, with the original model in the case of perfectly labeled data. To maintain the equivalence with the model introduced in [3] when dealing with perfectly labeled data, the proposition of using one γ for each neighbor according to their similarity is made. When it comes to imperfectly labeled data, γ is calculated in relation to the distance with the other samples and according to its resemblance with Jousselme distance introduced in [8]. The closer we get to perfect labeling, the closer we get to one γ per class:

$$\gamma_i = \frac{1}{d_i^\beta}, \qquad d_i = \frac{\displaystyle\sum_{\nu=0}^{N}\sum_{\mu=0}^{N}(1 - d_j^{i,\nu})(1 - d_j^{i,\mu})d^{\nu,\mu}}{[\displaystyle\sum_{\nu=0}^{N}(1 - d_j^{i,\nu})]^2 - \displaystyle\sum_{\nu=0}^{N}(1 - d_j^{i,\nu})^2}, \tag{11}$$

with N the total number of samples and $d_j^{i,\nu}$ Jousselme's distance between m^i and m^ν.

In order to study the relevance of using imperfect labels by comparing an eviendential and a non-evidential model, γ_i-EKNN will be used, at the cost of its complexity, as it maintains equivalence with the orginal model.

3.3   Labeling with Uncertainty and Imprecision

In order to work with imperfectly labeled data, we obtained a dataset from crowdsourcing campaigns using the model and the materials developed in [13].

Credal Bird-10 is a dataset composed of 200 pictures of birds imperfectly labeled. Each of these images belongs to a class corresponding to one of the 10 species of birds evenly distributed on the dataset. During crowdsourcing campaigns, the pictures are displayed and participants can choose multiple corresponding classes as well as the belief they have in their responses. The resulting dataset is a combination of pictures associated with BBAs, refer to [13] for construction of the dataset. When using non-evidential models, the class maximizing the pignistic probability is then chosen as the perfect label. Two datasets have been obtained, one on the Irisa laboratory (*Credal Bird-10 irisa*) and one on a non-specific crowd of paid contributors (*Credal Bird-10 public*).

Example for a picture of red/green/blue pixels corresponding to a marsh tit with y_1 the vector on 2^{Ω} which is the BBA describing its imperfect label:

	y_1
$m(\{Marsh\ tit\})$	0.2
$m(\{Marsh\ tit,\ Great\ tit\})$	0.5
$m(\Omega)$	0.3

4   Experiments

The following section presents several procedures for implementing the method. The interest is to show that allowing a source to provide imperfectly labeled data may be more realistic than perfectly labeled data and therefore yield better results. For each experiment 20% of the dataset is used as a test set and the remaining as a training set. The experiment of Sect. 4.1 is a brief comparison between the approaches discussed in Sect. 3.2. The experiments given in Sects. 4.2 and 4.3 are coupled to active learning in order to avoid expensive labeling. A comparison between K-NN and its evidential version is made to study the relevance of using imperfectly labeled data, other models are added for a general overview.

4.1   Different Approaches for γ Parameter

A comparison is made in Table 1 between K-NN and the approaches presented in Sect. 3.2. They are used with a K nearest neighbors value equal to 7 and the result is a mean accuracy over 100 iterations. The distance weighted K-NN is compared to the original version γ_q-EKNN, to the unique gamma γ-EKNN version using $\gamma = 1/d^{\beta}$, with d the mean distance between two training vectors, and to the proposed γ_i-EKNN. Datasets are split into two categories: perfectly labeled (Iris, Wine and Breast Cancer)[1] and imperfectly labeled (Credal Bird-10 public). The 95% confidence interval[2] is also given.

As it can be observed in Table 1, there is an equivalence between γ_q-EKNN and the proposed γ_i-EKNN on perfectly labeled datasets (Iris, Wine and Breast Cancer); this equivalence is discussed in Sect. 3.2. When dealing with imperfectly labeled data, the same γ_i-EKNN model is also competitive. Letting sources label imperfectly gave better results.

Table 1. Mean accuracy over 100 iterations on perfectly labeled (Iris, Wine, Breast Cancer) and imperfectly labeled (Credal Bird-10 public) datasets.

Dataset	K-NN	γ_q-EKNN	γ-EKNN	γ_i-EKNN
Iris	**0.965** $\pm$ 0.006	0.963 $\pm$ 0.006	0.964 $\pm$ 0.006	0.963 $\pm$ 0.006
Wine	**0.737** $\pm$ 0.013	0.696 $\pm$ 0.012	0.704 $\pm$ 0.012	0.696 $\pm$ 0.012
Breast Cancer	0.927 $\pm$ 0.004	**0.928** $\pm$ 0.004	**0.928** $\pm$ 0.004	**0.928** $\pm$ 0.004
Credal Bird-10 public	0.383 $\pm$ 0.015	0.389 $\pm$ 0.014	0.411 $\pm$ 0.014	**0.412** $\pm$ 0.014

4.2 Experiment on Noised Real World Datasets

In this experiment both Iris and Wine datasets have been noised as this is a common point in the literature. A noise parameter $\epsilon = [0, 1]$ is defined and the observations are randomly selected in order to have a proportion of noisy labels equal to ϵ. For each selected observation another singleton is randomly selected and a random mass assigned. The remaining mass is evenly distributed among all other elements. For non-evidential classifiers, the singleton which maximizes the pignistic probability of this new mass is the new label. The Iris and Wine datasets were altered with ϵ equal to 0.5. In this experiment, the mean accuracy of different models is compared using active learning. The models are as follows: K-NN based on a distance weight with 7 nearest neighbors, Logistic Regression with newton-cg for optimization and Random Forest, all used with the scikit-learn default parameters [9]. They are compared to γ_i-EKNN presented in this paper using 7 nearest neighbors. Both experiments used 8 randomly labeled instances (there must be more than K labeled instances) and 20 active learning queries were performed according to uncertainty sampling. The mean accuracy is calculated over 100 iterations.

Figure 1 shows that γ_i-EKNN achieves a mean accuracy of about 0.9 on Iris dataset with only 28 labeled instances, a 30% performance improvement over K-NN due to less significant alteration of the real labels. The distance between the mean accuracy of γ_i-EKNN and K-NN is also greater as the queries number

---

[1] https://archive.ics.uci.edu.
[2] Formula: $[\bar{x} - 1{,}96\frac{S}{\sqrt{n}}; \bar{x} + 1{,}96\frac{S}{\sqrt{n}}]$, with n the size of the sample, $\bar{x}$ its mean and S the standard deviation of the serie. This formula is used because it is a mean over 100 experiments and not a single proportion.

increases, which means that the model manages to select better instances to label while using the same uncertainty sampling. The same Fig. 1 shows less optimistic results on the Wine dataset, but still with a dominance of γ_i-EKNN over its non-evidential version. One must be careful with the results, even if the noisy data are distributed in the same way, the labels used for the non-evidential classifiers do not contain the same information as the labels used for γ_i-EKNN, making the comparison more difficult. Apart from the noise, one of the objectives of the paper is to find out whether by adding information during the labeling phase, interesting results can be obtained with a low labeling cost, which leads to the experiment presented in Sect. 4.3.

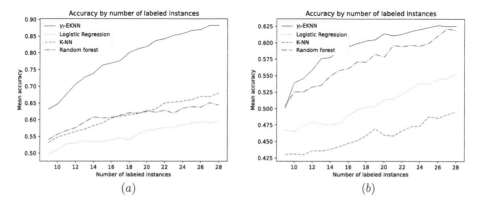

Fig. 1. Mean accuracy by number of labeled instances with 50% noise, on the Iris (a) and Wine (b) datasets.

4.3   Experiment on Imperfectly Labeled Datasets

So far, perfectly labeled datasets have been used for comparison. In this section, a procedure is proposed that can this time be fully compared to non-evidential methods as the labels are unchanged but fundamentally imperfect. To show a real application of the proposed method, we need to train on uncertain and imprecise labels. With the imperfectly labeled dataset of 10 classes introduced in Sect. 3.3 and dimensionally reduced to 512 variables on $\mathcal{X}$, the model is compared to its non-evidential version. The same models and active learning steps as in the Experiment 4.2 are used. Differences are present in the datasets, imperfectly labeled by the contributors.

Figure 2 represents the mean accuracy of 100 iterations on both *Credal Bird-10 irisa* and *Credal Bird-10 public* datasets. With 28 labeled instances, EK-NN performs better than its non-evidential version, with around 0.44 accuracy on *Credal Bird-10 irisa* and 0.48 on *Credal Bird-10 public* compared to 0.41 and 0.44 for K-NN respectively. The comparison between the two datasets also shows that the results may vary greatly depending on the labels, even with the same models. Here, two different populations labeled the same pictures, members of a laboratory and crowdsourcing contributors. This difference produces changes in the results with, in all cases, better results for EK-NN.

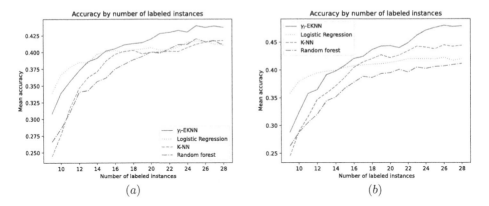

(a) (b)

Fig. 2. Mean accuracy by number of labeled instances, on the Credal Bird-10 irisa (a) and Credal Bird-10 public (b) datasets.

5   Conclusion

This study presents a model for efficient learning from a small amount of data derived from imperfect human contributions. It is proposed to couple the theory of belief functions, to model the uncertainty and imprecision of the data, with an active learning algorithm using only a fraction of the labeled data. In particular, our work focuses on the labeling method and how information can be added to allow the learning phase to work more efficiently and at lower cost. A version of the evidential K-nearest neighbors model is proposed, offering a new computation for the parameter γ and allowing to recover an equivalence with the original model in the case of imperfectly labeled data.

To validate this approach, experiments were first conducted on noisy data sets (Sect. 4.2). Very optimistic results are obtained with good performances of the credibility classifiers. However, as the nature of the noise makes it difficult to compare a credibility classifier with its classical version, two new imperfectly labeled datasets on bird species were produced via crowdsourcing to test the model on real data. Few labeled images are used in Sect. 4.3 for decent performance. The quality of the labeling, which depends on the oracle and the model used to represent the imperfection, has a strong influence on the final performance, and can make the results vary more significantly by improving the quality of the labels rather than the quality of the model itself. In future work, we plan to study how to maximize the quality of the imperfection contained in the labels, by working on its modelisation or on the interface allowing even an inexperienced user to give a relevant uncertain and imprecise answer. Improvement could also be done with active learning, taking into account at the sampling step, that the model can give an evidential answer.

References

1. Dempster, A.P.: Upper and lower probabilities induced by a multivalued mapping. Ann. Math. Stat. **38**(2), 325–339 (1967)
2. Denoeux, T., Kanjanatarakul, O., Sriboonchitta, S.: A new evidential k-nearest neighbor rule based on contextual discounting with partially supervised learning. Int. J. Approximate Reasoning **113**, 287–302 (2019)
3. Denæux, T.: A k-nearest neighbor classification rule based on dempster-shafer theory. IEEE Trans. Syst. Man Cybern. **219**(1995)
4. Denæux, T., Zouhal, L.: Handling possibilistic labels in pattern classification using evidential reasoning. Fuzzy Sets Syst. **122**, 409–424 (2001)
5. Dudani, S.A.: The distance-weighted k-nearest-neighbor rule. IEEE Trans. Syst. Man Cybern. SMC **6**(4), 325–327 (1976)
6. Fix, E., Hodges, J.L.: Discriminatory analysis. nonparametric discrimination: Consistency properties. Technical report 4. USAF, School of Aviation Medicine, Randolph Field (1951)
7. Fredriksson, T., Mattos, D.I., Bosch, J., Olsson, H.H.: Data labeling: an empirical investigation into industrial challenges and mitigation strategies. In: Morisio, M., Torchiano, M., Jedlitschka, A. (eds.) Product-Focused Software Process Improvement, pp. 202–216. Springer International Publishing, Cham (2020). https://doi.org/10.1007/978-3-030-64148-1_13
8. Jousselme, A.L., Grenier, D.: Éloi Bossé: a new distance between two bodies of evidence. Inf. Fusion **2**(2), 91–101 (2001)
9. Pedregosa, F., et al.: Scikit-learn: machine learning in python. J. Mach. Learn. Res. **12**, 2825–2830 (2011)
10. Roh, Y., Heo, G., Whang, S.E.: A survey on data collection for machine learning: a big data - AI integration perspective. IEEE Trans. Knowl. Data Eng. **33**(4), 1328–1347 (2021)
11. Settles, B.: Active learning literature survey. Computer Sciences Technical Report 1648, University of Wisconsin-Madison (2009)
12. Shafer, G.: A Mathematical Theory of Evidence. Princeton University Press, Princeton (1976)
13. Thierry, C., Hoarau, A., Martin, A., Dubois, J.C., Le Gall, Y.: Real bird dataset with imprecise and uncertain values. In: 7th International Conference on Belief Functions (2022)
14. Zhu, D., Martin, A., Le Gall, Y., Dubois, J.C., Lemaire, V.: Evidential nearest neighbours in active learning. In: Worksop on Interactive Adaptive Learning (IAL) - ECML-PKDD. Bilbao, Spain (2021)

Machine Learning and Pattern Recognition

An Evidential Neural Network Model for Regression Based on Random Fuzzy Numbers

Thierry Denœux[1,2]([✉]) [iD]

[1] Université de technologie de Compiègne, CNRS, UMR, 7253 Heudiasyc, France
[2] Institut Universitaire de France, Paris, France
tdenoeux@utc.fr

Abstract. We introduce a distance-based neural network model for regression, in which prediction uncertainty is quantified by a belief function on the real line. The model interprets the distances of the input vector to prototypes as pieces of evidence represented by Gaussian random fuzzy numbers (GRFN's) and combined by the generalized product intersection rule, an operator that extends Dempster's rule to random fuzzy sets. The network output is a GRFN that can be summarized by three numbers characterizing the most plausible predicted value, variability around this value, and epistemic uncertainty. Experiments with real datasets demonstrate the very good performance of the method as compared to state-of-the-art evidential and statistical learning algorithms.

Keywords: Evidence theory · Dempster-shafer theory · Belief functions · Machine learning · Random fuzzy sets

1 Introduction

The Dempster-Shafer (DS) theory of evidence is a general mathematical framework for reasoning and making decisions based on imprecise and uncertain information [7,14]. This framework is based on the representation of independent pieces of evidence by belief functions, and on their combination by a conjunctive operator called Dempster's rule. The greater number of degrees of freedom offered by belief functions, as compared to probabilities, makes it possible to distinguish between two situations of high uncertainty: equally supported hypotheses on the one hand, and total lack of support on the other hand, the latter situation characterizing complete ignorance.

In machine learning, DS theory has been mainly applied to classification and clustering tasks, in which the set of elementary hypotheses (or frame of discernment) is finite. In particular, several methods have been proposed to learn *evidential classifiers*, i.e., classifiers representing prediction uncertainty by belief functions. In the first such classifier, the evidential K-nearest neighbor (EKNN) rule [3], each neighbor of a feature vector to be classified is represented by a simple mass function, and the mass functions from the K nearest neighbors are

© The Author(s), under exclusive license to Springer Nature Switzerland AG 2022
S. Le Hégarat-Mascle et al. (Eds.): BELIEF 2022, LNAI 13506, pp. 57–66, 2022.
https://doi.org/10.1007/978-3-031-17801-6_6

combined by Dempster's rule. The evidential neural network (ENN) introduced in [5] is based on the same principle, the elementary mass functions being computed based on distances to prototypes. Recently, this distance-based approach has been extended to deep networks [9,15,16], by adding a DS layer to a deep architecture; output mass functions are then computed based on distances to prototypes in the space of high-level features extracted by convolutional layers.

Applying DS theory to regression is more challenging, because in regression tasks the frame of discernment is typically the real line or a real interval, whereas most tools of DS theory have been developed for finite frames. This difficulty can be circumvented by discretizing the response variable, as proposed in [4], in which a neural network model for regression directly extending the ENN model was introduced. The output of this model is a mass function with disjoint intervals and the whole frame as focal sets. Another approach, introduced in [12] [13], is to modify the EKNN rule by combining simple mass functions focussed either on a single real number, or on a fuzzy number in the case of learning data with fuzzy response variable. The output mass function then has a finite number of crisp or fuzzy focal sets. This method, called EVREG, was shown in [13] to yield good results in the case of crisp data, and to efficiently handle uncertain response data (such as provided by an unreliable sensor). However, the K nearest neighbor approach breaks down as dimension grows, and it cannot compete with state-of-the-art regression methods.

In this paper, we propose another evidential neural network model for regression inspired from the ENN model. This new model, called ENNreg, uses the formalism of Gaussian random fuzzy numbers (GRFN's) recently introduced in [8]. A GRFN is a random fuzzy subset of the real line, which can be described as a Gaussian possibility distribution whose mode is a Gaussian random variable. GRFN's induce belief functions and can be combined using a generalization of Dempster's rule. In ENNreg, GRFN's associated to each of the prototypes are combined to yield an output GRFN quantifying prediction uncertainty.

The rest of this paper is organized as follows. The general framework of epistemic random fuzzy sets and the GRFN model are first recalled in Sect. 2. The proposed ENNreg model is then introduced in Sect. 3. Experimental results are reported in Sect. 4, and Sect. 5 concludes the paper.

2   Epistemic Random Fuzzy Sets

The theory of epistemic random fuzzy sets (ERFS) was introduced in [6] and [8] as a general framework encompassing both DS theory and possibility theory. We first recall some important definitions in Sect. 2.1. Gaussian random fuzzy numbers, a parametric family of ERFS's on the real line are then described in Sect. 2.2.

2.1   General Framework

Let $(\Omega, \Sigma_\Omega, P)$ be a probability space and let (Θ, Σ_Θ) be a measurable space. Let $\widetilde{X}$ be a mapping from Ω to the set $[0,1]^\Theta$ of fuzzy subsets of Θ. For any

$\alpha \in [0,1]$, let ${}^{\alpha}\widetilde{X}$ be the mapping from Ω to 2^{Θ} defined as

$$^{\alpha}\widetilde{X}(\omega) = {}^{\alpha}[\widetilde{X}(\omega)],$$

where ${}^{\alpha}[\widetilde{X}(\omega)]$ is the weak α-cut of $\widetilde{X}(\omega)$. If for any $\alpha \in [0,1]$, ${}^{\alpha}\widetilde{X}$ is $\Sigma_{\Omega} - \Sigma_{\Theta}$ strongly measurable [11], the tuple $(\Omega, \Sigma_{\Omega}, P, \Theta, \Sigma_{\Theta}, \widetilde{X})$ is said to be a *random fuzzy set* (also called a *fuzzy random variable*) [2]. When there is no possible confusion about the domain and codomain, we will refer to mapping $\widetilde{X}$ itself as a random fuzzy set.

In ERFS theory, random fuzzy sets represent unreliable and fuzzy evidence. In this model, we see Ω as a *set of interpretations* of a piece of evidence about a variable $\boldsymbol{\theta}$ taking values in Θ. If interpretation $\omega \in \Omega$ holds, we know that "$\boldsymbol{\theta}$ is $\widetilde{X}(\omega)$", i.e., $\boldsymbol{\theta}$ is constrained by the possibility distribution defined by $\widetilde{X}(\omega)$. We qualify such random fuzzy sets as *epistemic*, because they encode a state of knowledge about some variable $\boldsymbol{\theta}$. If all images $\widetilde{X}(\omega)$ are crisp, then $\widetilde{X}$ defines an ordinary random set. If mapping $\widetilde{X}$ is constant, then it is equivalent to specifying a unique fuzzy subset of Θ, which defines a possibility distribution.

Belief and plausibility functions. In the following, we will assume that random fuzzy set $\widetilde{X}$ is *normalized*, i.e., that it verifies the following conditions: (1) For all $\omega \in \Omega$, $\widetilde{X}(\omega)$ is either the empty set, or a normal fuzzy set, i.e., $\mathsf{hgt}(\widetilde{X}(\omega)) = \sup_{\theta \in \Theta} \widetilde{X}(\omega)(\theta) \in \{0,1\}$; (2) $P(\{\omega \in \Omega : \widetilde{X}(\omega) = \emptyset\}) = 0$. For any $\omega \in \Omega$, let $\Pi_{\widetilde{X}}(\cdot \mid \omega)$ be the possibility measure on Θ induced by $\widetilde{X}(\omega)$:

$$\Pi_{\widetilde{X}}(B \mid \omega) = \sup_{\theta \in B} \widetilde{X}(\omega)(\theta), \tag{1}$$

and let $N_{\widetilde{X}}(\cdot \mid \omega)$ be the dual necessity measure:

$$N_{\widetilde{X}}(B \mid \omega) = \begin{cases} 1 - \Pi_{\widetilde{X}}(B^c \mid \omega) & \text{if } \widetilde{X}(\omega) \neq \emptyset \\ 0 & \text{otherwise,} \end{cases}$$

where B^c denotes the complement of B. The mappings $Bel_{\widetilde{X}}$ and $Pl_{\widetilde{X}}$ from Σ_{Θ} to $[0,1]$ defined as

$$Bel_{\widetilde{X}}(B) = \int_{\Omega} N_{\widetilde{X}}(B \mid \omega) dP(\omega) \tag{2}$$

and

$$Pl_{\widetilde{X}}(B) = \int_{\Omega} \Pi_{\widetilde{X}}(B \mid \omega) dP(\omega) = 1 - Bel_{\widetilde{X}}(B^c) \tag{3}$$

are, respectively, belief and plausibility functions.

Combination. Consider two epistemic random fuzzy sets $(\Omega_i, \Sigma_i, P_i, \Theta, \Sigma_{\Theta}, \widetilde{X}_i)$, $i = 1, 2$, encoding independent pieces of evidence. The independence assumption means here that the relevant probability measure on the joint measurable space $(\Omega_1 \times \Omega_2, \Sigma_1 \otimes \Sigma_2)$ is the product measure $P_1 \times P_2$. If interpretations $\omega_1 \in \Omega_1$ and

$\omega_2 \in \Omega_2$ both hold, we know that "$\boldsymbol{\theta}$ is $\widetilde{X}_1(\omega_1)$" and "$\boldsymbol{\theta}$ is $\widetilde{X}_2(\omega_2)$". It is then natural to combine the fuzzy sets $\widetilde{X}_1(\omega_1)$ and $\widetilde{X}_2(\omega_2)$ by an intersection operator. As discussed in [6], the normalized product intersection operator $\odot$ is suitable for combining fuzzy information from independent sources and it is associative. We thus consider the mapping $\widetilde{X}_\odot(\omega_1, \omega_2) = \widetilde{X}_1(\omega_1) \odot \widetilde{X}_2(\omega_2)$, assumed to be $\Sigma_1 \otimes \Sigma_2$-$\Sigma_\Theta$ strongly measurable.

If $\mathsf{hgt}(\widetilde{X}_1(\omega_1)\widetilde{X}_2(\omega_2)) = 0$, the two interpretations ω_1 and ω_2 are inconsistent and they must be discarded. If $\mathsf{hgt}(\widetilde{X}_1(\omega_1)\widetilde{X}_2(\omega_2)) = 1$, the two interpretations are fully consistent. If $0 < \mathsf{hgt}(\widetilde{X}_1(\omega_1)\widetilde{X}_2(\omega_2)) < 1$, ω_1 and ω_2 are *partially consistent*. The *soft normalization* proposed in [8] consists in conditioning the product probability $P_1 \times P_2$ by the fuzzy subset $\widetilde{\Theta}^*$ of consistent pairs of interpretations, with membership function $\widetilde{\Theta}^*(\omega_1, \omega_2) = \mathsf{hgt}\left(\widetilde{X}_1(\omega_1) \cdot \widetilde{X}_2(\omega_2)\right)$. Alternatively, we can use a *hard normalization* operation, which consists in conditioning $P_1 \times P_2$ by the crisp set Θ^* of interpretations that are not fully inconsistent, described formally as

$$\Theta^* = \{(\omega_1, \omega_2) \in \Omega_1 \times \Omega_2 : \mathsf{hgt}(\widetilde{X}_1(\omega_1)\widetilde{X}_2(\omega_2)) > 0\}.$$

Both combination rules, with soft or hard normalization, are commutative and associative, and both of them generalize Dempster's rule. In the following, we will use hard normalization as it leads to simpler calculations. This operation will be referred to as the *generalized product-intersection rule with hard normalization*, and the corresponding operator will be denoted by $\boxplus$.

2.2   Gaussian Random Fuzzy Numbers

A *Gaussian Fuzzy Number* (GFN) is a fuzzy subset of $\mathbb{R}$ with membership function

$$\varphi(x; m, h) = \exp\left(-\frac{h}{2}(x - m)^2\right),$$

where $m \in \mathbb{R}$ is the mode and $h \in [0, +\infty]$ is the precision. Such a fuzzy number will be denoted by $\mathsf{GFN}(m, h)$. The normalized product intersection of two GFN's $\mathsf{GFN}(m_1, h_1)$ and $\mathsf{GFN}(m_2, h_2)$ is a GFN $\mathsf{GFN}(m_{12}, h_{12}) = \mathsf{GFN}(m_1, h_1) \odot \mathsf{GFN}(m_2, h_2)$, with $m_{12} = (h_1 m_1 + h_2 m_2)/(h_1 + h_2)$ and $h_{12} = h_1 + h_2$.

Let $(\Omega, \Sigma_\Omega, P)$ be a probability space and let $M : \Omega \to \mathbb{R}$ be a Gaussian random variable (GRV) with mean μ and variance σ^2. The random fuzzy set $\widetilde{X} : \Omega \to [0, 1]^{\mathbb{R}}$ defined as

$$\widetilde{X}(\omega) = \mathsf{GFN}(M(\omega), h)$$

is called a *Gaussian random fuzzy number* (GRFN) with mean μ, variance σ^2 and precision h, which we write $\widetilde{X} \sim \widetilde{N}(\mu, \sigma^2, h)$. A GRFN is, thus, defined by a location parameter μ, and two parameters h and σ^2 corresponding, respectively, to possibilistic and probabilistic uncertainty.

A GRFN can be seen either as a generalized GRV with fuzzy mean, or as a generalized GFN with random mode. In particular, a GRFN $\widetilde{X}$ with infinite precision $h = +\infty$ is equivalent to a GRV with mean μ and variance σ^2, which we can write: $\widetilde{N}(\mu, \sigma^2, +\infty) = N(\mu, \sigma^2)$. If $\sigma^2 = 0$, M is a constant random variable taking value μ, and $\widetilde{X}$ is a possibilistic variable with possibility distribution $\mathrm{GFN}(\mu, h)$. Another case of interest is that where $h = 0$, in which case $\widetilde{X}(\omega)(x) = 1$ for all $\omega \in \Omega$ and each $x \in \mathbb{R}$, and the belief function induced by $\widetilde{X}$ is vacuous.

As shown in [8], the plausibility and belief of any real interval $[x, y]$ are given by the following formulas:

$$Pl_{\widetilde{X}}([x, y]) = \Phi\left(\frac{y - \mu}{\sigma}\right) - \Phi\left(\frac{x - \mu}{\sigma}\right) +$$
$$pl_{\widetilde{X}}(x)\Phi\left(\frac{x - \mu}{\sigma\sqrt{h\sigma^2 + 1}}\right) + pl_{\widetilde{X}}(y)\left[1 - \Phi\left(\frac{y - \mu}{\sigma\sqrt{h\sigma^2 + 1}}\right)\right], \quad (4a)$$

and

$$Bel_{\widetilde{X}}([x, y]) = Pl_{\widetilde{X}}([x, y]) - pl_{\widetilde{X}}(x)\Phi\left(\frac{(x + y)/2 - \mu}{\sigma\sqrt{h\sigma^2 + 1}}\right) -$$
$$pl_{\widetilde{X}}(y)\left[1 - \Phi\left(\frac{(x + y)/2 - \mu}{\sigma\sqrt{h\sigma^2 + 1}}\right)\right], \quad (4b)$$

where Φ is the standard normal cumulative distribution function (cdf), and

$$pl_{\widetilde{X}}(x) = \frac{1}{\sqrt{1 + h\sigma^2}}\exp\left(-\frac{h(x - \mu)^2}{2(1 + h\sigma^2)}\right) \quad (4c)$$

is the contour function. Denoting by $\underline{F}_{\widetilde{X}}(x) = Bel_{\widetilde{X}}((-\infty, x])$ and $\overline{F}_{\widetilde{X}}(x) = Pl_{\widetilde{X}}((-\infty, x])$, respectively, the lower and upper cdf's of $\widetilde{X}$, its lower and upper expectations are

$$\mathbb{E}_*(\widetilde{X}) = \int_{-\infty}^{+\infty} x\, d\overline{F}_{\widetilde{X}}(x) = \mu - \sqrt{\frac{\pi}{2h}}$$

and

$$\mathbb{E}^*(\widetilde{X}) = \int_{-\infty}^{+\infty} x\, d\underline{F}_{\widetilde{X}}(x) = \mu + \sqrt{\frac{\pi}{2h}}.$$

The usefulness of GRFN's as a model of uncertain information about a real quantity arises from the fact that GRFN's can easily be combined by the generalized product-intersection rule, with soft or hard normalization [8]. Here, we only consider hard normalization, which is used in the proposed regression model described in Sect. 3. Given two GRFN's $\widetilde{X}_1 \sim \widetilde{N}(\mu_1, \sigma_1^2, h_1)$ and $\widetilde{X}_2 \sim \widetilde{N}(\mu_2, \sigma_2^2, h_2)$, we have $\widetilde{X}_1 \boxplus \widetilde{X}_2 \sim \widetilde{N}(\mu_{12}, \sigma_{12}^2, h_{12})$, with

$$\mu_{12} = \frac{h_1\mu_1 + h_2\mu_2}{h_1 + h_2}, \quad \sigma_{12}^2 = \frac{h_1^2\sigma_1^2 + h_2^2\sigma_2^2}{(h_1 + h_2)^2}, \quad \text{and} \quad h_{12} = h_1 + h_2.$$

3   Neural Network Model

The ENN (Evidential Neural Network) model introduced in [5] for classification is based on prototypes in input space, each one having degrees of membership to the different classes. In this model, each prototype provides a piece of evidence regarding the class of a test instance. This evidence is represented by a DS mass function defined from the class membership degrees of the prototype and the distance from the input vector. The mass functions induced by the prototypes are then combined by Dempster's rule. Here, we propose a similar model for regression, called ENNreg. The propagation equations and the loss function are given, respectively, in Sects. 3.1 and 3.2.

3.1   Propagation Equations

We consider J prototypes $\boldsymbol{w}_j \in \mathbb{R}^p$, $j = 1, \ldots, J$, where p is the dimension of the input space. The activation of prototype j for input $\boldsymbol{x}$ is

$$a_j(\boldsymbol{x}) = \exp(-\gamma_j^2 \|\boldsymbol{x} - \boldsymbol{w}_j\|^2),$$

where γ_j is a positive scale parameter. The evidence of prototype j is represented by a GRFN $\widetilde{Y}_j(\boldsymbol{x}) \sim \widetilde{N}(\mu_j(\boldsymbol{x}), \sigma_j^2, a_j(\boldsymbol{x})h_j)$, where σ_j^2 and h_j are variance and precision parameters for prototype j; the mean $\mu_j(\boldsymbol{x})$ is defined as $\mu_j(\boldsymbol{x}) = \boldsymbol{\beta}_j^T \boldsymbol{x} + \alpha_j$, where $\boldsymbol{\beta}_j$ is a p-dimensional vector of coefficients, and α_j is a scalar parameter. The vector ψ_j of parameters associated to prototype j is, thus, $\psi_j = (\boldsymbol{w}_j, \gamma_j, \boldsymbol{\beta}_j, \alpha_j, \sigma_j^2, h_j)$.

The output $\widetilde{Y}(\boldsymbol{x})$ for input $\boldsymbol{x}$ is computed by combining the GRFN's $\widetilde{Y}_j(\boldsymbol{x})$, $j = 1, \ldots, J$ induced by the J prototypes using the $\boxplus$ operator. It is a GRFN $\widetilde{Y}(\boldsymbol{x}) \sim \widetilde{N}(\mu(\boldsymbol{x}), \sigma^2(\boldsymbol{x}), h(\boldsymbol{x}))$, with

$$\mu(\boldsymbol{x}) = \frac{\sum_{j=1}^{J} a_j(\boldsymbol{x})h_j\mu_j(\boldsymbol{x})}{\sum_{k=1}^{J} a_k(\boldsymbol{x})h_k}, \quad \sigma^2(\boldsymbol{x}) = \frac{\sum_{j=1}^{J} a_j^2(\boldsymbol{x})h_j^2\sigma_j^2}{\left(\sum_{k=1}^{J} a_k(\boldsymbol{x})h_k\right)^2},$$

and $h(\boldsymbol{x}) = \sum_{j=1}^{J} a_j(\boldsymbol{x})h_j$. Some special cases are of interest:

1. If $\boldsymbol{\beta}_j = 0$ for all j, then $\mu_j(\boldsymbol{x}) = \alpha_j$, and $\mu(\boldsymbol{x})$ is identical to the output of a radial basis function (RBF) neural network with hidden-to-output weights $h_j\alpha_j$ and normalized outputs;
2. If $J = 1$ and $\gamma_1 = 0$, $\mu(\boldsymbol{x}) = \boldsymbol{\beta}_1^T \boldsymbol{x} + \alpha_1$, $\sigma^2(\boldsymbol{x}) = h_1\sigma_1^2$ and $h(\boldsymbol{x}) = h_1$. We then have a linear model with constant variance.

3.2   Loss Function

We want to fit the model described in the previous section in such a way that the observed values of the response variable have a high degree of belief and a high plausibility. Because the degree of belief is zero for a single real value, we consider a small interval $[y - \epsilon, y + \epsilon]$, denoted as $[y]_\epsilon$, around each observed value y, and we define the following loss function:

$$\mathcal{L}_{\lambda,\epsilon}(y, \widetilde{Y}) = -\lambda \ln Bel_{\widetilde{Y}}([y]_\epsilon) - (1 - \lambda) \ln Pl_{\widetilde{Y}}([y]_\epsilon), \tag{5}$$

where y is the true response, $\widetilde{Y} \sim \widetilde{N}(\mu, \sigma^2, h)$, and $\lambda \in [0, 1]$ is a hyperparameter. This loss function is minimal for a perfect forecast, such that $\mu = y$, $h = +\infty$ and $\sigma^2 \to 0$. With a fixed variance σ^2, the term $Bel_{\widetilde{Y}}([y]_\epsilon)$ is maximized for $\mu = y$ and $h = +\infty$, while the term $Pl_{\widetilde{Y}}([y]_\epsilon)$ is maximized for $h = 0$, whatever μ. Hyperparameter λ thus determines the precision of the predictions. We can also remark that, when ϵ is small, we have, for a probabilistic GRFN $\widetilde{Y} \sim \widetilde{N}(\mu, \sigma^2, +\infty)$, $\mathcal{L}_{\lambda,\epsilon}(y, \widetilde{Y}) \approx -\ln \phi(\frac{y-\mu}{\sigma}) - \ln \epsilon$, where ϕ denotes the standard normal probability density function; loss function (5) then becomes equivalent to minus the log-likelihood.

Using a training set $\mathcal{T} = \{(\boldsymbol{x}_1, y_1), \ldots, (\boldsymbol{x}_n, y_n)\}$, we minimize the regularized average loss

$$C(\boldsymbol{\Psi}) = \frac{1}{n} \sum_{i=1}^{n} \mathcal{L}_{\lambda,\epsilon}(y_i, \widetilde{Y}(\boldsymbol{x}_i)) + \frac{\xi}{J} \sum_{j=1}^{J} h_j,$$

where $\boldsymbol{\Psi} = (\psi_1, \ldots, \psi_J)$ is the vector of all parameters, and ξ is a regularization coefficient. We note that setting $h_j = 0$ amounts to removing prototype j, as the GRFN $\widetilde{Y}_j(\boldsymbol{x})$ becomes vacuous for any input $\boldsymbol{x}$. Increasing ξ results in more cautious predictions and a more parsimonious model.

4 Experimental Results

We first give an illustrative example in Sect. 4.1. Results from a comparative experiment are then reported in Sect. 4.2.

4.1 Illustrative Example

As an illustrative example, we consider data with $p = 1$ input from the following distribution:

1. The input X is drawn from a mixture of two uniform distributions on $[-3, -1]$ and $[1, 4]$: $X \sim 0.5 \, \mathsf{Unif}(-3, -1) + 0.5 \, \mathsf{Unif}(1, 4)$;
2. Given $X = x$, $Y = x + \sin 3x + \eta$, where η is a Gaussian random variable with zero mean and variance $\sigma^2 = 0.01$ if $x < 0$ and $\sigma^2 = 0.3$ otherwise.

We generated a learning set of size $n = 200$ and a test set of size $n_t = 1000$ from that distribution. We trained a network with $J = 10$ prototypes initialized by the k-means algorithm, with $\lambda = 0.95$, $\xi = 10^{-3}$ and $\epsilon = 0.01$. Figure 1 shows the expected values $\mu(x)$, together with the lower and upper expectations $\mathbb{E}_*(\widetilde{Y}(x))$, $\mathbb{E}^*(\widetilde{Y}(x))$, as well as prediction intervals of the form

$$[\overline{F}_{\widetilde{Y}(x)}^{-1}(\alpha/2), \underline{F}_{\widetilde{Y}(x)}^{-1}(1 - \alpha/2)], \tag{6}$$

for $1 - \alpha \in \{0.5, 0.9, 0.99\}$. The estimated coverage probabilities of these intervals are, respectively, 0.76, 0.96 and 0.997, which suggests that expression (6) provides conservative prediction intervals. As can be seen from Fig. 1, the prediction intervals are wider when the variance of the data is larger, and in regions of the input space where there is no data.

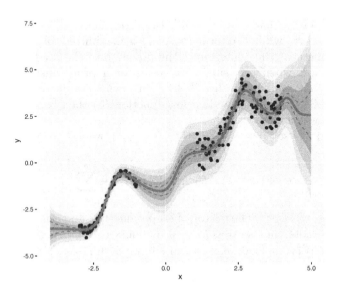

Fig. 1. Learning data and predictions for the illustrative example. The red solid and broken lines correspond, respectively, to the expected values $\mu(x)$, and to the lower and upper expectations $\mathbb{E}_*(\widetilde{Y}(x))$, $\mathbb{E}^*(\widetilde{Y}(x))$. Prediction intervals at levels $1 - \alpha \in \{0.5, 0.9, 0.99\}$ are shown as grey areas. (Color figure online)

4.2   Comparative Experiment

The performance of ENNreg model was compared to those of six alternative regression methods on four datasets from the UCI Machine Learning Repository[1]. The methods are:

- The two evidential regression algorithms published so far: the neural network model introduced in [4] (referred to as ENN97) and EVREG [13];
- Three state-of-the-art nonlinear regression algorithms with Radial Basis Function Kernel: Relevance Vector Machines (RVM), Support Vector Machines (SVM), and Gaussian Process (GP);
- The Random Forest (RF) algorithm, which is often considered as one of the best statistical learning procedures.

---

[1] Available at https://archive.ics.uci.edu/ml/.

For all methods, except ENN97 and EVREG, we used the implementation in the R package `caret` [10]. Each dataset was split randomly into a training set and a test set containing, respectively, 2/3 and 1/3 of the observations. All predictors were scaled to have zero mean and unit standard deviation. For each method, hyperparameters were tuned by 10-fold cross-validation. For ENN97, the number M of classes was set to 10. For ENNreg, we used $J = 30$, $\lambda = 0.9$ and $\epsilon = 0.01\sigma_y$ (where σ_y is the standard deviation of the response variable) for all the simulations; only ξ was tuned by cross-validation.

The results are reported in Table 1. We can see that ENNreg performs much better than the two other evidential methods on all datasets, and also better that the state-of-the-art methods for most datasets (it was only outperformed by RF on the Concrete dataset). From these results, it appears that ENNreg not only provides informative outputs with uncertainty quantification, but is also very competitive in terms of prediction accuracy.

Table 1. Test mean squared errors of ENNreg and six alternative algorithms on four UCI datasets. (See the description of the methods in the text).

	n	p	ENNreg	ENN97	EVREG	RVM	SVM	GP	RF
Boston	506	13	8.72	15.78	19.82	14.86	9.74	17.10	10.68
Energy	768	9	0.342	5.303	4.266	0.721	0.440	2.324	0.462
Concrete	1,030	8	28.32	62.0	71.4	42.3	29.6	52.7	25.6
Yacht	308	6	0.462	6.662	42.045	2.771	3.295	33.721	0.908

5 Conclusions

The evidential distance-based neural network described in this paper can be seen as regression counterpart of the ENN model introduced in [5] for classification. Both models are based on prototypes, and interpret the distances of the input vector to the prototypes as pieces of evidence. In ENN, pieces of evidence are represented by mass functions on the finite frame of discernment and are combined by Dempster's rule. In ENNreg, the frame of discernment is the real line; pieces of evidence are represented by GRFN's and combined by the generalized product-intersection rule with hard normalization, which generalizes Dempster's rule to random fuzzy sets.

The output of ENNreg for input vector x is a GRFN defined by three numbers: a point prediction $\mu(x)$, a variance $\sigma^2(x)$ measuring random uncertainty, and a precision $h(x)$, which can be seen as representing epistemic uncertainty. Experimental results show that the method outperforms previous evidential regression models in terms of mean squared error, and that it also performs better than or as well as some of the state-of-the-art nonlinear regression models. In future work, we will further investigate the calibration properties of the output belief functions, and study their potential to faithfully represent prediction uncertainty, particularly in information fusion contexts. We will also compare our approach to that of Cella and Martin [1], who propose a method, applicable to any regression algorithm, for constructing a predictive possibility distribution with some well-defined frequentist properties.

References

1. Cella, L., Martin, R.: Valid inferential models for prediction in supervised learning problems. Researchers. One (2021). https://researchers.one/articles/21.12.00002v2
2. Couso, I., Sánchez, L.: Upper and lower probabilities induced by a fuzzy random variable. Fuzzy Sets Syst. **165**(1), 1–23 (2011)
3. Denœux, T.: A k-nearest neighbor classification rule based on dempster-shafer theory. IEEE Trans. Syst. Man Cybern. **25**(05), 804–813 (1995)
4. Denœux, T.: Function approximation in the framework of evidence theory: a connectionist approach. In: Proceedings of the 1997 International Conference on Neural Networks (ICNN 1997), vol. 1, pp. 199–203, Houston, June 1997 . IEEE (1997)
5. Denœux, T.: A neural network classifier based on dempster-shafer theory. IEEE Trans. Syst. Man Cybern. A **30**(2), 131–150 (2000)
6. Denœux, T.: Belief functions induced by random fuzzy sets: A general framework for representing uncertain and fuzzy evidence. Fuzzy Sets Syst. **424**, 63–91 (2021)
7. Denœux, T., Dubois, D., Prade, H.: Representations of uncertainty in artificial intelligence: probability and possibility. In: Marquis, P., Papini, O., Prade, H. (eds.) A Guided Tour of Artificial Intelligence Research, pp. 69–117. Springer, Cham (2020). https://doi.org/10.1007/978-3-030-06164-7_3
8. Denœux, T.: Reasoning with fuzzy and uncertain evidence using epistemic random fuzzy sets: general framework and practical models. Fuzzy Sets Syst. (2022). https://doi.org/10.1016/j.fss.2022.06.004
9. Huang, L., Ruan, S., Decazes, P., Denoeux, T.: Lymphoma segmentation from 3D PET-CT images using a deep evidential network. Int. J. Approx. Reason. **149**, 39–60 (2022)
10. Kuhn, M.: Caret: classification and regression training (2021). R package version 6.0-90. https://CRAN.R-project.org/package=caret
11. Nguyen, H.T.: On random sets and belief functions. J. Math. Anal. Appl. **65**, 531–542 (1978)
12. Petit-Renaud, S., Denœux, T.: Handling different forms of uncertainty in regression analysis: a fuzzy belief structure approach. In: Hunter, A., Parsons, S. (eds.) ECSQARU 1999. LNCS (LNAI), vol. 1638, pp. 340–351. Springer, Heidelberg (1999). https://doi.org/10.1007/3-540-48747-6_31
13. Petit-Renaud, S., Denœux, T.: Nonparametric regression analysis of uncertain and imprecise data using belief functions. Int. J. Approximate Reasoning **35**(1), 1–28 (2004)
14. Shafer, G.: A Mathematical Theory of Evidence. Princeton University Press, Princeton (1976)
15. Tong, Z., Xu, P., Denœux, T.: An evidential classifier based on dempster-shafer theory and deep learning. Neurocomputing **450**, 275–293 (2021)
16. Tong, Z., Xu, P., Denœux, T.: Evidential fully convolutional network for semantic segmentation. Appl. Intell. **51**, 6376–6399 (2021)

Ordinal Classification Using Single-Model Evidential Extreme Learning Machine

Liyao Ma$^{(\boxtimes)}$, Peng Wei, and Bin Sun

School of Electrical Engineering, University of Jinan, Jinan 250022, China
cse_maly@ujn.edu.cn

Abstract. The extreme learning machine model for ordinal classification is extended to the uncertain case. Dealing with epistemic uncertainty by Dempster-Shafer theory, in this paper, the single-model multi-output extreme learning machine is learned from evidential training data. Taking both the uncertainty and the ordering relation of labels into consideration, given mass functions of training labels, different evidential encoding schemes for model output are proposed. On that basis, adopting the structure of a single extreme learning machine model with multiple output nodes, the construction procedure of evidential ordinal classification model is designed. According to the encoding mechanism and learning details, when there is no epistemic uncertainty in training labels, the proposed evidential ordinal method can be reduced to the traditional ordinal one. Experiments on artificial and UCI datasets illustrate the practical implementation and effectiveness of proposed evidential extreme learning machine for ordinal classification.

Keywords: Ordinal classification · Dempster-Shafer theory · Extreme learning machine

1 Introduction

Ordinal classification, also known as ordinal regression [1], is a special kind of supervised learning approach, with wide applications in various fields. It considers the ordinal variables appeared in various practical problems, such as the attitude degree in Likert Scale *Extremely* $\succ$ *Very Much* $\succ$ *Moderately* $\succ$ *A little* $\succ$ *Not at all*. Ordinal classification aims at learning a model from the training set to predict the output labels of new instances, with specifically the finite set of possible labels being naturally ordered. Due to the existence of the ordering relations of labels, ordinal classification bridges the classification and regression tasks. However, it differs from classification problem as the latter deals with nominal variables having no ordering among elements. On the other hand, since there is no metric on the ordinal output (*Extremely* is not five times more severe that *Not at all*), it also cannot be simply treated as a regression problem.

Supported by the Natural Science Foundation of Shandong Province ZR2021MF074 and the National Key R & D Program of China 2018AAA0101703

The origin of ordinal classification research can be traced back to the ordinal statistics methods in the 1980s.s. With the development of machine learning, more methods have been proposed by considering ordered labels within classification models, such as support vector machine [2], deep learning model [3] and so on. As supervised learning, the training set of ordinal classification model must be certain and precise, leading to high cost of preprocessing and loss of useful information. Also, there exists epistemic uncertainty in the model itself. Hence, considering uncertainty in data and model, within the framework of imprecise probability, Desdercke discussed the lower and upper medians of ordinal problems [4], and proposed to perform cautious ordinal classification by binary decomposition [5]. In this paper, we adopt Dempster-Shafer theory [6,7] (also called the theory of belief functions) to represent and deal with epistemic uncertainty. Although there are some researches focused on ordinal information, such as He discussed the entropy and evidence combination in ordinal environment [8,9], little work has been done on ordinal classification with evidential data.

This paper proceeds with extreme learning machine (ELM) that has a simple single-hidden-layer structure, and extends its model to learn directly from evidential data. The rest of this paper is organized as follows: Sect. 2 provides some brief descriptions and definitions of the evidential ordinal classification task. Section 3 is devoted to the evidential encoding scheme and the construction of proposed model. In Sect. 4, experiments on artificial and UCI datasets demonstrate the implementation and performance of proposed evidential ordinal ELM model. Finally, the main conclusions are summarized in Sect. 5.

2 Background

In this section, we recall some necessary definitions and descriptions from the Dempster-Shafer theory and the ordinal extreme learning machine approach.

2.1 Dempster-Shafer Theory

As a generalization of both set and probabilistic uncertainty, Dempster-Shafer theory provides a flexible framework for modelling and reasoning with uncertainty. Here only a few definitions needed in the rest of the paper are recalled. More complete descriptions can be found in Shafer's book [7] and in the recent survey [10]. Let finite set $\mathcal{K} = \{1, 2, \cdots, K\}$ be the *frame of discernment*, containing all the possible exclusive values that the ordinal variable can take. When the true value of the variable is ill-known, we can model its partial information by a *mass function* $m : 2^{\mathcal{K}} \rightarrow [0, 1]$ such that $m(\emptyset) = 0$ and

$$\sum_{A \subseteq \mathcal{K}} m(A) = 1. \tag{1}$$

A subset A of $\mathcal{K}$ with $m(A) > 0$ is called a *focal set* of m. We can interpret the quantity $m(A)$ as the amount of evidence indicating that the true value is specifically in A while in no strict subset. This formalism extends both probabilistic and set-valued uncertainty models. For example, the *vacuous* mass function verifies $m(\mathcal{K}) = 1$ and represents total ignorance. A *Bayesian* mass function is such that all its focal sets are singletons, being equivalent to a probability distribution. A mass function such that $m(A) = 1$ for some subset $A \subseteq \mathcal{K}$ is said to be *logical*, being equivalent to set A.

In the Transferable Belief Model, Smets proposed the *pignistic probability distribution* [11], which is obtained by distributing masses equally among the sets of A as

$$BetP(k) = \sum_{\{A \subseteq \mathcal{K}: k \in A\}} \frac{m(A)}{|A|}, \tag{2}$$

where $|A|$ denotes the cardinality of $A \subseteq \mathcal{K}$.

2.2   Ordinal Extreme Learning Machine

Developed from the single-hidden-layer feedforward neural network, Extreme Learning Machine (ELM) was proposed in [12] as a fast-learning algorithm, with input weights assigned randomly and output weights decided analytically. Deng [13] extended ELM to Ordinal Extreme Learning Machine (ORELM) by proposing different encoding schemes and multiple model structures.

Considering the ordering relation among labels, ORELM mainly includes the steps of encoding, learning and prediction. For the ORELM structure of one model with multiple output nodes, the target output is encoded in the form of $t_{1 \times K} = [1, ...1, -1, ..., -1]$, where label k and its lower-orders are set with value 1 and the others -1. ORELM can also be decomposed into multiple binary classifiers, such as $K-1$ binary ELM classifiers adopting 1-vs-all encoding scheme or $K(K-1)/2$ binary ones with 1-vs-1 scheme. Once the model structure and encoding strategy are decided, the learning procedure of ORELM basically follows that of traditional ELM. The weight vectors connecting the input nodes and hidden nodes, as well as the bias in the hidden nodes are randomly assigned according to certain probability distribution. The weight vectors connecting the hidden nodes and output nodes are obtained as the least square solution of a linear equation system. The learning procedure is similar to what described in Sect. 3.2. Due to length limitation, the calculation is not detailed here.

3   Single-Model Multi-output Evidential Ordinal Extreme Learning Machine

Inspired by the decision-making criteria in the Dempster-Shafer theory [14], several evidential encoding schemes are proposed to transfer mass functions of ordinal labels to output coding bits. Considering both the ordinal relationship among labels and the epistemic uncertainty in training instance labels, the learning procedure of single-model multi-output evidential ordinal extreme learning machine (EORELM) is proposed.

3.1 Evidential Encoding Schemes

The key to build the evidential ordinal ELM model is to achieve the output coding bits based on mass functions of training labels. For each evidential training instance $(\boldsymbol{x}_i, m_i)$ with inputs $\boldsymbol{x}_i$ and mass function m_i, $i = 1, \cdots, N$, the target output is encoded into K bits as $\boldsymbol{t}_i = [t_{i1}, \cdots, t_{iK}]$, where the total coding bit is the same as the number of ordinal classes $K = |\mathcal{K}|$.

Evidential encoding is based on the ordinal coding matrix generated in the following way. Taking the ordering relation of labels into consideration, when an instance belongs to the k-th class, it is also classified into its lower-order classes $\{1, 2, ..., k-1\}$. The target output is encoded in the formation of $\tilde{\boldsymbol{c}} = [1, ...1, 0, ..., 0]$, where

$$\tilde{c}_j = \begin{cases} 1, & 1 \leq j \leq k \\ 0, & k+1 \leq j \leq K \end{cases} \quad j = 1, \cdots, K. \tag{3}$$

Ordinal coding matrix is obtained by gathering the target outputs of all possible labels together. Taking $K=3$ for example, the $K \times K$ ordinal coding matrix C is shown in Table 1. The i-th column of C represents the target coding output of the i-th class. The matrix element c_{ij} represents the output coding bit t_i corresponding to output node i, given class j.

Table 1. Ordinal coding matrix C

Coding bits	True label		
	1	2	3
t_1	1	1	1
t_2	0	1	1
t_3	0	0	1

Given ordinal coding matrix $C_{K \times K}$ and mass function m of a training label, evidential encoding can be implemented with the following proposed schemes:

Generalized Maximin Encoding Scheme. When generating the evidential coding vector $\underline{\boldsymbol{t}} = [\underline{t}_1, \cdots, \underline{t}_K]$, the worst case of possible label is considered within each focal set B. Each coding bit of the evidential coding vector is calculated by the lower expected coding value

$$\underline{t}_i = \sum_{B \subseteq \mathcal{K}} m(B) \min_{j \in B} c_{ij}, \quad i = 1, \cdots, K. \tag{4}$$

Generalized Maximax Encoding Scheme. Taking an optimistic point of view, each coding bit of the evidential coding vector $\overline{\boldsymbol{t}} = [\overline{t}_1, \cdots, \overline{t}_K]$ is calculated by the upper expected coding value

$$\overline{t}_i = \sum_{B \subseteq \mathcal{K}} m(B) \max_{j \in B} c_{ij}, \quad i = 1, \cdots, K. \tag{5}$$

That is to say, when generating the evidential coding vector, the best case of possible label is considered within each focal set B.

Pignistic Encoding Scheme. In this scheme, each coding bit of the resulted coding vector $t^p = [t_1^p, \cdots, t_K^p]$ is computed as the weighted-sum of coding values for all labels, with Pignistic probabilities obtained from mass function as weights. It can also be seen from Equation (6) that actually coding values within the focal set are averaged.

$$t_i^p = \sum_{j=1}^{K} BetP(j)c_{ij} = \sum_{B \subseteq \mathcal{K}} m(B) \left(\frac{1}{|B|} \sum_{j \in B} c_{ij} \right), \quad i = 1, \cdots, K. \quad (6)$$

Generalized Hurwicz Encoding Scheme. To obtain the evidential coding vector, this scheme takes a convex combination of the minimum and maximum coding value. A *pessimism index* $\alpha \in [0, 1]$ is set to adjust the combination, with $\alpha = 1$ and $\alpha = 0$ corresponding, respectively, to the previously proposed generalized maximin and maximax encoding schemes. Each coding bit of $t^\alpha = [t_1^\alpha, \cdots, t_K^\alpha]$ is calculated by

$$t_i^\alpha = \sum_{B \subseteq \mathcal{K}} m(B) \left(\alpha \min_{j \in B} c_{ij} + (1 - \alpha) \max_{j \in B} c_{ij} \right) = \alpha \underline{t}_i + (1 - \alpha) \overline{t}_i. \quad (7)$$

Example 1. Suppose $\mathcal{K} = \{1, 2, 3\}$ and the mass function for the label of a training instance is $m(\{1\}) = 0.7$, $m(\{1, 2\}) = 0.2$, $m(\{1, 2, 3\}) = 0.1$, the evidential coding vectors obtained with different encoding schemes are listed as follows:

- Generalized maximin encoding scheme: $\underline{t} = [1, 0, 0]$
- Generalized maximax encoding scheme: $\overline{t} = [1, 0.3, 0.1]$
- Pignistic encoding scheme: $t^p = [1, 0.17, 0.03]$
- Generalized Hurwicz encoding scheme $(\alpha = 0.4)$: $t^\alpha = [1, 0.18, 0.06]$

3.2   Construction of Evidential Ordinal ELM Model

In this paper, the evidential ordinal classification model is constructed as the structure of one ELM model with multiple output nodes. The evidential training set contains N instances $\{(x_i, t_i)\}_{i=1}^N$, where $x_i = [x_{i1}, \cdots, x_{ip}] \in \mathcal{R}^p$ is the input of instance i consisting of p attributes, and $t_i = [t_{i1}, \cdots, t_{iK}]$ is the evidential coding vector encoded by mass function m_i using one of the proposed evidential encoding schemes.

As shown in Fig. 1, for the single-model multi-output EORELM, input and output layers respectively have p nodes (number of attributes) and K nodes (number of classes). There is one hidden layer with h hidden nodes, each of which is associated with bias b_i and activation function $g_i(x)$, $i = 1, \cdots, h$. With weights from input nodes to hidden nodes $w_i = [w_{1i}, \cdots, w_{pi}], i = 1, \cdots, h$, weights from hidden nodes to output nodes $\beta_{ij}, i = 1 \cdots, h, j = 1, \cdots, K$, biases and activation functions, the EORELM can be mathematically modelled as

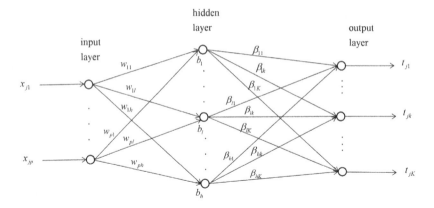

Fig. 1. Structure of single-model multi-output EORELM

$$\sum_{i=1}^{h} \beta_{ij} g_i(\boldsymbol{w}_i \boldsymbol{x}_j^T + b_i) = \boldsymbol{t}_j, \; j = 1, \cdots, N. \tag{8}$$

The N equations corresponding to the N training instances can be arranged into the matrix-vector format as

$$H\boldsymbol{\beta} = T, \tag{9}$$

in which $\boldsymbol{\beta} = (\beta_{ij})_{h \times K}$, $T = (t_{ij})_{N \times K}$ and

$$H_{N \times h} = \begin{bmatrix} g(\boldsymbol{w}_1 \boldsymbol{x}_1 + \boldsymbol{b}_1) & \cdots & g(\boldsymbol{w}_h \boldsymbol{x}_1 + \boldsymbol{b}_h) \\ \vdots & \cdots & \vdots \\ g(\boldsymbol{w}_1 \boldsymbol{x}_N + \boldsymbol{b}_1) & \cdots & g(\boldsymbol{w}_h \boldsymbol{x}_N + \boldsymbol{b}_h) \end{bmatrix}.$$

Being different from the weight-tuning methods, since ELM is a fast-learning method, the wights $\boldsymbol{w}_i$ and bias $\boldsymbol{b}_i$ are assigned randomly. Given $(\boldsymbol{w}_i, \boldsymbol{b}_i)$, the weights $\boldsymbol{\beta}$ from hidden layer to output layer are determined by solving the overdetermined system of linear equations. Analytically, it is the least-square solution of equation system

$$\hat{\boldsymbol{\beta}} = H^+ T, \tag{10}$$

where $H+$ is the Pseudo inverse of the hidden layer output matrix H.

Once the model is learned, given a new instance, the model output is compared with each column of ordinal coding matrix C by loss-based distance. The instance is then predicted as the label corresponding to the closest column. The learning procedure of single-model multi-output EORELM is summarised in Algorithm 1. It is notable that according to the mechanism of evidential encoding schemes, when there is no epistemic uncertainty in the training labels, the proposed EORELM is degraded to traditional ordinal ELM (ORELM).

Algorithm 1: Learning procedure of EORELM

Input: uncertain training set $\{(\boldsymbol{x}_i, m_i)\}_{i=1}^N$, new instance $\boldsymbol{x}$, type of encoding scheme TP, activation function g, number of hidden nodes h

Output: label prediction $\hat{y}$

1 % Learning;
2 C=CodingMatrixGeneration;
3 **for** $i=1$:N **do**
4   t_i=EvidentialEncoding(m_i,TP);

5 $(\boldsymbol{w}_i, \boldsymbol{b}_i)$=RandomGeneration;
6 **for** $i=1$:N **do**
7   **for** $i=1$:h **do**
8        $H_{ij} = g(\boldsymbol{w}_j \boldsymbol{x}_i + \boldsymbol{b}_j)$;

9 T=Gathering(t_i);
10 $\hat{\beta} = H^+ T$;
11 % Predicting;
12 $\hat{y}$=Prediction(EORELM,$\boldsymbol{x}$);
13 $\hat{y}$=DistanceJudgement($\tilde{y}$, C);

---

4 Experiments

4.1 Artificial Dataset

The proposed encoding schemes and learning procedure are illustrated by an artificial dataset firstly. A dataset of 1000 instances was generated following the same way of paper [13] (Section 4.1), with function $y = (0.5 - x_1)^2 + (2 - x_2)^2$.

For each instance in the data set $\{\boldsymbol{x}_i, y_i\}, i = 1, \cdots, 1000$, the two inputs were both randomly uniformly distributed on interval $[0, 1]$. To make the dataset closer to real applications, the random noise uniformly distributed on $[-0.1, 0.1]$ was added to y for all the training instances, while the test set remains unchanged. Set $\mathcal{K} = \{1, 2, 3\}$, the ordinal label for each instance is decided by value of y as

$$\tilde{t}_i = \begin{cases} 1, & y_i \leq 2.08 \\ 2, & 2.08 < y_i \leq 3.17 \\ 3, & y_i > 3.17 \end{cases} \tag{11}$$

ORELM can be learned from training instances with labels $\tilde{t}_i$. For EORELM, due to uncertainties in y and the ordinal label $\tilde{t}_i$, mass functions were generated for instances near the classification boundaries. In this paper, two focal sets were considered, being the singleton $\tilde{t}_i$, and the set consisting of $\tilde{t}_i$ and its adjacent class. Mass assigned to the set was computed as

$$m(set) = \begin{cases} \frac{y-a}{b-a}, & a < y \leq b \\ \frac{c-y}{c-b}, & b < y \leq c \\ 0, & others \end{cases} \tag{12}$$

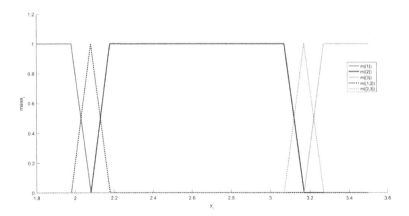

Fig. 2. Mass function generation

In detail, for the instances with y near boundary 2.08, $m(\{1,2\})$ is calculated with $a = 1.98$, $b = 2.08$, $c = 2.18$, while for the instances near boundary 3.17, $m(\{2,3\})$ is calculated with $a = 3.07$, $b = 3.17$, $c = 3.27$. The rest of the mass was assigned to the singleton. For example, when $y = 3.25$, the generated mass function is $m(\{3\}) = 0.8, m(\{2,3\}) = 0.2$. Fig. 2 shows the mass assigned to the focal sets with different y.

Table 2. Classification accuracies of different encoding schemes (artificial data)

Encoding scheme	Cross validation				
	1	2	3	4	5
Pignistic	0.4100	0.4200	0.3550	0.4200	0.3850
Maximax	0.9500	0.9550	0.9550	0.9650	0.9750
Maximin	0.9350	0.9800	0.9750	0.9700	0.9350
Hurwitz	0.9700	0.9900	0.9900	0.9850	0.9550

Mass functions were transformed into coding vectors with different encoding schemes, to obtain evidential training sets. Five-fold cross validation was implemented. The activation function was selected as the radial basis function $g(x) = exp(-x^2)$. Classification accuracies on the test sets are shown in Table 2. The Pignistic encoding scheme has a poor performance due to its average property. Meanwhile, the other three schemes can provide satisfying results, with Hurwitz encoding scheme ($\alpha = 0.3$) resulting in reletively higher accuracies.

4.2   UCI Datasets

In this section, the proposed method is tested on five UCI datasets [15], as detailed in Table 3. As regression datasets, they were turned into ordinal ones by output variable discretization, with discretizing points listed in the last column of Table 3. In the experiments, we tested the proposed EORELM models with different encoding schemes (Pignistic, Maximax, Maximin, Hurwitz with $\alpha = 0.3$), and compared them with traditional classification model (ELM) and ordinal classification model (ORELM).

Table 3. UCI datasets for validation

Dataset	# instances	# attributes	# classes	discretizing points
Boston housing	506	13	3	[15, 25]
Appliances energy prediction	19735	27	3	[55, 85]
Real estate valuation	414	27	5	[25, 34, 41, 49]
Parkinsons telemonitoring	5875	20	5	[15, 25, 35, 45]
Combined cycle power plant	9568	4	5	[428, 446, 459, 472]

The activation function in Sect. 4.1 was used. Following Equation (13), mass functions were generated according to the value of regression output variable. Let b be the discretizing points and $\delta = b - a = c - b$. The value of δ was set as 0.1 for Parkinsons and CCPP datasets, 0.5 for Bonston and Real state datasets, and 5 for Appliances dataset. Number of hidden nodes was set as 25 for Real state dataset, 200 for Parkinsons dataset, and 50 for the others. Accuracies were obtained from 5-fold cross validation, with each experiment repeated three times for average. Table 4 summarises the performances of different models. EORELM with Pignistic encoding scheme still performances poorly. Except for this one, all the models considering ordering relation outperform ELM. Compared to ORELM, EORELM models usually have a slightly performance improvement. One one hand, considering uncertainty in ordinal labels helps to make better use of information. On the other hand, as only instances near the classification boundaries are considered as uncertain, most instances actually obtain mass 1 for a certain singleton, providing little additional information.

Table 4. Classification accuracy comparisons on UCI datasets

Dataset	ELM	ORELM	Pignistic	Maximax	Maximin	Hurwitz
Boston	0.7746	0.8083	0.2450	0.8261	0.8261	0.8162
Appliances	0.4043	0.5655	0.3155	0.5648	0.5659	0.5672
Real state	0.3599	0.5606	0.1908	0.5630	0.5750	0.5653
Parkinsons	0.6708	0.7391	0.0756	0.7479	0.7457	0.7491
CCPP	0.7635	0.8305	0.1954	0.8311	0.8316	0.8328

5   Conclusion

Single-model multi-output EORELM is proposed in this paper to learn ordinal classification model directly from evidential data. The evidential encoding schemes of Maximax, Maximin, Hurwitz and Pignistic are proposed, with the former three have the value of practical application. Based on coding vectors transformed from mass functions, the learning procedure of EORELM is detailed. Experiments demonstrate the implementation and performances of proposed approach. The encoding-based framework can also be used in other ordinal classification models such as support vector machines. In future work, binary decomposition structure of EORELM, as well as other evidential ordinal classification models will be explored.

References

1. Gutierrez, P.A., Perez-Ortiz, M., Sanchez-Monedero, J., et al.: Ordinal regression methods: survey and experimental study. IEEE Trans. Knowl. Data Eng. **28**(1), 127–146 (2016)
2. Xiao, Y., Li, X., Liu, B., et al.: Multi-view support vector ordinal regression with data uncertainty. Inf. Sci. **589**, 516–530 (2022)
3. Lucas, K., Lisa, H., Torsten, H., et al.: Deep and interpretable regression models for ordinal outcomes. Pattern Recogn. **122**, 108263 (2022)
4. Destercke, S.: On the median in imprecise ordinal problems. Ann. Oper. Res. **256**(2), 375–392 (2017)
5. Destercke, S., Yang, G.: Cautious ordinal classification by binary decomposition. In: Calders, T., Esposito, F., Hüllermeier, E., Meo, R. (eds.) ECML PKDD 2014. LNCS (LNAI), vol. 8724, pp. 323–337. Springer, Heidelberg (2014). https://doi.org/10.1007/978-3-662-44848-9_21
6. Dempster, A.P.: Upper and lower probabilities induced by a multivalued mapping. In: Yager, R.R., Liu, L. (eds.) Classic Works of the Dempster-Shafer Theory of Belief Functions. Studies in Fuzziness and Soft Computing, vol. 219, pp. 57–72. Springer, Berlin, Heidelberg (2008). https://doi.org/10.1007/978-3-540-44792-4_3
7. Shafer, V.: A Mathematical Theory of Evidence. Princeton University Press, Princeton (1976)
8. He, Y.: An approach utilizing negation of extended-dimensional vector of disposing mass for ordinal evidences combination in a fuzzy environment. arXiv preprint arXiv:2104.05416 (2021)
9. He, Y.: Ordinal relative belief entropy. arXiv preprint arXiv:2102.12575 (2021)
10. Campagner, A., Ciucci, D., Denœux, T.: Belief functions and rough sets: survey and new insights. Int. J. Approximate Reasoning **143**, 192–215 (2022)
11. Smets, P.: Decision making in the TBM: the necessity of the pignistic transformation. Int. J. Approximate Reasoning **38**(2), 133–147 (2005)
12. Huang, G.B., Zhu, Q.Y., Siew, C.K.: Extreme learning machine: theory and applications. Neurocomputing **70**(1–3), 489–501 (2006)
13. Deng, W.Y., Zheng, Q.H., Lian, S., et al.: Ordinal extreme learning machine. Neurocomputing **74**(1–3), 447–456 (2010)
14. Ma, L., Denoeux, T.: Partial classification in the belief function framework. Knowl.-Based Syst. **214**, 106742 (2021)
15. Dua, D., Graff, C.: UCI Machine Learning Repository. Irvine, CA: University of California, School of Information and Computer Science (2019). http://archive.ics.uci.edu/ml

Reliability-Based Imbalanced Data Classification with Dempster-Shafer Theory

Hongpeng Tian[1], Zuowei Zhang[2,3(✉)], Arnaud Martin[3], and Zhunga Liu[2]

[1] School of Electrical Engineering, Zhengzhou University, Zhengzhou, China
[2] School of Automation, Northwestern Polytechnical University, Xi'an, China
`zuowei_zhang@mail.nwpu.edu.cn, liuzhunga@nwpu.edu.cn`
[3] Univ Rennes, CNRS, IRISA, Rue E. Branly, 22300 Lannion, France
`Arnaud.Martin@univ-rennes1.fr`

Abstract. The classification analysis of imbalanced data remains a challenging task since the base classifier usually focuses on the majority class and ignores the minority class. This paper proposes a reliability-based imbalanced data classification approach (RIC) with Dempster-Shafer theory to address this issue. First, based on the minority class, multiple under-sampling for the majority one are implemented to obtain the corresponding balanced training sets, which results in multiple globally optimal trained classifiers. Then, the neighbors are employed to evaluate the local reliability of different classifiers in classifying each test sample, making each global optimal classifier focus on the sample locally. Finally, the revised classification results based on various local reliability are fused by the Dempster-Shafer (DS) fusion rule. Doing so, the test sample can be directly classified if more than one classifier has high local reliability. Otherwise, the neighbors belonging to different classes are employed again as the additional knowledge to revise the fusion result. The effectiveness has been verified on synthetic and several real imbalanced datasets by comparison with other related approaches.

Keywords: Imbalanced data · Reliability · Dempster-Shafer theory

1 Introduction

Imbalanced data refers to the dataset has an unequal distribution between classes [1]. For a binary class problem, if the number of samples in the majority class is significant larger than that of the minority class, traditional classifiers, such as K-nearest neighbors (K-NN) [2], support vector machine classifier (SVM) [3], are dedicated to maximize the overall classification performance. In this case, most minority samples are assigned to majority class.

Increasingly works are emerged for classifying imbalanced data, and they can be roughly divided into three categories including sampling approaches [4], cost-sensitive learning [5] and ensemble learning [6]. Sampling approaches focus

© The Author(s), under exclusive license to Springer Nature Switzerland AG 2022
S. Le Hégarat-Mascle et al. (Eds.): BELIEF 2022, LNAI 13506, pp. 77–86, 2022.
https://doi.org/10.1007/978-3-031-17801-6_8

on preprocessing the input data to balance the classes. By doing this, the preprocessed data can be classified by basic classifiers. Cost-sensitive learning approaches assign relatively high weights to minority samples, which can reduce the misclassification of the minority class. Ensemble learning approaches combine different classifiers trained by various subsets, which supplies the complementarity information to improve the performance of classification with respect to an individual classifier. However, these imbalanced data classification approaches only consider the global optimum and are not suitable for each test sample. For instance, the samples lying in the overlap area of different classes are indistinguishable and easily misclassified. In this case, there are some uncertain information between between classes.

Dempster-Shafer theory (DST) [7,8], also known as the theory of belief functions, has the advantage of reasoning uncertain information, and has been widely used in classification [9–12]. Recently, a few works [13,14] have been proposed to deal with imbalanced data classification within the belief function theory. Although these approaches has the advantage of capturing uncertain information thanks to evidence reasoning, they fill to consider the local performance of classifiers for each test sample. In this paper, we propose a reliability-based imbalanced data classification approach with Dempster-Shafer theory. The contributions mainly include three aspects. 1) We design a reliability evaluation strategy to obtain local reliability of different classifiers for each test sample, which can characterize the local performance of classifiers. 2) We introduce a revision strategy to resubmit the samples with low local reliability of different classifiers according to neighbors from various classes. 3) We apply RIC to synthetic and several real imbalanced datasets to demonstrate the superiority.

The rest of this paper is organized as follows. The proposed approach is presented in detail in Sect. 2. Then, it is tested in Sect. 3 and compared with several other typical methods, followed by conclusions.

2   Reliability-Based Imbalanced Data Classification

In this section, a reliability-based imbalanced data classification approach is proposed in detail. Assume that a test set $\mathcal{X} = \{\mathbf{x}_1, ..., \mathbf{x}_N\}$ is classified under the frame of discernment $\Omega = \{\omega_{min}, \omega_{maj}\}$ according to a training set $\mathcal{Y} = \{\mathbf{y}_1, ..., \mathbf{y}_M\}$ on H different attribute spaces. $\mathcal{Y}_{min}$ and $\mathcal{Y}_{maj}$ represent the minority class and majority class, respectively.

2.1   Multiple Under-Sampling for Majority Class

In this subsection, we implement random under-sampling[1] for the majority class multiple times to obtain different training sets thereby training basic classifiers.

T subsets $\mathcal{Y}_{maj}^1, ..., \mathcal{Y}_{maj}^T$ are random sampled from the majority class $\mathcal{Y}_{maj}$. Each subset has the same number of samples as that of the minority class $\mathcal{Y}_{min}$,

---

[1] In applications, users can employ other appropriate under-sampling approaches according to the request of practice.

and is combined with $\mathcal{Y}_{min}$ to form a new training set. By doing this, we can obtain T training sets, named $\mathcal{Y}^1, ..., \mathcal{Y}^T$, and the number of T is denoted as:

$$T = [IR] \tag{1}$$

with

$$IR = \frac{|Y_{maj}|}{|Y_{min}|} \tag{2}$$

where IR, such that $IR \geq 1$, refers to the measurement for the imbalanced degree of the dataset, and $|.|$ represents the cardinality symbol. $[\cdot]$ is a rounding symbol that rounds the elements of IR to the nearest integers towards infinity.

Then, each training set can train a basic classifier that has high performance in classifying a balanced dataset. The classification result of $\mathbf{x}_i$ by t-th classifier is denoted as $\mathcal{P}_i^t = [p_i^t(\{\omega_{min}\}), p_i^t(\{\omega_{maj}\}], t = 1, ..., T$.

2.2 Evaluate the Local Reliability for Classifiers Fusion

In this subsection, we evaluate the local reliability of different classifiers for classifying each test sample. Then, we combine classifiers with various reliability by the original discounting fusion rule.

Here, we employ the neighbors $\mathbf{y}_1, ..., \mathbf{y}_K$ of the test sample $\mathbf{x}_i$ to evaluate the local reliability of different classifiers, since $\mathbf{x}_i$ has the similar data structure and distribution with respect to $\mathbf{y}_1, ..., \mathbf{y}_K$. The better the performance of the classifier to classify $\mathbf{y}_1, ..., \mathbf{y}_K$, the higher the reliability for classifying $\mathbf{x}_i$. Based on the above analysis, we define a rule to evaluate the degree of reliability of different classifiers, denoted as:

$$\xi_{it} = \frac{\exp(-\vartheta_{it})}{\sum_{t=1}^{T} \exp(-\vartheta_{it})} \tag{3}$$

with

$$\vartheta_{it} = \sum_{k=1}^{K} \sqrt{\sum_{\{\omega_c\} \in \Omega} [p_k^t(\{\omega_c\}) - l_k(\{\omega_c\})]^2} \tag{4}$$

where ξ_{it}, such that $0 < \xi_{it} < 1$, represents the reliability of the t-th classifier for classifying $\mathbf{x}_i$. $p_k^t(\{\omega_c\})$ refers to the probability of $\mathbf{y}_k$ belongs to $\{\omega_c\}$. The truth of classification of $\mathbf{y}_k$ is characterized by the binary vector $L_k = [l_k(\{\omega_{min}\}), l_k(\{\omega_{maj}\})]$. The $l_k(\{\omega_c\}) = 1$ if the true class of $\mathbf{y}_k$ is $\{\omega_c\}$. If not, $l_k(\{\omega_c\})$ is equal to 0. Ω is the frame of discernment, such that $\Omega = \{\{\omega_{min}\}, \{\omega_{maj}\}\}$. We can observe that the lower the deviation between classification results and truths, the higher reliability of the classifier.

Each classification result can be considered as a piece of evidence under the framework of DST, which is appealing to combine multi-source information. The reliability-based discount fusion method [8], is employed here for discounting and fusing pieces of evidence. The reliability ξ_{it} for different T classifiers can be considered as the discounting factors. The discounted masses of belief is denoted as:

$$\begin{cases} m_i^t(\{\omega_c\}) = \xi_{it} p_i^t(\{\omega_c\}), \{\omega_c\} \in \Omega \\ m_i^t(\Omega) = 1 - \xi_{it} \end{cases} \quad (5)$$

where $p_i^t(\{\omega_c\})$ represents the probability (Bayesian BBA) that the sample $\mathbf{x}_i$ belongs to $\{\omega_c\}$ under Bayesian framework, and Ω is the the total unknown class. We can find that the more important and reliable the classification result, the larger the corresponding discounting factor, and the less discounted information assigned to the total ignorance Ω. In particular, the degree of conflict between pieces of evidence is reduced since the conflict information is transferred into Ω that plays a particular neutral role in the fusion process. In this case, the global fusion result for T basic belief assignments (BBAs) of the sample $\mathbf{x}_i$ is denoted as:

$$\mathbf{m}_i = \mathbf{m}_i^1 \oplus \cdots \oplus \mathbf{m}_i^T \quad (6)$$

where $\mathbf{m}_i^t = [m_i^t(\{\omega_{min}\}), m_i^t(\{\omega_{maj}\})]$, and $\oplus$ refers to the DS fusion rule [7] for the combination of these T pieces of evidence. $\mathbf{m}_i$ represents the normalized combination result. As a result, the fused BBAs can be transferred into pignistic probability[15] for the preliminary decision-making.

2.3 Employ Neighbors for Final Decision

In this subsection, we employ neighbors from different classes as the additional information to make final decision.

For the test sample $\mathbf{x}_i$, it can be directly classified if there is more than one classifier that has high local reliability. In contrast, when all classifiers have low reliability to classify $\mathbf{x}_i$, which means it is hard to be correctly classified by different classifiers.

We evaluate the different degrees of local reliability of classifiers for classifying $\mathbf{x}_i$ before normalization, and obtain the max values of them, denoted as:

$$\widehat{\xi}_{i,\max} = \max\{\widehat{\xi}_{i1}, ..., \widehat{\xi}_{iT}\} \quad (7)$$

where $\widehat{\xi}_{it}$ represents the degree of local reliability of t-th classifier for classifying $\mathbf{x}_i$, such that $\widehat{\xi}_{it} = \exp(-\vartheta_{it})$. The higher the value of $\widehat{\xi}_{i,\max}$, the bigger the possibility of $\mathbf{x}_i$ being correctly classified. Thus, we define a threshold δ to distinguish whether $\mathbf{x}_i$ can be directly classified or not, given by:

$$\delta = \text{quantile}(\widehat{\Xi}_{\max}, \gamma) \quad (8)$$

with

$$\widehat{\Xi}_{\max} = \{\widehat{\xi}_{1,\max}, ..., \widehat{\xi}_{N,\max}\} \quad (9)$$

where γ is a quantile number such that $\gamma \in [0, 1]$. If $\widehat{\xi}_{i,\max} > \delta$, the test sample $\mathbf{x}_i$ can be classified directly according to the classification result obtained by discounting fusion. Otherwise, we need to mine some additional information by neighbors to revise classification results. We convert the distances between $\mathbf{x}_i$

and different classes into the mass of belief that it belongs to different classes, denoted as:

$$\widehat{m}_i(\{\omega_c\}) = \frac{\exp(-d(\mathbf{x}_i, \{\omega_c\}))}{\sum\limits_{\{\omega_c\}\in\Omega} \exp(-d(\mathbf{x}_i, \{\omega_c\}))}. \tag{10}$$

where $d(\mathbf{x}_i, \{\omega_c\})$ represent the mean Euclidean distance between $\mathbf{x}_i$ and its K neighbors in class $\{\omega_c\}$. We can observe that the larger the distance $d(\mathbf{x}_i, \{\omega_c\})$, the lower the possibility that $\mathbf{x}_i$ belongs to class $\{\omega_c\}$. Then, the BBA of $\mathbf{x}_i$, named $\widehat{m}_i$, is fused with $\mathbf{m}_i$ according to the DS fusion rule. Finally, the fused BBAs can be transferred into pignistic probability to make final decision.

3   Experiment Applications

In this section, the proposed RIC is compared with several typical approaches including ROS [4], RUS [16], SMOTE [17], CBU [18] and RUSBoost [6]. SVM [3] is taken as the basic classifier in different approaches. Two common indexes [1], *i.e.*, F-measure (FM) and G-mean (GM), widely used in imbalanced data classification, are employed to evaluate the performance of different approaches. The higher the values of FM and GM, the better the performance of the approach.

3.1   Benchmark Datasets

A 2-D dataset with two classes $\Omega = \{\omega_{min}, \omega_{maj}\}$ is given in Fig. 1(a)(b), where each sample denoted as a point has two dimensions of attributes corresponding to x-coordinate and y-coordinate. The minority class ω_{min} has 2000 samples and majority class ω_{maj} consists of 200 samples. All the samples are generated from two bivariate Gaussian densities and have the following means vectors and covariance matrices, denoted as: $\mu_{min} = (3.1, 5)$, $\Sigma_{min} = 0.01\mathbf{I}$, $\mu_{maj} = (4, 5)$, $\Sigma_{maj} = 0.1\mathbf{I}$, where $\mathbf{I}$ represents the 2×2 identity matrix. Half of the samples in each class are randomly selected as training samples and others are as test samples. ω^{tr}_{min} and ω^{te}_{maj} represent the minority class and majority class in the training set, respectively. The ground truth of test set is marked by different colors and represented by ω^{te}_{min} and ω^{te}_{maj}.

Ten generally used real imbalanced datasets from Keel repository[2] are employed to test and evaluate the performance of different approaches in classifying imbalanced data. Each dataset is partitioned using a five-folds stratified cross validation. The basic information of these datasets including the number of all samples (#Size.), majority class samples (#Maj.), minority class samples (#Min), attributes (#Attr.) and imbalance ratio (#IR.) are shown in Table 1.

---

[2] http://www.keel.es/.

Table 1. Basic information of the Keel datasets.

Data	#Size.	#Min.	#Maj.	#Attr.	#IR.
glass1	214	76	138	9	1.82
yeast1	1484	429	1055	8	2.46
vehicle2	846	218	628	18	2.88
ecoli2	336	52	284	7	5.46
page-blocks0	5472	559	4913	10	8.79
vowel0	998	90	898	13	9.98
led7digit	443	37	406	7	10.97
ecoli4	336	20	316	7	15.80
yeast5	1484	44	1440	8	32.73
shuttle2vs5	3316	49	3267	9	66.67

3.2   Performance Evaluation

1) RIC vs. comparisons in the synthetic dataset.

This experiment is designed to intuitively validate the effectiveness of RIC in classifying imbalanced data with the overlapped area between classes. We take $K = 7$ and $\gamma = 0.05$ in RIC, and the parameters in comparison approaches are as default. The training and test set of the synthetic dataset are reported in Fig. 1(a)(b). We can see that the majority class and minority class are partly overlapped on their borders. The classification results of comparison approaches are shown in Fig. 1(c)-(g), where most samples lying in the overlapped area are misclassified. Actually, these samples are hard to be correctly classified and there are uncertain information between classes in this case. We can observe from Fig. 1(h) that these samples marked by black points are correctly identified. As shown in Fig 1(i), RIC can correctly resubmit most of these samples according to neighbors from different classes. Moreover, an ablation study is carried out on this dataset to compare the contribution of each step in RIC, and the results are reported in Table 2. We can find that with the addition of each step in RIC, the performance continues to improve, which verifies the effectiveness of each step in RIC.

2) RIC vs. comparisons in Keel datasets.

In this experiment, ten Keel datasets are employed to further investigate the effectiveness of the proposed approach by comparing it to other comparison approaches in real word datasets. We take $K = 7$ and $\gamma = 0.15$ in RIC, and the parameters in comparison approaches are as default. The classification results of different approaches for classifying different approaches are reported in Table 3. We can observe that the proposed RIC generally provides better performances than comparison approaches in most datasets. The reason is that RIC evaluates the local reliability of different classifiers in classifying each test sample, making each global optimal classifier focus on the sample locally.

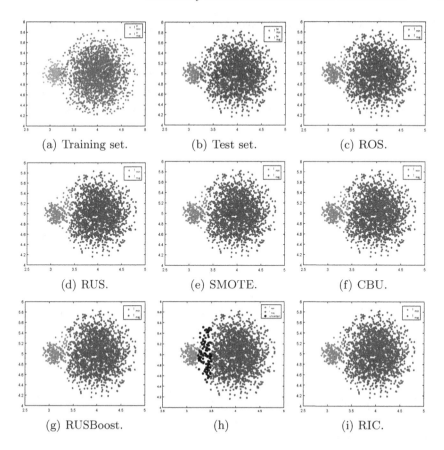

Fig. 1. Classification results of the synthetic dataset by different approaches.

Table 2. The values of FM and GM of different approaches for the synthetic dataset (IN %)

Indexes	ROS	RUS	SMOTE	CBU	RUSBoost	$RIC_{Step\ 1}$	$RIC_{Step\ 1-2}$	$RIC_{Step\ 1-3}$
FM	90.45	85.84	90.25	88.30	90.25	85.47	86.02	**94.06**
GM	98.72	98.34	98.70	98.67	98.70	98.29	98.36	**98.92**

3.3 Influence of K and γ

In this experiment, we employ the synthetic dataset to investigate the performance of RIC with various values of K and γ. The classification results of RIC with various parameters are shown in Fig. 2, where the x-coordinate denotes the value of K, ranging from 5 to 15, and the y-coordinate represents the value of γ, which is expressed in $[0, 1]$. The z-coordinate of Fig. 2(a) and (b) is the value of FM and GM, respectively. We can see that with the increase of K, the variations of the result are small, which verifies it is robust to the value of K. Of course,

Table 3. Classification results for Keel datasets by different approaches (IN %)

Indexes	Datasets	ROS	RUS	SMOTE	CBU	RUSBoost	RIC
FM	ecoli2	72.38	72.40	71.45	70.97	69.30	**83.70**
	ecoli4	73.21	60.64	74.35	63.69	68.55	**84.14**
	led7digit	61.99	58.16	64.44	60.15	60.16	**80.03**
	glass1	57.36	56.59	55.81	56.10	49.76	**66.26**
	page-blocks0	22.87	22.59	18.38	31.54	46.84	**80.54**
	shuttle2vs5	93.61	74.01	93.61	73.64	93.02	**97.89**
	vehicle2	92.04	91.90	92.79	89.42	80.80	**94.03**
	vowel0	78.23	72.45	81.58	72.29	75.25	**94.10**
	yeast1	58.15	58.29	58.57	57.24	57.87	**58.84**
	yeast5	46.69	38.65	46.96	37.78	56.96	**58.41**
GM	ecoli2	90.77	89.72	90.39	89.15	88.05	**93.36**
	ecoli4	95.28	95.08	92.92	93.79	87.00	**96.20**
	led7digit	88.15	89.66	87.51	88.82	87.95	**90.04**
	glass1	61.11	54.14	54.63	56.55	64.40	**71.87**
	page-blocks0	50.26	46.02	38.21	48.85	83.51	**87.94**
	shuttle-2vs5	96.69	99.46	96.69	99.43	**99.59**	98.96
	vehicle2	95.62	95.69	96.32	94.80	88.95	**96.93**
	vowel0	95.23	95.65	96.23	95.59	86.74	**97.31**
	yeast1	70.60	70.58	70.20	69.09	70.15	**71.18**
	yeast5	96.43	94.98	96.46	94.79	**96.49**	95.49

K is not the higher the better. The result in Fig. 2 reveals that the value of GM tends to decrease when the K is taken too large, which may be affected by noise samples. Thus, we recommend $K \in [5, 12]$ as the default in applications. Moreover, we can also observe that the value of γ should not taken too small, since RIC may fail to fully mine the uncertain information in such a case. Thus, $\gamma \in [0.02, 0.15]$ can be recommended in applications.

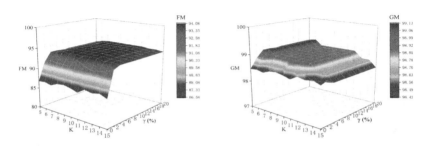

Fig. 2. Classification results of RIC with different parameters in the synthetic dataset.

3.4   Execution Time

The execution time in seconds of RIC and other comparison approaches on different datasets with SVM as shown in Table 4. We can see that the execution time of RIC is higher than other approaches since it needs to calculate a large number of distances between samples to obtain neighbors. In applications, the proposed RIC approach is more suitable for cases where high accuracy is required, whereas efficient computation is not a vital requirement.

Table 4. Execution time of different methods (In seconds)

Datasets	ROS	RUS	SMOTE	CBU	RUSBoost	RIC
ecoli2	0.1946	0.0512	0.1412	0.1800	2.9829	0.7967
ecoli4	0.1773	0.0589	0.1374	0.1389	2.7882	2.0548
led7digit	0.1885	0.0633	0.1487	0.1619	2.9539	1.5250
glass1	0.2001	0.0543	0.1380	0.1602	2.9692	0.4475
page-blocks0	216.1961	18.9831	186.3028	14.4645	3.0455	251.9016
shuttle-2vs5	71.1344	0.0591	70.4835	1.4666	2.2398	12.8517
vehicle2	14.9956	3.4566	14.5732	1.7628	3.0990	22.3558
vowel0	0.3269	0.0724	0.3118	0.2386	3.0903	1.7887
yeast1	0.2646	0.0633	0.2313	0.2744	3.0286	1.9574
yeast5	0.2919	0.0557	0.1942	0.2191	3.2067	4.5351

4   Conclusion

This paper proposes a reliability-based imbalanced data classification approach (RIC) with Dempster-Shafer theory. RIC considers not only the global optimization of different classifiers, but also the local optimization. Thus, we can obtain a more robust and reasonable performance for each test sample. The experiments on synthetic and several real imbalanced datasets have verified the effectiveness of RIC with respect to typical approaches. Moreover, we also investigate the influence of parameters on the classification performance of RIC. In the future, we will extend the application scope of RIC to other real-word tasks, such as network intrusion detection.

Acknowledgments. This work was supported by the National Natural Science Foundation of China under Grant U20B2067, Grant 61790552, and Grant 61790554; the Aeronautical Science Foundation of China under Grant 201920007001.

References

1. Zhu, Z., Wang, Z., Li, D., Du, W.: Globalized multiple balanced subsets with collaborative learning for imbalanced data. IEEE Trans. Cybern. **52**(4), 2407–2417 (2022)
2. Chaovalitwongse, W.A., Fan, Y.J., Sachdeo, R.C.: On the time series k-nearest neighbor classification of abnormal brain activity. IEEE Trans. Syst. Man Cybern. A Syst. Hum. **37**(6), 1005–1016 (2007)
3. Kashef, R.: A boosted SVM classifier trained by incremental learning and decremental unlearning approach. Expert Syst. Appl. **167**, 114154 (2021)
4. He, H., Garcia, E.A.: Learning from imbalanced data. IEEE Trans. Knowl. Data Eng. **21**(9), 1263–1284 (2009)
5. Wang, N., Liang, R., Zhao, X., Gao, Y.: Cost-sensitive hypergraph learning with f-measure optimization, IEEE Trans. Cybern. (2021)
6. Seiffert, C., Khoshgoftaar, T.M., van Hulse, J., et al.: RUSBoost: a hybrid approach to alleviating class imbalance. IEEE Trans. Syst. Man Cybern. A Syst. Hum. **40**(1), 185–197 (2009)
7. Dempster, A.P.: Upper and lower probabilities induced by a multivalued mapping. Ann. Statist. **83**, 325–339 (1967)
8. Shafer, G.: A Mathematical Theory of Evidence. Princeton Univ, Press (1976)
9. Liu, Z., Pan, Q., Dezert, J., Han, J.W., He, Y.: Classifier fusion with contextual reliability evaluation. IEEE Trans. Cybern. **48**(5), 1605–1618 (2018)
10. Zhang, Z.W., Tian, H.P., Yan, L.Z., Martin, A., Zhou, K.: Learning a credal classifier with optimized and adaptive multiestimation for missing data imputation, IEEE Trans. Syst. Man Cybern. (2021)
11. Niu, J., Liu, Z., Lu, Y., Wen, Z.: Evidential combination of classifiers for imbalanced data. IEEE Trans. Syst. Man Cybern. (2022)
12. Zhang, Z.W., Ye, S.T., Zhang, Y.R., Ding, W.P., Wang, H.: Belief combination of classifiers for incomplete data. IEEE-CAA J. Automatica Sin. **9**(4), 652–667 (2022)
13. Grina, Fares, Elouedi, Zied, Lefevre, Eric: A preprocessing approach for class-imbalanced data using SMOTE and belief function theory. In: Analide, Cesar, Novais, Paulo, Camacho, David, Yin, Hujun (eds.) IDEAL 2020. LNCS, vol. 12490, pp. 3–11. Springer, Cham (2020). https://doi.org/10.1007/978-3-030-62365-4_1
14. Niu, Jiawei, Liu, Zhunga: Imbalance data classification based on belief function theory. In: Denœux, Thierry, Lefèvre, Eric, Liu, Zhunga, Pichon, Frédéric. (eds.) BELIEF 2021. LNCS (LNAI), vol. 12915, pp. 96–104. Springer, Cham (2021). https://doi.org/10.1007/978-3-030-88601-1_10
15. Smets, P.: Constructing the pignistic probability function in a context of uncertainty. Uncertainty Artif. Intell. **5**, 29–39 (1990)
16. Zhang, Y., Liu, G., Luan, W.: An approach to class imbalance problem based on stacking and inverse random under sampling methods, In: 2018 IEEE 15th International Conference on Networking, Sensing and Control (ICNSC), pp. 1–6. IEEE (2018)
17. Chawla, N.V., Bowyer, K.W., Hall, L.O., et al.: SMOTE: synthetic minority oversampling technique. J. Artif. Intell. Res. **16**, 321–357 (2002)
18. Lin, W.C., Tsai, C.F., Hu, Y.H., et al.: Clustering-based undersampling in class-imbalanced data. Inf. Sci. **409**, 17–26 (2017)

Evidential Regression by Synthesizing Feature Selection and Parameters Learning

Chao Liu, Zhi-gang Su[✉], and Gang Zhao

National Engineering Research Center of Power Generation Control and Safety,
School of Energy and Environment, Southeast University, Nanjing 210096, China
zhigangsu@seu.edu.cn

Abstract. In this paper, we propose an evidential regression (EVREG) that can simultaneously perform feature selection and model parameters learning. To achieve these two functions, an evaluation function for the significance of the features to be selected is proposed, which contains two terms, one for evaluating the impact of the feature to be selected on the prediction accuracy and one for evaluating the redundancy between it and the already selected features, respectively. According to the sequentially forward selection strategy, the features with high degree of significance are selected one by one, while irrelevant or redundant features are eliminated. In contrast to traditional EVREG, instead of performing feature selection as a separate data pre-processing, this method combines it with the parameters learning in model training, thus improving the accuracy of regression prediction. The effectiveness of this method was verified by implementing it on Friedman dataset.

Keywords: Evidential regression · Feature selection · Parameters learning · Mutal information

1 Introduction

Regression is a statistical technique that learns functional relationship among variables based on a learning set $\{(\boldsymbol{x}_i, y) \mid i = 1 \text{ to } n\}$, and uses the learned relationship to make predictions. Many methods have been proposed to estimate the regression function, such as k-nearest neighbor regression, SVM regression, decision tree regression, in particular, the evidential regression (EVREG) proposed by Petit-Renaud and Denoeux [1] takes into account the imprecision and uncertainty of observations, and does not need to assume perfect knowledge of the value of the response y for given sample set. It has received a lot of attention in the field of machine learning.

This work is supported by the National Natural Science Foundation of China under Grants 52076037 and 51976032.

Irrespective of the regressor used, superfluous information in the training set have a serious impact on parameters learning and prediction effects. In order to simplify and clean up the training set, there are usually two typical approaches: instance selection [2,3] and feature selection, with the latter being the focus of this study. The purpose of feature selection is to eliminate irrelevant and redundant parts of the original features and to select a smaller subset of features that are sufficient for prediction. In general, the widely used feature selection methods can be divided into three categories: filter, wrapper and embedded methods. For the filter method, feature selection can be regarded as a data pre-processing process before the model parameters are learned. The importance of each feature to the target variable is evaluated based on the inherent nature of the data or statistical criteria, the features most relevant to the target variable can be selected by setting a threshold value. This method is fast and simple to calculate, however, feature selection is conducted without considering its effect on regression prediction accuracy, which resulted in parameters learning and feature selection being independent of each other, and thus the selected features may have less impact on improving prediction ability. While the wrapper and embedded methods perform feature selection and model parameters learning simultaneously, such as sequential selection algorithms [4] and direct objective optimization methods [5], where the selected feature subsets are determined by the predictive effect.

For EVREG, feature selection is more often used as a data preprocessing, just like the filter method, such as Chen et al. [6] used the frequency ratio method to give the coefficient weights of each feature and implemented the selection with SWM method. However, for the evidential classifier, Su et al. [7] proposed a REK-NN classification rule that can perform feature selection and classification simultaneously, and inspired by this approach, this paper aims to propose a method for evidential regression that can perform feature selection and parameters learning simultaneously. More precisely, we want to define a feature evaluation criterion to measure the significance of a feature to be selected, taking into account its contribution to the predictive effect and the redundancy between the already selected features. There are various methods that can be used to evaluate the relationship between features, such as Pearson's correlation coefficient, Euclidean distance, and mutual information. The mutual information based on Shannon's information theory [8], which is not affected by whether the variables are linearly related to each other, has received wide attention in the field of feature selection. Therefore, we choose mutual information to describe the redundancy between features, and based on that we propose an evidential regression with feature selection and parameters learning functions.

The rest of this paper is organized as follows. In Sect. 2, the concept of Dempster-Shafer theory and EVREG is briefly introduced. Section 3 introduces the EVREG with feature selection and parameters learning functions. In Sect. 4, the validity of this method is verified with a synthetic dataset. Conclusion is drawn in Sect. 5.

2 Preliminaries

2.1 Dempster-Shafer Theory

In this section, we will introduce the basic concepts of belief function theory. More material on belief function can be found for instance in [9]. Let Ω be a finite set called the frame of discernment, which contains all possible answers to a given question of interest Q. And let 2^Ω be the set of all subsets of Ω. Then a *mass function* can be defined as a mapping from 2^Ω to $[0, 1]$, verifying:

$$\sum_{A \subseteq \Omega} m_y(A) = 1. \tag{1}$$

Mass function, also called basic belief assignment(BBA), represents the uncertainty about y when an evidential corpus EC is given. Each number $m_y(A)$ represents the belief assigned to the hypothesis that "$y \in A$", and that cannot be assigned to any more restrictive hypothesis, given the available knowledge. Any subset of Ω such as $m(A) > 0$ is called a focal element of m, denoted by $\mathscr{F}(m)$.

Let m_1 and m_2 be two mass functions induced by two distinct sources separately. The conjunctive combination of m_1 and m_2, denoted by $\bigcirc$, yields the following unnormalized mass function:

$$m_{1\bigcirc2}(A) = \sum_{B \cap C = A} m_1(B)m_2(C), \forall A \subseteq \Omega. \tag{2}$$

The normality condition $m(\phi) = 0$ can be recovered by dividing each mass $m_{1\bigcirc2}(A)$ by $1 - m_{1\bigcirc2}(\phi)$. This operation is noted $\oplus$ and is called Dempster's rule of combination:

$$m_{1\oplus2}(A) = \frac{m_{1\bigcirc2}(A)}{1 - m_{1\bigcirc2}(\phi)}, \phi \neq A \subseteq \Omega. \tag{3}$$

Both operations are associative, commutative and admit the vacuous BBA as a unique neutral element.

It may occur that we have some doubt about the reliability of the source of information inducing m. To solve this problem, the discount of BBA is proposed:

$$m^\alpha(A) = \begin{cases} (1 - \alpha)m(A), & \forall A \subset \Omega, \\ \alpha + (1 - \alpha)m(A), & A = \Omega. \end{cases} \tag{4}$$

The discount rate $\alpha \in [0, 1]$ characterizes the degree of reliability of information provided by the source. If $\alpha = 0$, it means the information is absolutely reliable. On the contrary, the information is absolutely unreliable.

The above three functions m, bel, pl describes a belief state, but not suitable for decision making. For decision making from mass functions, the pignistic transformation is proposed [9]:

$$BetP(\omega) = \sum_{\{A \subseteq \Omega, \omega \in A\}} \frac{m^*(A)}{|A|}, \tag{5}$$

where $|\cdot|$ denotes the cardinality of a focal element.

2.2   EVREG: Evidential Regression

For evidential regression problem, the frame of discernment Ω represents all possible values of the output variable y, usually taken as the output in training set T and denoting as $U_{(y_{min}, y_{max})}$. U represents the uniform distribution, and y_{min}, y_{max} represents the minimum and maximum values of the output in T. Let x be an arbitrary vector to be predicted, y be the corresponding unknown output and $\mathcal{N}_K(x)$ be the set of K nearest neighbors of x in T. The information on y can be deduced from the neighbors. Each neighbor x_i in $\mathcal{N}_K(x)$ with output y_i can provide a piece of evidence about the possible value of y, which can also be represented by the following mass function:

$$m_i(y = A) \mid x_i) = \begin{cases} \phi(d_i), & A = y_i, \\ 1 - \phi(d_i), & A = \Omega, \end{cases} \tag{6}$$

where ϕ is a decreasing distance function from $\mathbb{R}^+$ to $[0, 1]$ verifying $\phi(0) \in (0, 1]$ and $\lim_{d \to \infty} \phi(d) = 0$. Function ϕ is called a discounting function. In (9), $1 - \phi(d_i)$ is defined as the discount rate α_i, which determines the influence of x_i on x. α_i is close to 0 if x_i is close to x. The distance used here is the Euclidean distance: $d_i = \|x - x_i\|^{1/2}$. So a natural choice for ϕ is:

$$\phi(d) = \theta \exp(-\gamma d^2), \tag{7}$$

where $\theta \in (0, 1)$ is a constant parameter, in this paper, we set θ to 0.95. And $\gamma > 0$ is an important structure parameter that controls the decay gradient of the distance function.

DS rule is used to combine the information provided by each neighbor in $\mathcal{N}_K(x)$, then the final BBA is:

$$m = \oplus_{i=1}^{K} m_i(y = A) \mid x_i). \tag{8}$$

After the final BBA is calculated, various forms of output can be obtained. For example, the probability density function about the value of y, $BetP(y)$ can be obtained through the evidential decision-making process. Based on that, the point prediction $\widehat{y}$ can be calculated:

$$\widehat{y} = \sum_{i=1}^{K} m(y = y_i) \cdot y_i + m(y = \Omega) \cdot \frac{y_{min} + y_{max}}{2}. \tag{9}$$

γ is an important structural parameter that directly determines the prediction effect of EVREG. To complete the model training of EVREG, γ needs to be identified and optimized in order to complete the model training of Evreg. K-fold cross-validation is applied to optimize γ by minimizing the following criterion:

$$CV(\gamma) = \frac{1}{k} \sum_{j=1}^{k} (\frac{k}{N} \sum_{i=1}^{\frac{N}{k}} (y_i - \widehat{y}_i[x_i, T - T_j, \gamma])^2, \tag{10}$$

where $T - T_j$ is the training set without the validation sets $T_j = \{x_j | j = 1, 2, \ldots, \frac{N}{k}\}$. The estimator $\widehat{\gamma}$ of parameter γ is then obtained by minimizing this criterion:

$$\widehat{\gamma} = \arg\min_{\gamma} CV(\gamma). \tag{11}$$

3   Proposed Method

In this section, we will propose an evidential regression with feature selection and parameters learning functions.

3.1   Construction of Evaluation Function

This method can be formulated as an optimization problem through evaluating the significance of features, searching optimal neighborhood size K^* and its corresponding γ^*, and minimal feature subset $\mathcal{B}^*$. Formally speaking, we want to solve

$$(K^*, \mathcal{B}, \gamma^*) = \arg\min_{K, \mathcal{B}, \gamma} \mathcal{J}(K, \mathcal{B}, \gamma) \tag{12}$$

with an objective function $\mathcal{J}$ defined as follows:

$$\mathcal{J}_{(K, \mathcal{B}, \gamma)} = CV_{(K, \mathcal{B}, \gamma)} + \lambda \frac{\sum_{b_i, b_j \in \mathcal{B}} NI(b_j, b_i)}{|\mathcal{B}|}, i = 1, \ldots, |\mathcal{B}| - 1, j > i. \tag{13}$$

The first term CV reflects the fitting ability on the training set and can approximate the prediction accuracy of the model. The second term is a penalty item used to describe the redundancy between features, where NI represents the normalized mutual information, its specific calculation method can be referred to [10]. As described in Sect. 1, mutual information (MI) can describe the amount of information shared between variables, and the larger the MI, the greater the redundancy. The hyperparameters λ is a penalty factor, which controls the tradeoff between the two terms in (14), in this paper, it is set to 0.01.

Hence, the significance of one feature a to be selected relative to the feature subset $\mathcal{B}$, denoted by SIG is defined as follows:

$$SIG(a, K, \mathcal{B}, \gamma) = \mathcal{J}_{(K, \mathcal{B}, \gamma)} - \mathcal{J}_{(K, \mathcal{B} \cup a, \gamma)}. \tag{14}$$

The evaluation function (15) indicates that significance will increase while adding an informative feature. For a given K, the minimal feature subset $\mathcal{B}^*$ of the whole feature set $\mathcal{A}$ can be found when the condition $\forall a \in \mathcal{A} - \mathcal{B}$, $SIG(a, \mathcal{B}, \gamma) < 0$ can be satisfied.

3.2   Feature Selection and Parameters Learning

As previously stated, our purpose is to propose evidential regression with feature selection function, the search strategy is important to select the optimal feature subsets. There are numbers of candidate search strategies to select a minimal feature subset, e.g., the greedy search strategy such as sequentially forward selection (SFS) and sequentially backward selection (SBS), B&B search strategy, and GA-based feature selection. SFS is adopted for convenience. Firstly, for a given K, using only one feature for model training, optimize γ according to (12), iterate through all the features and select a_k satisfying the following criterion to add it to $\mathcal{B}$:

$$SIG(a_k, K, \mathcal{B}) = \max\{SIG(a_i, \mathcal{B}, K)\}. \tag{15}$$

Repeat this operation, increasing selected features one by one until SIG is no longer increasing to end the feature selection and optimization of γ for the current K. According to above interpretations, our method can be realized as follows.

Algorithm 1. Procedure of the proposed method

Input: traing data (x, y), feature set $\mathcal{A}$, bound $[\underline{K}, \overline{K}]$ for K, λ
          and testing data $x_t, t = 1, 2, \ldots, n_t$
Output: Optimal K^*, γ^*, selected feature subset $\mathcal{B}^*$, estimations $\widehat{y}_t$ of x_t
1: Calculate NI between each feature in $\mathcal{A}$
2: $K^* = \underline{K}$, $\gamma^* = 0$, $\mathcal{B}^* \leftarrow \emptyset$, $\mathcal{J}^* = \inf$
3: **For** $K = \underline{K}$ *to* $\overline{K}$ **do**
4:     $\mathcal{B} \leftarrow \emptyset$, $\gamma = 0$
5:     **while** $\mathcal{A} - \mathcal{B} \neq \emptyset$ **do**
6:         **For** each $a_i \in \mathcal{A} - \mathcal{B}$ **do**
7:             Compute $\widehat{\gamma}_i$, $SIG(a, K, \mathcal{B})$
8:         **end For**
9:         Select a_k: $SIG(a_k, K, \mathcal{B}) = \max\{SIG(a_i, \mathcal{B}, K)\}$
10:        **if** $SIG(a_k, \mathcal{B}, K) > 0$
11:            $\mathcal{B} \leftarrow \mathcal{B} \bigcup a_k$, $\gamma = \widehat{\gamma}_k$
12:        **else**
13:            break
14:        **end if**
15:    **end while**
16:    **if** $\mathcal{J}_{(K, \mathcal{B})} < \mathcal{J}^*$
17:            $\mathcal{B}^* \leftarrow \mathcal{B}$, $K^* \leftarrow K$, $\gamma^* = \gamma$, $\mathcal{J}^* \leftarrow \mathcal{J}$
18:    **end if**
19: **end For**

---

4   Numerical Experiment

In this section, we use synthetic datasets to show the prediction capabilities and feature selection effect of our method. The experiments were conducted on the Friedman dataset. Friedman regression problem is a synthetic dataset proposed in [11], which has 5 relevant variables. The formula is shown below:

$$f(x) = 10\sin(\pi x_1 x_2) + 20(x_3 - 0.5)^2 + 10x_4 + 5x_5 + \varepsilon, \tag{16}$$

where the inputs x_1–x_5 and irrelevant variables x_6–x_{10} are uniformly distributed over the interval $[0, 1]$ and a Gaussian white noise ε with variance of 0.1 is applied. In addition, to verify the ability to select redundant features by adding the NI penalty term, eight redundant variables x_{11}–x_{18} are applied: x_{11}–x_{16} are generated by linear combinations of the original five variables, x_{17}-x_{18} are generated by nonlinear transformations of the original variables, and x_{19}–x_{20} are generated by random permutations of the original two variables x_1 and x_2.

In this paper, we randomly generated 500 sets of data and then randomly divided them into the training data of 350 samples with the remainder making up the test data. All the data was normalized into interval $[0, 1]$. Model was trained on the training dataset with increasing K from 3 to 13. When optimizing for γ^*, five-fold cross-validation was used according to (10) and (11).

As described in Sect. 3, our method has two functions: searching optimal neighborhood size K^* and its corresponding γ^*, and performing feature selection to reduce the dimensionality of input, thus improving the prediction accuracy of the regressor. To verify its validity, predictions were made on the test data using the optimized K^*, γ^* and selected features, and the expected output $\hat{y}_t$ was calculated using (6)–(9). The mean square error (MSE) between the expected and true values is used as the evaluation metric for the prediction effectiveness.

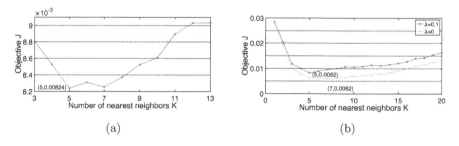

Fig. 1. Variation of objective $\mathcal{J}$ with the number of nearest neighbors K when λ is 0.01 (a) and with the order of selected variables (b).

Figure 1a shows the variation of $\mathcal{J}$ with the number of nearest neighbors K. It can be seen that as K increases from 3 to 13, $\mathcal{J}$ first decreases and then increases, and when K is taken as 5, $\mathcal{J}$ has the minimum value, which means the optimal value $K^*$is 5. The selected features $\mathcal{B}^*$ in this case is $\{4, 1, 2, 5, 3\}$, and the corresponding γ^* is 1.0390.

The hyperparameters λ are taken as 0 and 0.01 to illustrate the effect of adding the NI penalty term on the selection of redundant variables, respectively. The optimal K^* and selected variable subset for the two cases are shown in Table 1. The objective $\mathcal{J}$ varying with the order of selected variables when K takes 5 are shown in Fig. 1b.

Table 1. Optimized parameters and selected variables with different λ

λ	K^*	Selected variables $\mathcal{B}^*$	γ^*	$\mathcal{J}^*_{(K^*, \mathcal{B}^*)}$
0	5	4, 1, 2, 5, 3, 11, 16	0.9755	0.0062
0.01	5	4, 1, 2, 5, 3	1.0390	0.0082

From Table 1 and Fig. 1b, we can find that when the number of selected variables increases from 5 to 6, the objective $\mathcal{J}$ stops its decreasing trend and starts to gradually increase, that can avoid redundant variables from being selected, thus proving the validity of the penalty term on the selection of redundant variables.

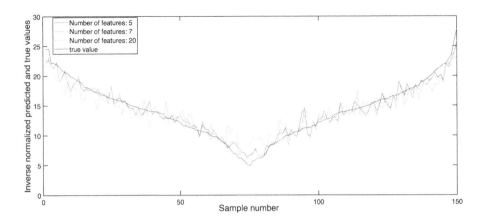

Fig. 2. Predicted values after inverse normalization against the true values.

To illustrate the effect of feature selection on prediction accuracy, predictions were made on the test data using the subset of variables $\mathcal{B}_1^*$ $\{4, 1, 2, 5, 3, 11, 16\}$ and $\mathcal{B}_2^*$ $\{4, 1, 2, 5, 3\}$, which are selected when λ is 0 and 0.01, and the unselected 20 original variables $\mathcal{A}$, respectively. MSE for the three cases are presented in Table 2, and for ease of observation, Fig. 2 plots the predicted values after inverse normalization against the true values.

Table 2. MSE on the test data when using different sets of variables for prediction

Variables set	MSE (normalized)	MSE (inverse normalized)
$\mathcal{B}_1^*$	0.0060	3.7058
$\mathcal{B}_2^*$	0.0057	3.5533
$\mathcal{A}$	0.0142	8.7938

It can be found that the prediction results after performing feature selection are significantly better than those using all variables for prediction. The prediction results using $\mathcal{B}_1^*$ and $\mathcal{B}_2^*$ are close in Fig. 2, from Table 2 we can clearly find that the prediction accuracy after adding the penalty term is also improved compared to when it is not added.

5   Conclusion

In this paper, an improved evidential regression is proposed, which can synchronously accomplish feature selection and parameters learning: the determination of the optimal number of nearest neighbors K and the corresponding structural parameters γ. By adding the penalty term about mutual information in the objective function, the redundant features are effectively reduced and the accuracy of regression prediction is improved. A synthetic dataset is used to show the performance of this method. Some practical issues and more experimental validation are not presented and will be discussed in our subsequent study.

References

1. Petit-Renaud, S., Denoeux, T.: Nonparametric regression analysis of uncertain and imprecise data using belief functions. Int. J. Approximate Reasoning **35**(1), 1–2 (2004)
2. Arnaiz-Gonzalez, A., Diez-Pastor, J.F., Garcia-Osorio, C., Rodriguez, J.J.: Instance selection for regression: adapting drop. Neurocomputing **201**, 66–81 (2016)
3. Gong, Chaoyu, Wang, Pei-hong, Su, Zhi-gang: An interactive nonparametric evidential regression algorithm with instance selection. Soft Comput. **24**(5), 3125–3140 (2020). https://doi.org/10.1007/s00500-020-04667-4
4. Nakariyakul, S., Casasent, D.P.: An improvement on floating search algorithms for feature subset selection. Pattern Recogn. **42**(9), 1932–1940 (2009)
5. Guyon, I., Weston, J., Barnhill, S., Vapnik, V.: Gene selection for cancer classification using support vector machines. Mach. Learn. **46**(1–3), 389–422 (2002)
6. Chen, W., Li, Y., Tsangaratos, P., Shahabi, H., Bian, H.: Groundwater spring potential mapping using artificial intelligence approach based on kernel logistic regression, random forest, and alternating decision tree models. Appl. Sci. **10**(2), 425 (2020)
7. Su, Z., Hu, Q., Denaeux, T.: A distributed rough evidential k-nn classifier: integrating feature reduction and classification. IEEE Trans. Fuzzy Syst. **PP**(99), 1 (2020)
8. Shannon, C.E.: A mathematical theory of communication, 1948. Bell Syst. Techn. J. **27**(3), 3–55 (1948)
9. Denoeux, T.: Conjunctive and disjunctive combination of belief functions induced by nondistinct bodies of evidence. Artif. Intell. **172**(2–3), 234–264 (2008)
10. Kraskov, A., Stoegbauer, H., Grassberger, P.: Estimating mutual information (2003)
11. Friedman, J.H.: Multivariate adaptive regression splines. Annal. Stat. **19**(1), 1–67 (1991)

Algorithms and Evidential Operators

Distributed EK-NN Classification

Chaoyu Gong[1,2,4(✉)], Zhi-gang Su[1,2(✉)], Qian Wang[3], and Yang You[4]

[1] National Engineering Research Center of Power Generation Control and Safety,
Nanjing 210096, China

[2] School of Energy and Environment, Southeast University, Nanjing 210096, China
{chaoyugong,zhigangsu}@seu.edu.cn

[3] School of Energy and Power, Jiangsu University of Science and Technology,
Zhenjiang 212003, China

[4] Schoold of Computing, National University of Singapore,
Singapore 119077, Singapore

Abstract. For big data, the Evidential K-Nearest Neighbor (EK-NN) classifier is still impractical due to the restrictions of time and memory. In both the training and testing stage, searching for K closest neighbors requires intensive quadratic computation and has to be repeated for each input sample. To address this issue, we propose a distributed EK-NN classifier, named Global Exact EK-NN, for fast processing with Apache Spark. We compare the proposed classifier, which can be scaled to 48 nodes (2688 cores) at a cluster named the Texas Advanced Computing Center Frontera, with several other parallel K-NN based algorithms over 4 large datasets. Our method is able to achieve state-of-the-art scaling efficiency and accuracy on the large datasets having more than 10 million samples.

Keywords: EK-NN · Apache spark · Distributed calculation

1 Introduction

As a powerful nonparametric learning method, the K-nearest neighbor (K-NN) technique has been widely used for performing classification tasks. K-NN adopts a distance measure to search K nearest neighbors (KNNs) of a testing sample among training set, then assigns it with a label voted by a majority of the KNNs. Different from the eager learners that establish a mathematical model, K-NN classifier belongs to the family of lazy learners. To better describe the imperfect knowledge in class labels, the evidential K-NN (EK-NN) classifier is proposed in [1] under the framework of Dempster-Shafer (DS) theory [2] and proved to usually outperform K-NN and fuzzy weighted K-NN (FK-NN) classifier [3].

In EK-NN, KNNs of one testing sample provide pieces of evidence supporting hypotheses regarding its class membership. Each evidence is formulated as a mass function concerning the dissimilarity between one neighbor and such a

This paper is supported by the National Natural Science Foundation of China (52076037, 51976032).

sample. By pooling those K pieces of evidence based on the Dempster's rule [4], a decision about the predicted label (single) of a testing sample can be then made. In this way, the EK-NN classifier describes the ambiguous and uncertain class membership and has been used in a variety of fields [5].

As far as we know, there are generally three schemes to enhance the quality of EK-NN classifier. The first scheme is to improve information transmission and fusion between KNNs and their corresponding samples. In [6], a class of parametric t-norm based combination rule [7] is proposed and used to substitute for the Dempster's rule in EK-NN. In this way, the hypothesis of evidence independence is relaxed, which is often not be guaranteed in practice. By maximizing an evidential version of likelihood [8], the classical discounting operation of evidence is replaced by the contextual discounting. The second scheme is to expand the representation of class labels. For example, methods introduced in [9] allow to classify samples in some specific classes, such as meta-classes defined by a union of several specific classes and an ignorant class for outlier detection. The last scheme is utilizing the ensemble classifier system [10] to adapt EK-NN for solving complex classification, consisting of mainly two distinct levels: generation of base classifiers and combination of their output predictions. More recently, we can mention a rough EK-NN [11] integrating feature reduction and classification to deal with the curse of dimensionality. Nevertheless, none of these methods is focused on the classification of big data.

Over the two decades, technology improvements result in a relatively inexpensive and automatic way to gather data. This brings a significant overhead for machine learning tasks in terms of processing huge datasets [12]. For EK-NN, the large number of samples is particularly a crucial factor affecting its performance, although it has become a widely used method in machine learning applications. That is to say, KNN search has to traverse all training/testing samples. Due to the computation of pairwise dissimilarity and corresponding sorting operation, EK-NN becomes further impractical in big data applications. Therefore, how to improve the performance of EK-NN classifier on huge datasets is still an open problem.

Motivated by above discussion, we propose a distributed EK-NN classifier for big data on Apache Spark [13–15] that is arguably considered as a more flexible engine, by the name of Global Exact EK-NN (GE^{2}K-NN), to remove the bottleneck of computation. For GE^{2}K-NN, the training set is partitioned into several splits and candidate KNNs in the corresponding split of each testing sample are found by map operations. Then, all these candidate KNNs are merged to obtain exact KNNs and output the classification result after combining K pieces of evidence. The novelties of this work are summarized as follows:

- EK-NN classifier is implemented on Apache Spark as a distributed version for the first time to conduct classification for huge datasets;
- GE^{2}K-NN shows great scalability in experiments on both a single computer and supercomputer cluster without losing prediction accuracy.

2   Preliminaries

2.1   EK-NN: Evidential K-NN Classifier

Consider the problem of classifying samples into c classes, set of which is denoted by a frame of discernment $\Omega = \{\omega_1, \omega_2, \cdots, \omega_c\}$. The training set $\mathcal{T} = \{(\mathbf{x}_1, \omega_1), (\mathbf{x}_2, \omega_2), \cdots, (\mathbf{x}_n, \omega_n)\}$ consists of n samples, where $\mathbf{x}_i \in \mathbb{R}^{p \times 1} = (x_{i1}, x_{i2}, \cdots, x_{ip}), i = 1, 2, \cdots, n$, is a p-dimensional vector, p is the number of dimensions and $\omega_i \in \Omega$ is the label of $\mathbf{x}_i$.

Denote a testing sample to be classified as $\mathbf{x}_t$ and the KNNs of it in $\mathcal{T}$ as $\mathcal{N}_K(\mathbf{x}_t)$. K distinct pieces of evidence regarding the class membership of $\mathbf{x}_t$ are provided by neighbors $\mathbf{x}_{(j)} \in \mathcal{N}_K(\mathbf{x}_t)$ with $\omega_{(j)} = \{\omega_q\}, q = 1, 2, \cdots, c$. A small $d(\mathbf{x}_{(j)}, \mathbf{x}_t)$ means both patterns belong to the same class, and a large $d(\mathbf{x}_{(j)}, \mathbf{x}_t)$ means otherwise. Each item of evidence can be formalized to a basic belief assignment (BBA) over Ω defined by

$$\begin{cases} m_t(\{\omega_q\}|\mathbf{x}_{(j)}) = \alpha_0 \exp(-\gamma_q d(\mathbf{x}_{(j)}, \mathbf{x}_t)^2), \\ m_t(\Omega|\mathbf{x}_{(j)}) = 1 - \alpha_0 \exp(-\gamma_q d(\mathbf{x}_{(j)}, \mathbf{x}_t)^2), \end{cases} \tag{1}$$

where $m_t(\{\omega_q\}|\mathbf{x}_{(j)})$ is a *mass* function [4], α_0 is a constant such that $0 < \alpha_0 \leq 1$ and $\gamma_q > 0$ is a discounting parameter. Usually, $\alpha_0 = 0.95$ and γ_q is adaptively fixed according to data structure.

After obtaining all K items of evidence provided by neighbors in $\mathcal{N}_K(\mathbf{x}_t)$, the final mass function m_t regarding class of $\mathbf{x}_t$ can be synthesized,

$$m_t = \bigoplus_{\mathbf{x}_{(j)} \in \mathcal{N}_K(x_t)} m_t(\cdot|\mathbf{x}_{(j)}), \tag{2}$$

where the symbol $\bigoplus$ denotes the fusion of evidences by using Dempster' rule. More details can be found in [4].

Converting the mass function to the *pignistic probability distribution* [16]

$$BetP_t(\{\omega_q\}) = m_t(\{\omega_q\}) + \frac{m_t(\Omega)}{c}, \tag{3}$$

a decision regarding the assignment of $\mathbf{x}_t$ to a class can be made by

$$\hat{\omega}_t = \arg\max_{\omega_q \in \Omega} BetP_t(\{\omega_q\}). \tag{4}$$

Especially, when we have $m_t(\Omega) \approx 1$, i.e., $BetP_t(\omega_q) = \frac{1}{c}$ for all q, the class of $\mathbf{x}_t$ is unknown and it may be assigned to Ω.

2.2   Apache Spark

Apache spark [13], an open source processing engine for big data analysis, is originally designed under the MapReduce programming paradigm and can well coexist with the Hadoop [17] ecosystem. Spark copies the data from distributed

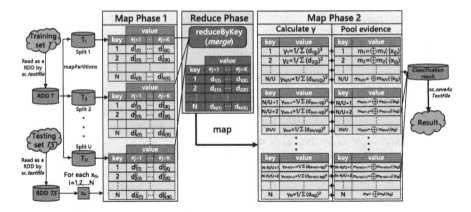

Fig. 1. Workflow of GE^{2}K-NN. Blue and red blocks respectively represent the map and reduce phase. (Color figure online)

physical storage to RAM memory, i.e., does in-memory computation, and thus processes data 100 times faster than Hadoop. The data structure used in Spark is named *Resilient Distributed Dataset* (RDD) [13]. Two kinds of distributed operation can be performed on RDD: transformations (a RDD is transformed into another RDD by it) and actions (one answer has to be given on a question). Based on these two operations, Spark follows a "lazy" approach such that a RDD is loaded into RAM only when an action rather than a transformation operation must be performed. Spark also has supplied a scalable machine learning library named MLlib on top of it.

3   GE^{2}K-NN: Global Exact EK-NN

The basic idea of GE^{2}K-NN can be stated as follows. After partitioning the training set $\mathcal{T}$ into U splits, the candidate KNNs of each testing sample $\mathbf{x}_t$ are firstly searched against the samples in local splits. That is to say, total $U \times K$ neighbors of $\mathbf{x}_t$ are determined. Then, these candidate neighbors are sorted according to the distance from them to $\mathbf{x}_t$ to find exact KNNs. Considering the information provided by KNNs, parameter γ defined in (1) is individually calculated. Ultimately, the classification rule (4) is conducted for every testing sample.

Denote testing set and the number of testing samples as TS and N. As shown in the left part of Fig. 1, the above idea of finding KNNs can be firstly realized using a divided-and-conquer approach under Spark:

– Training set $\mathcal{T}$ is firstly read as a RDD using a spark function *sc.textfile*, after declaring the SparkContext as *sc*. Using another spark function *mapPartitions*, RDD $\mathcal{T}$ is divided into U splits $\mathcal{T}_1, \mathcal{T}_2, \cdots, \mathcal{T}_U$ with approximately the same number of samples and being distributed among different computation cores. Along with these U splits, vectors of all $\mathbf{x}_t$ are regarded as input to map phase 1;

Algorithm 1: Function *reduceByKey(Merge)*

Input: U key-value pairs $(i, < d_{i(1)}^z, d_{i(2)}^z, \cdots, d_{i(K)}^z >)$ of $\mathbf{x}_{ti}$.

Output: The *value* array $out = < d_{i(1)}, \cdots, d_{i(K)} >$ denoting the distances between $\mathbf{x}_{ti}$ and its exact KNNs.

1 $z \leftarrow 1$, $mout_1 = < d_{i(1)}^1, \cdots, d_{i(K)}^1 >$

2 $mout_2 = < d_{i(1)}^2, \cdots, d_{i(K)}^2 >$

3 **while** $z \leq U - 1$ **do**

4   **if** $z \geq 2$ **then**

5          $\lfloor$ $mout_2 = < d_{i(1)}^{z+1}, \cdots, d_{i(K)}^{z+1} >$

6      $id_1 = 1$, $id_2 = 1$, $t \leftarrow 1$

7      $out = < \cdot, \cdots, \cdot >$, where $< \cdot, \cdots, \cdot >$ is an empty array with length K

8   **while** $t \leq K$ **do**

9   **if** $mout_1(id_1) \leq mout_2(id_2)$ **then**

10              $out(t) == mout_1(id_1)$

11   **if** $mout_1(id_1) == mout_2(id_2)$ **then**

12                  $\lfloor$ $t \leftarrow t + 1$, $id_2 \leftarrow id_2 + 1$

13              $id_1 \leftarrow id_1 + 1$

14   **else**

15              $\lfloor$ $out(t) = mout_2(id_2)$, $id_2 \leftarrow id_2 + 1$

16          $t \leftarrow t + 1$

17      $mout_1 = out$, $z \leftarrow z + 1$

- For each $\mathbf{x}_{ti}$, the candidate KNNs of it are searched in local split. Then, a *key-value* pair, where the *key* denotes the index i of $\mathbf{x}_{ti}$ and the *value* is an distances array, is obtained. The *value* array contains distances $<d_{i(1)}^z, d_{i(2)}^z, \cdots, d_{i(K)}^z>$ between $\mathbf{x}_{ti}$ and its candidate KNNs in zth, $z = 1, 2, \cdots, U$, split. Every *value* array is sorted in ascending order regarding the distance to x_i, i.e., $d_{i(1)}^z < d_{i(2)}^z < \cdots, d_{i(K)}^z$;
- After receiving all the candidate KNNs from U splits, these constructed U key-value pairs of $\mathbf{x}_{ti}$ are input into the reduce phase, i.e., a spark function *reduceByKey(merge)* that is summarized in Algorithm 1.

This function combines two sorted value array assigned to *mout1* and *mout2* to obtain the *out* array that is closer to the exact KNNs. The distance value $d_{i(t)}^z$ of each neighbor is compared one by one in each circle, starting with the closest neighbor. Concretely, if the $d_{i(t)}^z$ is smaller than the current value in one position, value of this position is updated with $d_{i(t)}^z$, otherwise we proceed with the following $d_{i(t)}^{z+1}$. After this circle, *out* is assigned to *mout1* and *mout2* is given to a new value array. Until all the candidate KNNs are combined, we can obtain the final distances between $\mathbf{x}_{ti}$ and its exact KNNs.

Then, all N testing samples associated with corresponding $<d_{i(1)}, d_{i(2)}, \cdots, d_{i(K)}>$ are input to map phase 2 by a *map* operation. As shown in the right part of Fig. 1, γ_i are parallel calculated as $\gamma_i = 1/\sum d_{i(j)}$, i.e., γ_i of each $\mathbf{x}_{ti}$ is determined adaptively according to $<d_{i(1)}, d_{i(2)}, \cdots, d_{i(K)}>$ rather than a learned value. After pooling K pieces of evidence, the classification results of testing samples are obtained based on (4) and collected from different computation cores. To visualize the result in RDD form, the $sc.saveAsTextFile$ function is used to transform it into a $.txt$ file.

The computational complexity of GE^{2}K-NN can be analyzed as follows. First, the computation cost to search local KNNs of testing samples is $O(Nnp + Nn \log \frac{n}{U})$, where the first term is to calculate distances between testing and training samples and the second one is sorting those obtained distances on U computation cores. Then, the reduce phase needs $O(NK \log NK)$ operations. Finally, in map phase 2 of Fig. 1, $O(NKp)$ operations need to be performed to calculate individual γ and make prediction by pooling evidences for each testing sample. The total computational complexity of GE^{2}K-NN is $O(Nnp + Nn \log \frac{n}{U} + NK \log NK + NKp) \sim O(Nnp + Nn \log \frac{n}{U})$.

4   Experiments

In this section, the performances of GE^{2}K-NN are statistically compared with that of other state-of-the-art distributed KNN-based algorithms on 4 big datasets (as summarized in Table 1) from UCI database [18]. The running time (RT) in seconds and speedup defined as speedup $= \frac{RT_{\min U}}{RT_U}$ are used to check the acceleration achieved when using more computation cores, where RT_U and $RT_{\min U}$ are running time executed respectively by U and the minimal number $minU$ of computation cores. The efficiency defined as efficiency $= \frac{\text{speedup} \cdot \min U}{U}$ is also used to check the acceleration efficiency by using U computation cores. The experiments of Sect. 4.1 are carried out on a server with Intel(R) Xeon(R) Gold 6230 CPU @ 2.10 GHz 160 cores 256 GB RAM, and a 10-fold cross-validation (10-fcv) procedure is used. We show experiments run on a supercomputer cluster and discuss the communication cost between worker nodes and master nodes in Sect. 4.2.

Table 1. Dataset description.

Dataset	#Samples	#Dimensions	#Classes
Covtype	581,012	54	7
Poker	1,025,010	11	10
Susy	5,000,000	19	2
Higgs	11,000,000	29	2

4.1   Performance Evaluation

We compare GE^{2}K-NN with other 3 state-of-the-art parallel KNN-based algorithms: an exact K-NN classifier, named as K Nearest Neighbor-Iterative Spark (KNN-IS) [19]; an approximate FK-NN classifier based on local approximate nearest neighbor (ANN) search, named Local Hybrid Spill Tree FK-NN (LHS-FKNN) [20]; an approximate FK-NN classifier using hybrid tree to find ANNs globally, named Global Approximate Hybrid Spill Tree FK-NN (GAHS-FKNN) [20], which are also scaled up under Apache Spark. Hyper-parameters of KNN-IS, LHS-FKNN and GAHS-FKNN are set to recommended values according to corresponding literature.

Table 2. Comparative results among different algorithms.

Algorithm	K	Covtype	Poker	Susy	Higgs
KNN-IS	3	0.9363	0.5180	0.7114	0.6402
	5	0.9326	0.5314	0.7262	0.6421
	7	0.9291	0.5408	0.7303	0.6420
GAHS-FKNN	3	0.9415	0.5270	0.7210	0.6401
	5	0.9397	0.5379	0.7266	0.6384
	7	0.9391	0.5487	0.7218	0.6593
LHS-FKNN	3	0.9385	0.5365	0.7253	0.6496
	5	0.9337	0.5489	0.7254	**0.6615**
	7	0.9303	0.5526	0.7295	0.6586
GE^{2}K-NN	3	0.9465	0.5571	0.7187	0.6413
	5	**0.9470**	0.5998	0.7346	0.6448
	7	0.9462	**0.6107**	**0.7388**	0.6588

The comparative experiment results are reported in Table 2, where the bolded and underlined value(s) indicates the best performance. As can be seen, the proposed GE^{2}K-NN has the best performance in most cases. The biggest gaps between GE^{2}K-NN and remaining algorithms are reflected on Poker dataset, on which GE^{2}K-NN achieves the accuracy of 0.6107 whereas the accuracy of other 3 algorithms are lower than 0.5526.

Then, we set $K = 7$ and analyze the scalability of different algorithms. The running time with different numbers of computation cores is shown in Fig. 2. Comparing the KNN-IS and GE^{2}K-NN without a training stage, we can find that GE^{2}K-NN and KNN-IS almost cost the same amount of time, and both of them show well scalability. Note that the evidence combination in GE^{2}K-NN can be completed by some matrix operations, whereas the voting process in KNN-IS is done by a loop that traverses all the samples on local cores. This is why GE^{2}K-NN is not slower than KNN-IS even though it has an additional step of calculating the adaptive γ. As U increases, the time consumed by the GE^{2}K-NN decreases significantly. When $U = 320$, the GE^{2}K-NN consumes less time than algorithms GAHS-FKNN and LHS-FKNN on the data sets Covtype, Poker and higgs.

The speedup values of the distributed algorithms are shown in Table 3. We set the maximum value of U to 320 that is greater than 160 (the number of available cores of the used server), to achieve faster running speed of the algorithms.

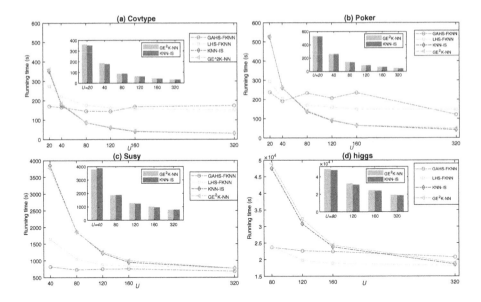

Fig. 2. Running time vs. number of cores with $K = 7$.

This is because that Apache Spark can queue the remaining map tasks and dispatch them as soon as anyone has been finished [13]. From this, we can see that GE^{2}K-NN can be linearly scaled up when the numbers of used cores are less than 160. On every dataset, GE^{2}K-NN has comparable scalability with KNN-IS. When U is increased from 160 to 320, the speedup can not continue to increase linearly, because the available computing resources are only 160 cores and there are remaining tasks waiting to be executed in a queue. Focusing on the GAHS-FKNN and LHS-FKNN, they do not show good scalability, because the HS tree method is used in their testing stage and this method can not be significantly accelerated by adopting more computing resources [20]. It indicates that GE^{2}K-NN will cost less running time when there are enough computation cores, compared with the GAHS-FKNN and LHS-FKNN.

Table 3. Speedup values of different distributed algorithms. The minimal number of computation cores are set to 80 on the largest Higgs dataset, and are set to 40 on the Susy.

Covtype	$U = 20$	40	80	120	160	320	Poker	$U = 20$	40	80	120	160	320
KNN-IS	1	1.98	4.06	5.88	8.65	11.24	KNN-IS	1	2.01	3.87	5.87	8.16	12.65
GAHS-FKNN	1	1.03	1.17	1.18	1.01	1.03	GAHS-FKNN	1	1.25	1.02	1.15	1.01	1.98
LHS-FKNN	1	1.24	1.56	1.68	1.75	1.81	LHS-FKNN	1	1.42	1.74	1.85	1.98	2.01
GE^{2}K-NN	1	1.95	4.17	5.76	8.42	11.52	GE^{2}K-NN	1	2.03	3.76	5.74	8.13	10.83
Susy	$U = 20$	40	80	120	160	320	Higgs	$U = 20$	40	80	120	160	320
KNN-IS	–	1	2.06	3.15	4.05	5.01	KNN-IS	–	–	1	1.54	1.99	2.56
GAHS-FKNN	–	1	1.12	1.09	1.08	1.2	GAHS-FKNN	–	–	1	1.05	1.06	1.14
LHS-FKNN	–	–	1	1.21	1.27	1.32	LHS-FKNN	–	–	1	1.21	1.27	1.32
GE^{2}K-NN	–	1	2.03	2.99	3.76	4.89	GE^{2}K-NN	–	–	1	1.52	2.01	2.57

4.2   Multi-node Experiments on TACC Frontera

Our methods are scaled up on the supercomputer cluster named TACC Frontera in the section. The CPU specifications of each node are Model Intel Xeon Platinum 8280 ("Cascade Lake") 56 cores (partitions) on two sockets (28 cores/socket) 192 GB (2933 MT/s) RAM. We set $K = 3$ for all experiments and use P to signify the number of nodes in this section. For example, if we use $P = 10$ nodes, then there are $U = P * 56 = 560$ cores being adopted.

Table 4. Results of GE^{2}K-NN on TACC Frontera for Higgs dataset

	$P = 8$	16	32	40	48
Time (s)	4243.9	1920.0	921.1	700.7	618.1
Speedup	1	2.210	4.607	6.057	5.121
Efficiency	100%	110.5%	115.2%	121.1%	114.4%
Accuracy	0.6413	0.6413	0.6413	0.6413	0.6413

Table 4 shows the results of Higgs dataset, which illustrates the excellent scalability of GE^{2}K-NN a High Performance Computing (HPC) system. As the number of nodes is increased from 8 to 48, we are able to get a strong scaling efficiency of $\{110.5\%, 115.2\%, 121.1\%, 114.4\%\}$. Due to the complexity $O(n \log \frac{n}{U})$ of sorting operations for local KNNs on each split, we can achieve an efficiency higher than 100%. However, the communication overhead becomes larger with further increase of the node number, leading to a decline in efficiency ($114.4\% < 121.1\%$) when we using 48 nodes. Besides, more candidate KNNs from worker nodes need to be merged by the reducers on the master node, relatively increasing the running time and lowering the efficiency. With different number of nodes, GE^{2}K-NN has the same accuracy (0.6413).

5   Conclusions

In this paper, we a distributed EK-NN classifier, GE^{2}K-NN, to tackle the big data problem. With the Apache Spark framework, our proposed classifier can be executed with large datasets up to 11 million samples on either a single computer or an HPC cluster. Compared to previous algorithms, experimental results show that GE^{2}K-NN achieve great performance. On the one hand, GE^{2}K-NN have similar scalability as KNN-IS but better prediction accuracy than it. On the other hand, GE^{2}K-NN is much more scalable than GAHS-FKNN and LHS-FKNN and achieve higher prediction accuracy in most cases. Besides, by using more computation cores, GE^{2}K-NN does not lose prediction accuracy and achieve high speedup and efficiency.

References

1. Denoeux, T.: A k-nearest neighbor classification rule based on Dempster-Shafer theory. IEEE Trans. Syst. Man Cybern. **25**(5), 804–813 (1995)
2. Denoeux, T.: 40 years of Dempster-Shafer theory. Int. J. Approx. Reason. **79**(C), 1–6 (2016)
3. Gong, C., Su, Z.-G., Wang, P.-H., Wang, Q., You, Y.: A sparse reconstructive evidential-nearest neighbor classifier for high-dimensional data. IEEE Trans. Knowl. Data Eng. (2022)
4. Shafer, G.: A Mathematical Theory of Evidence, vol. 42. University Press (1976)
5. Chen, X.-L., Wang, P.-H., Hao, Y.-S., Zhao, M.: Evidential KNN-based condition monitoring and early warning method with applications in power plant. Neurocomputing **315**, 18–32 (2018)
6. Su, Z.-G., Denoeux, T., Hao, Y.-S., Zhao, M.: Evidential K-NN classification with enhanced performance via optimizing a class of parametric conjunctive t-rules. Knowl.-Based Syst. **142**, 7–16 (2018)
7. Pichon, F., Denoeux, T.: T-norm and uninorm-based combination of belief functions. In: NAFIPS 2008–2008 Annual Meeting of the North American Fuzzy Information Processing Society, pp. 1–6. IEEE (2008)
8. Denoeux, T., Kanjanatarakul, O., Sriboonchitta, S.: A new evidential k-nearest neighbor rule based on contextual discounting with partially supervised learning. Int. J. Approx. Reason. **113**, 287–302 (2019)
9. Liu, Z.-G., Pan, Q., Dezert, J.: Evidential classifier for imprecise data based on belief functions. Knowl.-Based Syst. **52**, 246–257 (2013)
10. Trabelsi, A., Elouedi, Z., Lefevre, E.: Ensemble enhanced evidential k-NN classifier through random subspaces. In: Antonucci, A., Cholvy, L., Papini, O. (eds.) ECSQARU 2017. LNCS (LNAI), vol. 10369, pp. 212–221. Springer, Cham (2017). https://doi.org/10.1007/978-3-319-61581-3_20
11. Su, Z.-G., Hu, Q., Denoeux, T.: A distributed rough evidential K-NN classifier: integrating feature reduction and classification. IEEE Trans. Fuzzy Syst. **29**(8), 2322–2335 (2020)
12. Ghosh, D., Cabrera, J., Adam, T.N., Levounis, P., Adam, N.R.: Comorbidity patterns and its impact on health outcomes: two-way clustering analysis. IEEE Trans. Big Data **6**(2), 359–368 (2016)
13. Zaharia, M., et al.: Resilient distributed datasets: a fault-tolerant abstraction for in-memory cluster computing. In: 9th USENIX Symposium on Networked Systems Design and Implementation (NSDI 2012), pp. 15–28 (2012)
14. Gong, C., Su, Z.-G., Wang, P.-H., You, Y.: Distributed evidential clustering toward time series with big data issue. Expert Syst. Appl. **191**, 116279 (2022)
15. Gong, C., Su, Z.-G., Wang, P.-H., Wang, Q., You, Y.: Evidential instance selection for K-nearest neighbor classification of big data. Int. J. Approx. Reason. **138**, 123–144 (2021)
16. Smets, P.: The combination of evidence in the transferable belief model. IEEE Trans. Pattern Anal. Mach. Intell. **12**(5), 447–458 (1990)
17. Apache Hadoop (2011). http://hadoop.apache.org
18. Asuncion, A., Newman, D.: UCI machine learning repository (2007)
19. Maillo, J., Ramírez, S., Triguero, I., Herrera, F.: kNN-IS: an iterative spark-based design of the k-nearest neighbors classifier for big data. Knowl.-Based Syst. **117**, 3–15 (2017)
20. Maillo, J., García, S., Luengo, J., Herrera, F., Triguero, I.: Fast and scalable approaches to accelerate the fuzzy k-nearest neighbors classifier for big data. IEEE Trans. Fuzzy Syst. **28**(5), 874–886 (2019)

On Improving a Group of Evidential Sources with Different Contextual Corrections

Siti Mutmainah[1](✉) ⓘ, Samir Hachour[2](✉) ⓘ, Frédéric Pichon[2](✉) ⓘ,
and David Mercier[2](✉) ⓘ

[1] UIN Sunan Kalijaga, Yogyakarta, Indonesia
`siti.mutmainah@uin-suka.ac.id`
[2] Univ. Artois, UR 3926 LGI2A, 62400 Béthune, France
{`samir.hachour,frederic.pichon,david.mercier`}`@univ-artois.fr`

Abstract. In this paper, we investigate the interest of learning a group of evidential sources using contextual corrections, which is equivalent to directly learning an optimized conjunctive combination instead of optimizing each source individually. Several experiments on synthetic and real UCI data demonstrates the interest of the approach.

Keywords: Information fusion · Belief functions · Group of sources · Contextual corrections · Optimization

1 Introduction

Information fusion [1,11] allows one, by combining different heterogeneous sources of information, to obtain a better understanding (possibly more complete, more precise) of the situation under evaluation.

The Dempster-Shafer theory of belief functions [2,17,18], being able to represent the imprecision and uncertainty of a piece of information, is an interesting and already widely used framework for modeling a fusion scheme [9,16]. One classical evidential fusion scheme consists in modeling the individual outputs of the sources as finely as possible to make independent and reliable pieces information so that they can be combined using the conjunctive rule of combination (meaning the unnormalized Dempster's rule). The reliability of the outputs of the sources can be ensured using the discounting operation [15,15,18] or more refined corrections such that contextual corrections [13,14]. For instance, we can use the contextual discounting (CD), allowing one to weaken a piece of information and which generalizes the discounting, or the contextual reinforcement (CR), which can reinforce the output of a source, or the contextual negating (CN), able to negate what a source indicates.

In the discounting operation [18], the reliability of the source, providing a mass function m, is taken into account using a real $\beta \in [0, 1]$ quantifying the degree of belief in the fact that the source is reliable, and the corrected mass

© The Author(s), under exclusive license to Springer Nature Switzerland AG 2022
S. Le Hégarat-Mascle et al. (Eds.): BELIEF 2022, LNAI 13506, pp. 109–118, 2022.
https://doi.org/10.1007/978-3-031-17801-6_11

function is denoted by $^\beta m$. In the contextual correction mechanisms (CD, CR and CN), the imperfection of the source, its bias in a broad sense, is modeled using a vector $\beta \in [0,1]^C$, with $C \leq 2^K$ and K the number of elements in the universe (more specific details can be found in [14]). The resulting corrected mass function is also denoted by $^\beta m$ for simplicity.

If moreover, a learning set composed of the outputs of a source, expressed in the form of mass functions, are available regarding the classes of n objects o_i, $i \in \{1, \ldots, n\}$ the true class (belonging to the universe) of each object being known, then it is possible [10,13] to find optimal parameters β, *i.e.*, to learn the parameters β minimizing a discrepancy measure between the corrected outputs and the ground truths.

This classical information fusion scheme is illustrated in Fig. 1.

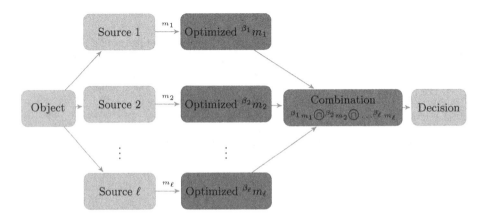

Fig. 1. Fusion scheme using individual corrections (Scheme 1).

Another idea, illustrated in Fig. 2, consists in learning directly an optimized conjunctive combination instead of optimizing each source individually. This idea has been mentioned in [10] for the discounting operation and in [13] for a particular CD.

In this paper, we use classifiers as sources of information, and we explore this idea of optimizing directly the performance of the combination using possibly different corrections among CD, CR and CN.

This paper is organized as follows. The notations and evidential concepts used are recalled in Sect. 2. The learning of contextual corrections for a group of evidential classifiers is presented in Sect. 3. Experiments on synthetic and real data demonstrating the interest of the approach are exposed in Sect. 4. Finally, a conclusion is given in Sect. 5.

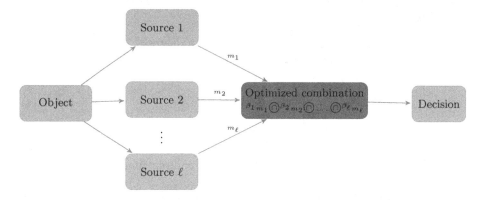

Fig. 2. Fusion scheme using global corrections (Scheme 2).

2   Belief Functions: Notations and Concepts Used

2.1   Basic Concepts

Basic concepts are briefly recalled. Details of the theory can be found for example in [5,15,18].

The universe Ω, a finite set, is composed of K elements ω_1, ..., ω_K. We consider a question of interest Q whose answer lies in Ω. A piece of information regarding this answer can be represented by a mass function (MF) m defined from 2^Ω to $[0,1]$ verifying s.t. $\sum_{A \subseteq \Omega} m(A) = 1$. The real $m(A)$ represents the part of belief allocated to the fact that the true searched value belongs to A and nothing more. A subset $A \subseteq \Omega$ s.t. $m(A) > 0$ is called a focal element of m. A categorical MF has only one focal element $A \subseteq \Omega$ and is denoted by m_A. We then have $m_A(A) = 1$. In particular, m_Ω represents the total ignorance.

A MF m is in one-to-one correspondence with a belief function Bel and a plausibility function Pl respectively defined for all $A \subseteq \Omega$ by:

$$Bel(A) = \sum_{\emptyset \neq B \subseteq A} m(B) , \qquad (1)$$

$$Pl(A) = \sum_{A \cap B \neq \emptyset} m(B) = Bel(\Omega) - Bel(\overline{A}) \qquad (2)$$

with $\overline{A} = \Omega \setminus A$.

The *contour function* pl corresponds to the restriction of the plausibility function to the singletons of Ω, it is defined for all $\omega \in \Omega$ by $pl(\omega) = Pl(\{\omega\})$.

Two reliable and independent MFs m_1 and m_2 defined on the same universe Ω can be combined using the conjunctive rule of combination (CRC) (or unnormalized Dempster's rule) defined by

$$(m_1 \!\bigcirc\!\!\!\!\cap\, m_2)(A) = m_{1 \bigcirc\!\!\!\!\cap 2}(A) = \sum_{B \cap C = A} m_1(B) \cdot m_2(C), \quad \forall A \subseteq \Omega . \qquad (3)$$

2.2   Corrections

A source providing a MF m and only reliable at a degree $\beta = 1 - \alpha \in [0, 1]$ can be discounted using the following operation

$$
\begin{aligned}
{}^{\beta}m &= \beta\,m + \alpha\,m_{\Omega} \\
&= \begin{cases} A \mapsto \beta\,m(A) & \forall A \subset \Omega \\ \Omega \mapsto \beta\,m(\Omega) + \alpha \end{cases}
\end{aligned}
\tag{4}
$$

The contour function associated with the discounted MF ${}^{\beta}m$ (4) verifies for all $w \in \Omega$, ${}^{\beta}pl(w) = 1 - (1 - pl(w))\beta$, with pl the contour function of m (Details can be found for example in [13–15]).

In Table 1, we summarize the contour functions of the contextual discounting (CD), contextual reinforcement (CR) and contextual negating (CN) of a MF m that can be obtained by a specific choice of $C = K$ parameters $\beta_w \in [0, 1]$, for each contextual corrections; the reasons for limiting ourselves to $C = K$ parameters and the definitions of these K parameters for each contextual corrections can be found in [14, Section 8].

Table 1. Contour functions of each contextual correction of a MF m given for any $w \in \Omega$. Each parameter β_w may vary in $[0, 1]$.

Corrections	Contour functions
CD	${}^{\beta}pl(w) = 1 - (1 - pl(w))\beta_w$
CR	${}^{\beta}pl(w) = pl(w)\beta_w$
CN	${}^{\beta}pl(w) = 0.5 + (pl(w) - 0.5)(2\beta_w - 1)$

As recalled in the introduction, if for a source we have a learning set containing its outputs, meaning MF $m\{o_i\}$, regarding the classes of n objects o_i, $i \in \{1, \ldots, n\}$, the true classes are known, we can then compute the CD, CR and CN parameters β optimizing the following measure of discrepancy between the corrected outputs and the true classes of the objects

$$
E_{pl}(\beta) = \sum_{i=1}^{n} \sum_{k=1}^{K} ({}^{\beta}pl\{o_i\}(\{w_k\}) - \delta_{i,k})^2 \ ,
\tag{5}
$$

where ${}^{\beta}pl\{o_i\}$ is the contour function regarding the class of the object o_i corrected with a vector $\beta = (\beta_w \in [0, 1], w \in \Omega)$ and $\delta_{i,k}$ is the indicator function of the truths of the objects o_i, $i \in \{1, \ldots, n\}$, meaning $\delta_{i,k} = 1$ if the class of the object o_i is w_k, otherwise $\delta_{i,k} = 0$.

The measure E_{pl} yields, for each correction (CD, CR, and CN), a constrained linear least-squares optimization problem which can be efficiently solved.

3   Learning a Group of Evidential Sources

When several sources are available, instead of learning the best correction parameters individually for each source knowing that these adjusted MFs are going to be next combined, it is possible to directly optimize the combination of the adjusted MFs.

With ℓ sources to be combined, ℓ vectors $\beta_1, \ldots, \beta_\ell$, each one associated with either CD or CR or CN, can be obtained by minimizing the following measure

$$E_{pl}(\beta_1, \ldots, \beta_\ell) = \sum_{i=1}^{n} \sum_{k=1}^{K} (^{\beta_1}pl_1\{o_i\}(\{\omega_k\}) \times \ldots \times^{\beta_\ell} pl_\ell\{o_i\}(\{\omega_k\}) - \delta_{i,k})^2 \quad (6)$$

Indeed, after the conjunctive combination, the plausibility of each singleton is equal to the product of the plausibilities given by the ℓ sources to this singleton.

Optimizing (5) for each classifier or (6) is not the same thing as a classifier can be used in a different manner if it is used alone or through a collective.

One drawback, however, of this approach, is that the optimization of (6) is no more a linear least-squares optimization problem, it can be minimized using a standard constrained nonlinear optimization procedure reaching to a possible local minimum.

Another critical point concerns the number of optimizations to undertake in each scenario. With three possible mechanisms (CD, CR and CN), which can be applied on each source, and ℓ sources, we have for the first scheme using individual corrections (cf Fig. 1) $3 \times \ell$ possible corrections to test, while for this second scheme optimizing the combination (cf Fig. 2), we have 3^ℓ possible corrections to test.

As an example, let us consider the case of two sources S_1 and S_2 ($\ell = 2$). For the individual optimizations, we have for each source three optimisations to undertake, using (5), to know what correction between CD, CR and CN to keep for each source, and thus finally 6 optimisations in total of (5). While, for the direct optimization of the combination using (6), we have to compare all the possible associations of corrections for sources S_1 and S_2 (CD-CD, CD-CR, CD-CN, CR-CD, CR-CR, CR-CN, CN-CD, CN-CR and CN-CN) leading then to a richer frame of possible corrections, but with more comparisons to do, $3^2 = 9$ in this scenario.

In the following section, we show with several experiments both on synthetic and real data, that this second scheme can have an interest due to its performances.

4   Experiments

To test these schemes (individual corrections - Fig. 1 - vs global correction - Fig. 2), several numerical experiments conducted on synthetic and real data sets using two evidential classifiers are exposed in this Section.

The first classifier is the evidential k-nearest neighbor (EkNN) [3,6] with $k = 5$. The second chosen classifier is the evidential neural networks (ENN) [4,6] with number of prototypes $np = 5$.

For each data set, the following experiment was repeated 10 times:

– One half of the data ($\mathcal{L}_1$) is used to learn the classifier (EkNN or ENN);
– A 10-fold cross validation is then performed on the second half of the data with 9 folds ($\mathcal{L}_2$) to learn the best correction, and 1 fold for testing.

The synthetic data set, illustrated in Fig. 3, has been generated by multivariate normal distribution composed of 2 features, 900 objects and 3 classes with the means $\mu_1 = (0, 2), \mu_2 = (1, 3), \mu_3 = (2, 2)$ and the following covariance matrices for each class: $\Sigma_1 = 0.1I$, $\Sigma_2 = 0.5I$ and $\Sigma_3 = \begin{bmatrix} 0.3 & -0.15 \\ -0.15 & 0.3 \end{bmatrix}$, where I is the 2×2 identity matrix.

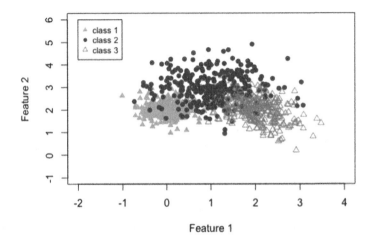

Fig. 3. Generated data set.

The real data sets used were taken from UCI [8]. Their descriptions can be seen in Table 2.

The results are summed up in Table 3 using as a measure of performance E_{pl} (5), meaning the squared error between the contour function resulting from the combination and the indicator function of the truths of the objects in the test set.

It can be seen from Table 3 that the second scheme optimizing the combination reaches better performances according to E_{pl} (5) than the first scheme combining individual optimizations.

Table 2. Description of the UCI data sets used [8]

Data sets	#Instances	#Features	#Classes
Haberman	306	3	2
Iris	150	4	3
Glass	214	10	6
Ionosphere	350	34	2
Lymphography	140	18	3
Liver	345	6	2
Pima	768	8	2
Sonar	208	60	2
Transfusion	748	3	2
Vehicle	846	19	4
Vertebral	310	6	3

Table 3. Performances (Average E_{pl} (5) values), the lower the better, obtained from two sources (EkNN and ENN) for a conjunctive combination without correction (No correction), for scheme 1 (best individual corrections), for scheme 2 (best parameterized combination), for scheme 2 with only CD, only CR and only CN to highlight the interest of possibly using multiple distinct corrections. Standard deviations are indicated in parentheses. In bold the best performance for each data set.

Data	No correction CC	Scheme 1	Scheme 2	Scheme 2 only CD	Scheme 2 only CR	Scheme 2 only CN
Synthetic	10.953 (3.094)	11.690 (2.748)	9.496 (2.492)	**9.491 (2.525)**	10.961 (3.076)	10.882 (2.913)
Haberman	6.054 (2.381)	6.742 (1.664)	**5.515 (1.725)**	5.886 (2.096)	5.717 (2.070)	5.577 (1.849)
Iris	0.503 (0.619)	0.569 (0.608)	**0.467 (0.715)**	0.471 (0.712)	0.503 (0.619)	0.503 (0.619)
Glass	5.016 (1.765)	6.007 (0.778)	**4.703 (1.340)**	4.965 (1.737)	4.771 (1.408)	4.763 (1.267)
Ionosphere	2.411 (0.882)	2.874 (0.831)	**2.057 (0.958)**	**2.057 (0.958)**	2.411 (0.882)	2.411 (0.882)
Lympho	2.305 (1.077)	2.748 (0.929)	2.253 (1.058)	**2.239 (1.045)**	2.322 (1.086)	2.322 (1.059)
Liver	7.937 (1.778)	9.481 (0.942)	**7.515 (1.242)**	7.728 (1.564)	7.848 (1.565)	7.743 (1.405)
Pima	13.123 (2.770)	16.408 (1.970)	**12.455 (2.415)**	12.489 (2.490)	13.130 (2.738)	13.107 (2.681)
Sonar	3.489 (1.071)	4.365 (0.834)	3.164 (0.918)	**3.160 (0.969)**	3.491 (1.037)	3.492 (0.992)
Transfusion	17.018 (3.798)	16.393 (2.240)	**13.218 (2.230)**	15.561 (3.111)	15.096 (2.944)	13.528 (2.190)
Vehicles	23.277 (3.161)	30.886 (1.220)	**21.741 (2.209)**	23.106 (3.193)	22.949 (2.687)	21.851 (2.166)
Vertebral	4.632 (1.777)	5.173 (1.600)	**3.979 (1.512)**	3.995 (1.525)	4.645 (1.763)	4.573 (1.576)

We also wanted to highlight the possible interest of taking advantage of using possibly several different corrections and so the performances of scheme 2 with only CD, only CR and only CN were also exposed for comparisons. Using only one kind of correction can limit the performances.

It can be observed that it happens that scheme 2 with only CD (Scheme 2 only CD) obtains slightly better performances (on Iris, Lympho and Sonar data) than scheme 2 testing all combinations including CD-CD. Several non-exclusive explanations may be given: first, the best configuration on the training set is not necessarily the best one on the test set; second, the optimization on the learning set is only local; and at last, the performance measure E_{pl} (5) is somewhat favorable to CD (Details in [14, Section 8.5.1]).

As expected, the drawback to reach these performances is a longer time to learn the parameters as shown in Table 4. With only two sources, this time remains reasonable. If the number of sources were to become too large, it would certainly be necessary to see if Scheme 2 is still applicable within a reasonable time.

Table 4. Time consumption in seconds on a macbook Air M1 3.2 GHz 8 GB RAM for the learning phase for Scheme 1 and Scheme 2. Standard deviations are indicated in parentheses.

Data	Scheme 1	Scheme 2
Synthetic	0.0477 (0.0116)	42.7417 (16.4333)
Haberman	0.0120 (0.0079)	11.4312 (1.3702)
Iris	0.0059 (0.0010)	5.9721 (1.1307)
Glass	0.0084 (0.0012)	18.5167 (3.9630)
Ionosphere	0.0120 (0.0020)	7.9370 (0.4878)
Lympho	0.0074 (0.0110)	6.5583 (0.7286)
Liver	0.0125 (0.0027)	13.5109 (1.0809)
Pima	0.0300 (0.0184)	22.2145 (3.2915)
Sonar	0.0070 (0.0013)	4.8880 (0.5885)
Transfusion	0.0281 (0.0078)	27.4938 (7.9787)
Vehicles	0.0596 (0.0223)	1.4006 (0.0770)
Vertebral	0.0124 (0.0017)	15.9031 (2.0436)

We now give the results according to another performance measure, and to consider the interest of belief function modeling, we look at partial decisions (meaning decision possibly in favor of a group of classes) [7], and we consider that the set of possible decisions (or acts) is equal to Ω, so we can use [12][Page 6, Strong dominance criterion with $0 - 1$ utilities and pieces of information represented by belief functions] the following relation of dominance between the singletons of Ω:

$$\omega \succeq \omega' \iff Bel(\{\omega\}) \geq Pl(\{\omega'\}) , \tag{7}$$

and make a partial decision composed of the non dominated singletons according to relation (7).

The results are then exposed in Table 5 using the u_{65} utility measure. This measure, introduced by Zaffalon et al. [20], allows one to take into account the interest of partial decisions for preferring the imprecision to being randomly correct.

The U_{65} value of a partial decision d, possibly in favor a set of singletons, is formally defined by

$$U_{65}(x) = 1.6x - 0.6x^2 \tag{8}$$

with x the so called discounted accuracy of d defined by $\frac{\mathbb{I}(\omega \in d)}{|d|}$, with $\mathbb{I}$ the indicator function, ω the true class of the instance, and $|d|$ the number of elements in d. The u_{65} utility measure gives a greater utility to imprecise but correct partial decisions of size n (meaning decisions equal to a set of n singletons one of them being the true class) than precise decisions (in favor of one singleton) only randomly correct with probability $\frac{1}{n}$.

Table 5. Performances (Average U_{65} values), the higher the better, obtained from two sources (EkNN and ENN) for a conjunctive combination without correction (No correction), for scheme 1 (best individual corrections), for scheme 2 (best parameterized combination), for scheme 2 with only CD, only CR and only CN to highlight the interest of possibly using multiple distinct corrections. Standard deviations are indicated in parentheses. In bold the best performance for each data set.

Data	No correction	Scheme 1	Scheme 2	Scheme 2 only CD	Scheme 2 only CR	Scheme 2 only CN
Synthetic	0.850 (0.052)	0.857 (0.046)	0.857 (0.046)	**0.858 (0.046)**	0.850 (0.052)	0.850 (0.052)
Haberman	0.755 (0.103)	0.747 (0.112)	0.758 (0.096)	**0.762 (0.097)**	0.756 (0.103)	0.759 (0.104)
Iris	**0.971 (0.062)**	**0.971 (0.062)**	0.968 (0.064)	0.969 (0.063)	**0.971 (0.062)**	**0.971 (0.062)**
Glass	0.677 (0.151)	0.679 (0.148)	0.681 (0.149)	0.684 (0.144)	**0.688 (0.150)**	0.683 (0.146)
Ionosphere	**0.938 (0.047)**	0.938 (0.046)	0.933 (0.043)	0.933 (0.043)	**0.938 (0.047)**	**0.938 (0.047)**
Lympho	**0.815 (0.139)**	0.805 (0.149)	0.805 (0.143)	0.806 (0.143)	0.814 (0.139)	**0.815 (0.135)**
Liver	0.697 (0.102)	0.684 (0.110)	0.703 (0.087)	**0.704 (0.092)**	0.700 (0.104)	0.698 (0.099)
Pima	0.769 (0.067)	0.767 (0.068)	**0.777 (0.066)**	0.775 (0.066)	0.769 (0.067)	0.771 (0.067)
Sonar	0.789 (0.132)	0.783 (0.132)	**0.812 (0.096)**	0.810 (0.096)	0.788 (0.135)	0.791 (0.131)
Transfusion	0.746 (0.060)	0.756 (0.064)	**0.768 (0.056)**	0.762 (0.053)	0.752 (0.062)	0.763 (0.057)
Vehicles	0.623 (0.062)	0.614 (0.064)	**0.635 (0.058)**	0.631 (0.059)	0.621 (0.063)	0.633 (0.058)
Vertebral	0.809 (0.106)	0.808 (0.104)	**0.833 (0.094)**	0.833 (0.093)	0.809 (0.106)	0.816 (0.105)

At last, with a classical error rate with decisions made for example by choosing the class maximizing the pignistic probability [17], we obtain results very similar with or without correction (Due to the page limit, it is difficult to put all results). The results on this point are preliminary, we will explore other experiments with surely more classifiers to see a possible interest of this approach.

5   Conclusion

In this paper, we have illustrated through experiments the interests of using different contextual corrections to optimize the conjunctive combination of the outputs of a group of evidential sources. We have also given elements of the possible limitations of this strategy when the number of sources to be combined increases, and according to classical error rate, these limitations remaining to be more clarified. A perspective of interest and topicality will be to study the possibility of using these schemes in an end-to-end learning of a group of deep evidential classifiers in the lines of the works of Tong et al. for example [19].

Acknowledgements. The authors would like to thank the anonymous reviewers for their very helpful and relevant comments.

References

1. Bloch, I., et al.: Fusion: general concepts and characteristics. Int. J. Intell. Syst. **16**(10), 1107–1136 (2001)
2. Dempster, A.P.: Upper and lower probabilities induced by a multiple valued mapping. Ann. Math. Stat. **38**, 325–339 (1967)
3. Denœux, T.: A k-nearest neighbor classification rule based on dempster-shafer theory. IEEE Trans. Syst. Man Cybern. **25**(5), 804–813 (1995)
4. Denœux, T.: A neural network classifier based on dempster-shafer theory. IEEE Trans. Syst. Man Cybern. **30**(2), 131–150 (2000)
5. Denœux, T.: Conjunctive and disjunctive combination of belief functions induced by nondistinct bodies of evidence. Artif. Intell. **172**, 234–264 (2008)
6. Denœux, T.: Evclass: evidential distance-based classification. [https://cran.r-project.org/web/packages/evclass/index.html], R package version 1.1.1. (2017)
7. Denœux, T.: Decision-making with belief functions: a review. Int. J. of Approx. Reason. **109**, 87–110 (2019)
8. Dua, D., Graff, C.: UCI Machine Learning Repository [http://archive.ics.uci.edu/ml]. Irvine, CA: University of California (2019)
9. Dubois, D., Liu, W., Ma, J., Prade, H.: The basic principles of uncertain information fusion. An organised review of merging rules in different representation frameworks. Inf. Fusion **32**(A), 12–39 (2016)
10. Elouedi, Z., Mellouli, K., Smets, P.: Assessing sensor reliability for multisensor data fusion within the transferable belief model. IEEE Trans. Syst. Man Cybern. B **34**(1), 782–787 (2004)
11. Kuncheva, L.I.: Combining Pattern Classifiers: Methods and Algorithms, 2nd edn. Wiley, Hoboken (2014)
12. Ma, L., Denœux, T.: Partial classification in the belief function framework. Knowl. Based Syst. **214**, 106742 (2021)
13. Mercier, D., Quost, B., Denœux, T.: Refined modeling of sensor reliability in the belief function framework using contextual discounting. Inf. Fusion **9**(2), 246–258 (2008)
14. Pichon, F., Mercier, D., Lefèvre, E., Delmotte, F.: Proposition and learning of some belief function contextual correction mechanisms. Int. J. Approx. Reason. **72**, 4–42 (2016)
15. Smets, P.: Belief functions: the disjunctive rule of combination and the generalized Bayesian theorem. Int. J. Approx. Reason. **9**(1), 1–35 (1993)
16. Smets, P.: Analyzing the combination of conflicting belief functions. Inf. Fusion **8**(4), 387–412 (2007)
17. Smets, P., Kennes, R.: The transferable belief model. Artif. Intell. **66**(2), 191–234 (1994)
18. Shafer, G.: A Mathematical Theory of Evidence. Princeton University Press, Princeton (1976)
19. Tong, Z., Xu, P., Denœux, T.: Fusion of evidential CNN classifiers for image classification. In: Denœux, T., Lefèvre, E., Liu, Z., Pichon, F. (eds.) BELIEF 2021. LNCS (LNAI), vol. 12915, pp. 168–176. Springer, Cham (2021). https://doi.org/10.1007/978-3-030-88601-1_17
20. Zafallon, M., Corani, G., Mauá, D.-D.: Evaluating credal classifiers by utility-discounted predictive accuracy. Int. J. Approx. Reason. **53**(8), 1282–1301 (2012)

Measure of Information Content of Basic Belief Assignments

Jean Dezert[1]([⊠]) , Albena Tchamova[2] , and Deqiang Han[3]

[1] The French Aerospace Lab, Chemin de la Hunière, 91761 Palaiseau, France
`jean.dezert@onera.fr`
[2] Institute for IC & Technologies, Bulgarian Academy of Sciences, Sofia, Bulgaria
`tchamova@bas.bg`
[3] Institute of Integrated Automation, Xi'an Jiaotong University, Xi'an 710049, China
`deqhan@xjtu.edu.cn`

Abstract. In this paper, we present a measure of Information Content (IC) of Basic Belief Assignments (BBAs), and we show how it can be easily calculated. This new IC measure is interpreted as the dual of the effective measure of uncertainty (i.e. generalized entropy) of BBAs developed recently.

Keywords: Belief functions · Information content · Generalized entropy

1 Introduction

Information quality (IQ) evaluation is of major importance for information processing and for helping the decision-making under uncertainty. In [1], the authors introduced the Accessibility, Interpretability, Relevance, and Integrity concepts as main attributes to describe the information quality in the context of assurance and belief networks, but unfortunately they present only general concepts without explicit formulas to evaluate quantitatively these attributes. In several recent books devoted to IQ [2–5], the authors proposed different models and methods of IQ evaluations. Recently in [6], Bouhamed et al. proposed a quantitative IQ evaluation using the possibility theory framework, which could be extended to the belief functions theory framework with further investigations. In this latter work, the information quantity component being necessary for the IQ evaluation is based on Gini's entropy rather than classical Shannon entropy. From the examination of these aforementioned references (and some references therein), it is far from obvious to make a clear justified choice among all these methods, especially when we model the uncertain information by belief functions (BF). What is clear however is that several distinct factors (or components) must be taken into account in the IQ evaluation mechanism. In this paper we focus on one of these components which is the Information Content (IC) component that we consider as the very (if not the most) essential component for IQ evaluation

S. Le Hégarat-Mascle et al. (Eds.): BELIEF 2022, LNAI 13506, pp. 119–128, 2022.
https://doi.org/10.1007/978-3-031-17801-6_12

and indispensable for developing an effective IQ evaluation method in future research works.

It is worth noting that we do not address directly the whole IQ evaluation problem in this work but to provide a mathematical solution for measuring the IC of any Basic Belief Assignments (BBA) in the belief functions (BF) framework. Our new IC measure is interpreted as the dual of an effective Measure of Uncertainty (MoU) developed recently [7]. We show how to calculate the IC of a BBA, and we also discuss the notion of information gain and information loss in the BF context. In our opinion, we cannot define a measure of Information Content independently of a Measure of Uncertainty (MoU) because they must be strongly related to each other. Actually these measures are two different sides of a same *abstract coin* we would say. On one side (the uncertainty side), more uncertainty content we have harder is the decision or choice to make, and on the other side (the information side) more information content we have easier and stronger is the decision or choice to make. This very simple and natural basic principle will be clarified mathematically next. So, the measure of information content of a BBA must reflect somehow the easiness and strength in the choice of an element of the frame of discernment drawn from the BBA (i.e. in the decision-making). This paper is organized as follows. After a brief recall of basics of belief functions in Sect. 2, we recall the effective MoU adopted in this work in Sect. 3. Section 4 defines the measure of information content of a BBA and the information granules vector. Section 5 introduces the notions of information gain and information loss. Conclusions and perspectives appear in the last section.

2   Belief Functions

The belief functions (BF) were introduced by Shafer [8] for modeling epistemic uncertainty, reasoning about uncertainty and combining distinct sources of evidence. The answer of the problem under concern is assumed to belong to a known finite discrete frame of discernement (FoD) $\Theta = \{\theta_1, \ldots, \theta_N\}$ where all elements (i.e. members) of Θ are exhaustive and mutually exclusive. The set of all subsets of Θ (including empty set $\emptyset$, and Θ) is the power-set of Θ denoted by 2^Θ. The number of elements (i.e. the cardinality) of the power-set is $2^{|\Theta|}$. A (normalized) basic belief assignment (BBA) associated with a given source of evidence is a mapping $m^\Theta(\cdot) : 2^\Theta \to [0,1]$ such that $m^\Theta(\emptyset) = 0$ and $\sum_{X \in 2^\Theta} m^\Theta(X) = 1$. A BBA $m^\Theta(\cdot)$ characterizes a source of evidence related with a FoD Θ. For notation shorthand, we can omit the superscript Θ in $m^\Theta(\cdot)$ notation if there is no ambiguity on the FoD we work with[1]. The quantity $m(X)$ is called the mass of belief for X. The element $X \in 2^\Theta$ is called a focal element (FE) of $m(\cdot)$ if $m(X) > 0$. The set of all focal elements of $m(\cdot)$ is denoted[2] by $\mathcal{F}_\Theta(m) \triangleq \{X \in 2^\Theta | m(X) > 0\}$.

The belief and the plausibility of X are defined for any $X \in 2^\Theta$ by [8]

$$Bel(X) = \sum_{Y \in 2^\Theta | Y \subseteq X} m(Y) \tag{1}$$

---

[1] However, we will keep $m^\Theta(\cdot)$ notation when very necessary.
[2] $\triangleq$ means *equal by definition*.

$$Pl(X) = \sum_{Y \in 2^{\Theta} | X \cap Y \neq \emptyset} m(Y) = 1 - Bel(\bar{X}). \tag{2}$$

where $\bar{X} \triangleq \Theta \setminus \{X\}$ is the complement of X in Θ.

One has always $0 \leq Bel(X) \leq Pl(X) \leq 1$, see [8]. For $X = \emptyset$, $Bel(\emptyset) = 0$ and $Pl(\emptyset) = 0$, and for $X = \Theta$ one has $Bel(\Theta) = 1$ and $Pl(\Theta) = 1$. $Bel(X)$ and $Pl(X)$ are often interpreted as the lower and upper bounds of unknown probability $P(X)$ of X, that is $Bel(X) \leq P(X) \leq Pl(X)$. To quantify the uncertainty (i.e. the imprecision) of $P(X) \in [Bel(X), Pl(X)]$, we use the notation $u(X) \in [0, 1]$ defined by

$$u(X) \triangleq Pl(X) - Bel(X) \tag{3}$$

The quantity $u(X) = 0$ if $Bel(X) = Pl(X)$ which means that $P(X)$ is known precisely, and one has $P(X) = Bel(X) = Pl(X)$. One has $u(\emptyset) = 0$ because $Bel(\emptyset) = Pl(\emptyset) = 0$, and one has $u(\Theta) = 0$ because $Bel(\Theta) = Pl(\Theta) = 1$. If all focal elements of $m(\cdot)$ are singletons of 2^{Θ} the BBA $m(\cdot)$ is a Bayesian BBA because $\forall X \in 2^{\Theta}$ one has $Bel(X) = Pl(X) = P(X)$ and $u(X) = 0$. Hence the belief and plausibility of X coincide with a probability measure $P(X)$ defined on the FoD Θ. The vacuous BBA characterizing a totally ignorant source of evidence is defined by $m_v(X) = 1$ for $X = \Theta$, and $m_v(X) = 0$ for all $X \in 2^{\Theta}$ different from Θ. This particular BBA has played a major role in the establishment of a new effective measure of uncertainty of BBA defined in [7].

3   Generalized Entropy of a BBA

In [9] we did analyze in details forty-eight measures of uncertainty (MoU) of BBAs by covering 40 years of research works on this topic. Some of these MoUs capture only a particular aspect of the uncertainty inherent to a BBA (typically, the non-specificity and the conflict). Other MoUs propose a total uncertainty measure to capture jointly several aspects of the uncertainty. Unfortunately, most of these MoUs fail to satisfy four very simple reasonable and essential desiderata, and so they cannot be considered as really effective and useful. Actually only six MoUs can be considered as effective from the mathematical sense presented next, but unfortunately they appear as conceptually defective and disputable, see discussions in [9]. That is why, a better effective measure of uncertainty (MoU), i.e. generalized entropy of BBAs has been developed and presented in [7]. The mathematical definition of this new effective entropy is given by

$$U(m) = \sum_{X \in 2^{\Theta}} s(X) \tag{4}$$

with

$$s(X) \triangleq -m(X)(1 - u(X)) \log(m(X)) + u(X)(1 - m(X)) \tag{5}$$

The quantity $-(1 - u(X))\log(m(X)) = (1 - u(X))\log(1/m(X))$ entering in $s(X)$ in (5) is the surprisal[3] $\log(1/m(X))$ of X discounted by the confidence $(1 - u(X))$ one has on the precision of $P(X)$. The term $-m(X)(1 - u(X))\log(m(X))$ is the weighted discounted surprisal of X. The term $u(X)(1 - m(X))$ entering in (5) corresponds to the imprecision of $P(X)$ discounted by $(1 - m(X))$ because the greater $m(X)$ the less one should take into account the imprecision $u(X)$ in the MoU. The quantity $s(X)$ is the uncertainty contribution related to element X (named the *entropiece* of X) in the MoU $U(m)$. This entropiece $s(X)$ involves $m(X)$ and the imprecision $u(X) = Pl(X) - Bel(X)$ about the unknown probability of X in a subtle interwoven manner. The cardinality of X is indirectly taken into account in the derivation of $s(X)$ thanks to $u(X)$ which requires the derivation of $Pl(X)$ and $Bel(X)$ functions depending on the cardinality of X. Because $u(X) \in [0,1]$ and $m(X) \in [0,1]$ one has $s(X) \geq 0$, and $U(m) \geq 0$. The quantity $U(m)$ is expressed in *nats* because we use the natural logarithm. $U(m)$ can be expressed in *bits* by dividing the $U(m)$ value in *nats* by $\log(2) = 0.69314718....$ This measure of uncertainty $U(m)$ is a continuous function in its basic belief mass arguments because it is a summation of continuous functions. In formula (5), we always take $m(X)\log(m(X)) = 0$ when $m(X) = 0$ because $\lim_{m(X)\to 0^+} m(X)\log(m(X)) = 0$ which can be proved using L'Hôpital rule [11]. Note that for any BBA m, one has always $s(\emptyset) = 0$ because $m(\emptyset) = 0$ and $u(\emptyset) = Pl(\emptyset) - Bel(\emptyset) = 0 - 0 = 0$. For the vacuous BBA, one has $s(\Theta) = 0$ because $m_v(\Theta) = 1$ and $u(\Theta) = Pl(\Theta) - Bel(\Theta) = 1 - 1 = 0$.

The set $\{s(X), X \in 2^\Theta\}$ of the entropieces values $s(X)$ can be represented by an entropiece vector $\mathbf{s}(m^\Theta) = [s(X), X \in 2^\Theta]^T$, where any order of elements X of the power set 2^Θ can be chosen. For simplicity, we suggest to use the classical N-bits representation (if $|\Theta| = N$) with the increasing order - see the next example.

This measure of uncertainty $U(m)$ is effective because it can be proved (see proofs in [7]) that it satisfies the following four essential properties:

1. $U(m) = 0$ for any BBA $m(\cdot)$ focused on a singleton X of 2^Θ.
2. $U(m_v^\Theta) < U(m_v^{\Theta'})$ if $|\Theta| < |\Theta'|$.
3. $U(m) = -\sum_{X \in \Theta} m(X)\log(m(X))$ if the BBA $m(\cdot)$ is a Bayesian BBA. Hence, $U(m)$ reduces to Shannon entropy [12] in this case.
4. $U(m) < U(m_v)$ for any non-vacuous BBA $m(\cdot)$ and for the vacuous BBA $m_v(\cdot)$ defined with respect to the same FoD.

The proof of the three first properties is quite simple to make. The proof of the last property is much more difficult. As explained in [7], we do not consider that the sub-additivity property [13] of $U(m)$ is a fundamental desideratum that an effective MoU must satisfy in general. In fact the sub-additivity desideratum is incompatible with the fourth important property $U(m) < U(m_v)$ above which stipulates that none non-vacuous BBA can be more uncertain (i.e. more ignorant

---

[3] This terminology is not used by Shannon in his original paper but it has been introduced by Tribus in [10] in the probabilistic context, and by analogy we adopt Tribus' terminology also for BBAs.

about the problem under consideration) than the vacuous BBA. Actually, it does not make sense to have the entropy $U(m_v^{\Theta \times \Theta'})$ of the vacuous joint BBA $m_v^{\Theta \times \Theta'}$ defined on the cartesian product space $\Theta \times \Theta'$ smaller than (or equal to) the sum $U(m_v^{\Theta}) + U(m_v^{\Theta'})$ of entropies of vacuous BBAs m_v^{Θ} and $m_v^{\Theta'}$ defined respectively on Θ and Θ'. There is no theoretical justification, nor intuitive reason for this sub-additivity desideratum in the context of non-bayesian BBAs. Of course for Bayesian BBAs, $U(m)$ is equivalent to Shannon entropy which is in this case sub-additive.

It can be also proved, see [7] for details, that the entropy of the vacuous BBA m_v related to a FoD Θ is equal to

$$U(m_v^{\Theta}) = 2^{|\Theta|} - 2 \tag{6}$$

This maximum entropy value $U(m_v)$ makes perfect sense because for this very particular BBA there is no information at all about the conflicts between the elements of the FoD. Actually for all $X \in 2^{\Theta} \setminus \{\emptyset, \Theta\}$ one has $u(X) = 1$ because $[Bel(X), Pl(X)] = [0, 1]$, and one has $u(\emptyset) = 0$ and $u(\Theta) = 0$. Hence, the sum of all imprecisions of $P(X)$ for all $X \in 2^{\Theta}$ is exactly equal to $2^{|\Theta|} - 2$ which corresponds to $U(m_v^{\Theta})$ as expected. Moreover, one has always $U(m_v^{\Theta}) > \log(|\Theta|)$ which means that the vacuous BBA has always an entropy greater than the maximum of Shannon entropy $\log(|\Theta|)$ obtained with the uniform probability mass function distributed on Θ.

Example 1 of Entropy Calculation: consider $\Theta = \{\theta_1, \theta_2\}$ and the BBA $m^{\Theta}(\theta_1) = 0.5$, $m^{\Theta}(\theta_2) = 0.3$ and $m^{\Theta}(\theta_1 \cup \theta_2) = 0.2$, then one has $[Bel(\emptyset), Pl(\emptyset)] = [0, 1]$ and $u(\emptyset) = 0$, $[Bel(\theta_1), Pl(\theta_1)] = [0.5, 0.7]$, $[Bel(\theta_2), Pl(\theta_2)] = [0.3, 0.5]$, and $[Bel(\Theta), Pl(\Theta)] = [1, 1]$. Hence, $u(\theta_1) = 0.2$, $u(\theta_2) = 0.2$ and $u(\Theta) = 0$. Applying (5), one gets $s(\emptyset) = 0$, $s(\theta_1) \approx 0.377258$, $s(\theta_2) \approx 0.428953$ and $s(\Theta) \approx 0.321887$. Using the 2-bits representation with increasing ordering[4], we encode the elements of the power set as $\emptyset = 00$, $\theta_1 = 01$, $\theta_2 = 10$ and $\theta_1 \cup \theta_2 = 11$. The entropiece vector for this simple example is

$$\mathbf{s}(m^{\Theta}) = \begin{bmatrix} s(\emptyset) \\ s(\theta_1) \\ s(\theta_2) \\ s(\theta_1 \cup \theta_2) \end{bmatrix} \approx \begin{bmatrix} 0 \\ 0.377258 \\ 0.428953 \\ 0.321887 \end{bmatrix} \tag{7}$$

If we use the classical N-bits (here $N = 2$) representation with increasing ordering (as we recommand) the first component of entropiece vector $\mathbf{s}(m^{\Theta})$ will be $s(\emptyset)$ which is always equal to zero for any BBA m, hence the first component of $\mathbf{s}(m^{\Theta})$ is always zero. By summing all the components of the entropiece vector $\mathbf{s}(m^{\Theta})$ we obtain the entropy $U(m^{\Theta}) \approx 1.128098$ nats of the BBA $m^{\Theta}(\cdot)$. Note that the components $s(X)$ (for $X \neq \emptyset$) of the entropieces vector $\mathbf{s}(m^{\Theta})$ are not independent because they are linked to each other through the calculation of $Bel(X)$ and $Pl(X)$ values entering in $u(X)$.

---

[4] Once the binary values are converted into their digit value with the most significant bit on the left (i.e. the least significant bit on the right).

Example 2 of Entropy Calculation: for the vacuous BBA m_v^Θ, and when using the binary increasing encoding of elements of 2^Θ, the first component $s(\emptyset)$ and the last component $s(\Theta)$ of entropiece vector $\mathbf{s}(m_v^\Theta)$ will always be equal to zero, and all other components of $\mathbf{s}(m_v^\Theta)$ will be equal to one. For instance, if we consider $\Theta = \{\theta_1, \theta_2\}$ and the vacuous BBA $m_v^\Theta(\theta_1) = 0$, $m_v^\Theta(\theta_2) = 0$ and $m_v^\Theta(\theta_1 \cup \theta_2) = 1$, the corresponding entropiece vector $\mathbf{s}(m_v^\Theta)$ is

$$\mathbf{s}(m_v^\Theta) = \begin{bmatrix} s(\emptyset) \\ s(\theta_1) \\ s(\theta_2) \\ s(\theta_1 \cup \theta_2) \end{bmatrix} = \begin{bmatrix} 0 \\ 1 \\ 1 \\ 0 \end{bmatrix} \tag{8}$$

By summing all the components of the entropiece vector $\mathbf{s}(m_v^\Theta)$ we obtain the entropy value $U(m_v^\Theta) = 2$ nats for this vacuous BBA $m_v^\Theta(\cdot)$, which is of course in agreement with the formula (6).

4   Information Content of a BBA

We consider a (non-empty) FoD of cardinality $|\Theta| = N$, and we model our state of knowledge about the problem under consideration by a BBA defined on 2^Θ. Without more knowledge than the FoD itself (and its cardinality N), we are totally ignorant about the solution of the problem we want to solve, and of course we have no clue for making a decision/choice among the elements of the FoD. The BBA reflecting this total ignorant situation is the vacuous BBA $m_v(\cdot)$, whose maximal entropy is $U(m_v) = 2^N - 2$. In such case, we naturally expect that the information content we have[5] is zero when the uncertainty measure is maximal. In the very opposite case, it is very natural to consider that the information content of a BBA is maximal if the entropy value (the MoU value) of a BBA $m(\cdot)$ is zero, meaning that we make a choice of one element of the FoD without hesitation. Based on these very simple ideas, we propose to define the information content of any BBA $m(\cdot)$ as the dual of the effective measure of uncertainty, more precisely by

$$IC(m^\Theta) \triangleq U(m_v^\Theta) - U(m^\Theta) = (2^{|\Theta|} - 2) - \sum_{X \in 2^\Theta} s(X) \tag{9}$$

where $s(X)$ is the *entropiece* of the element $X \in 2^\Theta$ given by (5), that is

$$s(X) \triangleq -(1 - u(X))m^\Theta(X)\log(m^\Theta(X)) + u(X)(1 - m^\Theta(X))$$

and where $u(X)$ is the level of imprecision of the probability $P(X)$ given by

$$u(X) = Pl^\Theta(X) - Bel^\Theta(X) = \sum_{Y \in 2^\Theta | X \cap Y \neq \emptyset} m^\Theta(Y) - \sum_{Y \in 2^\Theta | Y \subseteq X} m^\Theta(Y) \tag{10}$$

---

[5] aside of the value of N of course.

From the definition (9), one sees that for $m^\Theta \neq m_v^\Theta$ one has $IC(m^\Theta) > 0$ because $U(m^\Theta) < U(m_v^\Theta)$, and for $m^\Theta = m_v^\Theta$ one has $IC(m_v^\Theta) = 0$, which is what we naturally expect.

It is worth mentioning that the information content $IC(m^\Theta)$ of a BBA depends not only of the BBA $m(.)$ itself but also on the cardinality of the frame of discernment[6] Θ because $IC(m^\Theta)$ requires the knowledge of $|\Theta|$ to calculate the max entropy value $U(m_v^\Theta) = 2^{|\Theta|} - 2$ entering in (9). This remark is very important to understand that even if two BBAs (defined on different FoDs) focus entirely on a same focal element, their information contents are necessarily different. For instance, if we consider the Bayesian BBA with $m^\Theta(\theta_1) = 1$ defined on the FoD $\Theta = \{\theta_1, \theta_2\}$, then

$$IC(m^\Theta) = U(m_v^\Theta) - U(m^\Theta) = (2^{|\Theta|} - 2) - 0 = 2 \text{ (nats)}$$

whereas if we consider the Bayesian BBA with $m^{\Theta'}(\theta_1) = 1$ defined on the larger FoD $\Theta' = \{\theta_1, \theta_2, \theta_3\}$ (for instance), then

$$IC(m^{\Theta'}) = U(m_v^{\Theta'}) - U(m^{\Theta'}) = (2^{|\Theta'|} - 2) - 0 = 6 \text{ (nats)}$$

So even if the decision θ_1 that we would make based either on m^Θ or on $m^{\Theta'}$ is the same, these decisions must not be considered actually with the same strength, and this is what reflects our information content measure.

From this very simple definition of information content, we can also define the Normalized Information Content (NIC) (if needed later in some applications), denoted by $NIC(m^\Theta)$ by normalizing $IC(m^\Theta)$ with respect to the maximal value of entropy $U(m_v^\Theta)$ as

$$NIC(m^\Theta) \triangleq \frac{U(m_v^\Theta) - U(m^\Theta)}{U(m_v^\Theta)} = 1 - \frac{U(m^\Theta)}{U(m_v^\Theta)} \tag{11}$$

Hence we will have $NIC(m^\Theta) \in [0, 1]$ and $NIC(m^\Theta) = 0$ for $m = m_v$, and $NIC(m^\Theta) = 1$ for $U(m) = 0$ which is obtained when $m(\cdot)$ is entirely focused on a singleton $\theta_i \in \Theta$, that is $m^\Theta(\theta_i) = 1$ for some $i \in \{1, 2, \ldots, |\Theta|\}$.

In fact, the (total) information content of a BBA $IC(m^\Theta)$ is the sum of all the *information granules* $IG(X|m^\Theta)$ of elements $X \in 2^\Theta$ carried by a BBA m^Θ, that is

$$IC(m^\Theta) = \sum_{X \in 2^\Theta} IG(X|m^\Theta) \tag{12}$$

where

$$IG(X|m^\Theta) \triangleq \begin{cases} 0, \text{if } X = \emptyset \\ -s(X), \text{if } X = \Theta \\ 1 - s(X) \text{ otherwise} \end{cases} \tag{13}$$

---

[6] That is why it is better, we think, to use the notation $IC(m^\Theta)$ instead of $IC(m)$.

We can define the information granules vector[7] $\mathbf{IG}(m) = [IG(X|m^\Theta), X \in 2^\Theta]^T$ by

$$\mathbf{IG}(m^\Theta) \triangleq \mathbf{s}(m_v^\Theta) - \mathbf{s}(m^\Theta) \tag{14}$$

One sees that the (total) information content $IC(m^\Theta)$ of a BBA m^Θ is just the sum of all components $IG(X|m^\Theta)$ of the information granules vector $\mathbf{IG}(m)$. The information granules vector $\mathbf{IG}(m)$ is interesting and useful because it helps to see the contribution of each element X in the whole measure of the information content $IC(m^\Theta)$ of a BBA m^Θ.

Example 1 (continued): consider $\Theta = \{\theta_1, \theta_2\}$ and the BBA $m^\Theta(\theta_1) = 0.5$, $m^\Theta(\theta_2) = 0.3$ and $m^\Theta(\theta_1 \cup \theta_2) = 0.2$. The information granules vector $\mathbf{IG}(m^\Theta)$ is given by

$$\mathbf{IG}(m^\Theta) = \mathbf{s}(m_v^\Theta) - \mathbf{s}(m^\Theta) = \begin{bmatrix} 0 \\ 1 \\ 1 \\ 0 \end{bmatrix} - \begin{bmatrix} 0 \\ 0.377258 \\ 0.428953 \\ 0.321887 \end{bmatrix} = \begin{bmatrix} 0 \\ 0.622742 \\ 0.571047 \\ -0.321887 \end{bmatrix} \tag{15}$$

By summing all the components of the information granules vector $\mathbf{IG}(m^\Theta)$ we obtain the (total) information content $IC(m^\Theta) = 0.871902$ nats of the BBA m^Θ, which can of course be calculated directly also as

$$IC(m^\Theta) = U(m_v^\Theta) - U(m^\Theta) = 2 - 1.128098 = 0.871902$$

However, the information granules vector $\mathbf{IG}(m^\Theta)$ is interesting to identify the contribution of each element X in the whole measure of the information content.

5   Information Gain and Information Loss

Once the IC measure is defined for a BBA, it is rather simple to define the information gain and information loss of a BBA with respect to another one, both defined on a same FoD Θ. Suppose that we have a first BBA m_1^Θ and a second BBA m_2^Θ, then we can calculate by formula (9) their respective information contents $IC(m_1^\Theta)$ and $IC(m_2^\Theta)$. The difference of information content measure of m_2^Θ with respect to m_1^Θ is defined by[8]

$$\Delta_{IC}(m_2|m_1) \triangleq IC(m_2^\Theta) - IC(m_1^\Theta) \tag{16}$$

If we replace $IC(m_2^\Theta)$ and $IC(m_1^\Theta)$ by their expressions according to (9), it comes

$$\Delta_{IC}(m_2|m_1) = [U(m_v^\Theta) - U(m_2^\Theta)] - [U(m_v^\Theta) - U(m_1^\Theta)] = U(m_1^\Theta) - U(m_2^\Theta) \tag{17}$$

---

[7] We suppose for convenience that the elements $X \in 2^\Theta$ are listed in increasing order using the classical $|\Theta|$-bits representation with the least significant bit on the right.
[8] Similarly, we can define $\Delta_{IC}(m_1|m_2) \triangleq IC(m_1^\Theta) - IC(m_2^\Theta) = -\Delta_{IC}(m_2|m_1)$.

If $\Delta_{IC}(m_2|m_1) = 0$, the BBAs m_1^Θ and m_2^Θ have same measure of information content. So, there is no gain and no loss in information content if one switches from m_1^Θ to m_2^Θ or vice versa. $\Delta_{IC}(m_2|m_1) = 0$ does not mean that the decisions based on m_1^Θ and on m_2^Θ are the same. It does only means that the decision based on m_1^Θ must be as easy as the decision made based on m_2^Θ. It means that they have the same informational strength. That's it. If $\Delta_{IC}(m_2|m_1) > 0$, one has $IC(m_2^\Theta) > IC(m_1^\Theta)$, i.e. the BBA m_2^Θ is more informative than m_1^Θ. In this case we get an information gain if one switches from m_1^Θ to m_2^Θ, and by duality we get an uncertainty reduction by switching from m_1^Θ to m_2^Θ. It means that it must be easier to make a decision based on m_2^Θ rather on m_1^Θ. If $\Delta_{IC}(m_2|m_1) < 0$, one has $IC(m_2^\Theta) < IC(m_1^\Theta)$, i.e. the BBA m_2^Θ is less informative than m_1^Θ. In this case we get an information loss if one switches from m_1^Θ to m_2^Θ, and by duality we get an uncertainty raise by switching from m_1^Θ to m_2^Θ. It means that it must be easier to make a decision based on m_1^Θ rather on m_2^Θ.

As simple example, consider $\Theta = \{\theta_1, \theta_2, \theta_3\}$. For the vacuous BBA one has $U(m_v^\Theta) = 2^3 - 2 = 6$ nats. Suppose at time $k = 1$ one has the BBA $m_1^\Theta(\theta_1 \cup \theta_2) = 0.2$, $m_1^\Theta(\theta_1 \cup \theta_3) = 0.3$, $m_1^\Theta(\theta_1 \cup \theta_2 \cup \theta_3) = 0.5$, then $U(m_1^\Theta) \approx 5.1493$ nats, and $IC(m_1^\Theta) = U(m_v^\Theta) - U(m_1^\Theta) \approx 0.8507$ nats. Suppose that after some information processing (belief revision, or fusion, etc.) we come up with the BBA m_2^Θ at time $k = 2$ defined by $m_2^\Theta(\theta_1) = 0.2$ and $m_2^\Theta(\theta_1 \cup \theta_3) = 0.8$, then $U(m_2^\Theta) \approx 0.5004$ nats and $IC(m_2^\Theta) = U(m_v^\Theta) - U(m_2^\Theta) \approx 5.4996$ nats. In this case, we get $\Delta_{IC}(m_2|m_1) = 5.4996 - 0.8507 = 4.6489$ which is positive. Hence we get an information gain by switching from m_1^Θ to m_2^Θ thanks to the information processing applied.

6   Conclusions

In this paper we have introduced a measure of information content (IC) for any basic belief assignment (BBA). This IC measure based on an effective measure of uncertainty of BBAs is quite simple to calculate, and it reflects somehow the informational strength and easiness ability to make a decision based on any belief mass function. We have also shown how it is possible to identify the contribution of each focal element of the BBA to this information content measure thanks to the information granule vector. This new IC measure is also interesting because it allows to well quantify the information loss or gain between two BBAs, and thus as perspectives we could use it to quantify precisely and compare the performances of information processing using belief functions (like fusion rules, belief conditioning, etc.). We hope that this new theoretical IC measure will open interesting tracks for forthcoming research works on reasoning about uncertainty with belief functions.

References

1. Bovee, M., Srivastava, R.S.: A conceptual framework and belief-function approach to assessing overall information quality. Int. J. Intell. Syst. **18**(1), 51–74 (2003)

2. Floridi, L., Illari, P. (eds.): The Philosophy of Information Quality. Springer International Publishing, Switzerland (2014)

3. Batini, C., Scannapieco, M.: Data and Information Quality: Dimensions, Principles and Techniques. Springer International Publishing, Switzerland (2016)

4. Kenett, R.S., Shmueli, G.: Information Quality. Wiley & Sons, Hoboken (2017)

5. Bossé, E., Rogova, G.L. (eds.): Information Quality in Information Fusion and Decision Making, Information Fusion and Data Science. Springer Nature, Switzerland (2019)

6. Bouhamed, S.A., Kalle, I.K., Yager, R.R., Bossé, E., Solaiman, B.: An intelligent quality-based approach to fusing multi-source possibilistic information. Inf. Fusion **55**, 68–90 (2020)

7. Dezert, J.: An effective measure of uncertainty of basic belief assignments, fusion. In: 2022 International Conference, Linköping, Sweden, pp. 1–10 (2022)

8. Shafer, G.: A Mathematical Theory of Evidence. Princeton University Press, Princeton (1976)

9. Dezert, J., Tchamova A.: On effectiveness of measures of uncertainty of basic belief assignments. Inf. Secur. J.: Int. J. (ISIJ)**52**, 9–36 (2022)

10. Tribus, M.: Rational Descriptions, Decisions and Designs. Pergamon Press, New York, pp. 26–28 (1969)

11. Bradley, R.E., Petrilli, S.J., Sandifer, C.E.: L'Hôpital's analyse des infiniments petits (An annoted translation with source material by Johann Bernoulli), Birkhäuser, p. 311 (2015)

12. Shannon, C.E.: A mathematical theory of communication. Bell Syst. Tech. J. **27**, 379–423 & 623–656 (1948)

13. Klir, G.J.: Principles of uncertainty: what are they? why do we need them? Fuzzy Sets Syst. **74**, 15–31 (1995)

Belief Functions on Ordered Frames
of Discernment

Arnaud Martin[✉]

Univ Rennes, CNS, IRISA, DRUID, IUT de Lannion, Lannion, France
arnaud.martin@irisa.fr
http://people.irisa.fr/Arnaud.Martin

Abstract. Most questionnaires offer ordered responses whose order is poorly studied via belief functions. In this paper, we study the consequences of a frame of discernment consisting of ordered elements on belief functions. This leads us to redefine the power space and the union of ordered elements for the disjunctive combination. We also study distances on ordered elements and their use. In particular, from a membership function, we redefine the cardinality of the intersection of ordered elements, considering them fuzzy.

Keywords: Ordinal variable · Ordered frame of discernment ·
Ordered and fuzzy elements · Ordered power set · Distance

1 Introduction

The theory of belief functions is used in more and more applications such as machine learning, pattern recognition, clustering, etc. To apply the theory of belief functions, we consider the frame of discernment given by $\Omega = \{\omega_1, \omega_2, \ldots, \omega_n\}$ and the basic belief assignments are defined as a mapping from the power set of Ω, noted 2^Ω, to $[0,1]$ [13]. The elements ω_i of Ω are considered exclusive and exhaustive. The assumption of the exclusivity can be lifted, considering the hyper power set D^Ω [9].

However, we may be confronted with applications where a semantic or proximity link exists between the elements of the frame of discernment. For example, in questionnaires graduated or ordered answers can be proposed:

What is the distance between Lannion and Paris?:
314 km, *414* km, *514* km, *614* km

Other possible answers use the Likert scale such as:

Not happy, neutral, happy
Strongly disagree; Disagree; Neither agree nor disagree; Agree; Strongly agree

These answers can be represented by an ordinal variable. In these cases, there is no way to deal with this link between the elements of the frame of discernment. Forcing these elements to have a zero mass is not stable in the combination and

S. Le Hégarat-Mascle et al. (Eds.): BELIEF 2022, LNAI 13506, pp. 129–138, 2022.
https://doi.org/10.1007/978-3-031-17801-6_13

decision processes. In this paper, we explore the possibilities of considering an order between the elements of the frame of discernment with the theory of belief functions. First in Sect. 2 we introduce the ordered power set, then we show how we can combine mass functions defined on an ordered power set in Sect. 3. In Sect. 4, we propose the definition of a distance between ordered elements used to define a distance between belief functions, that can be used for conflict measure and decision in Sect. 5. In Sect. 6, we consider ordered fuzzy elements to redefine the cardinality of the intersection of elements before concluding.

2   Power Set of Ordered Elements

Let us take again the example of the question about the distance between Paris and Lannion. The frame of discernment is given by $\Omega = \{\omega_1, \omega_2, \omega_3, \omega_4\}$, with the answers $\omega_1 = 314$ km, $\omega_2 = 414$ km, $\omega_3 = 514$ km, $\omega_4 = 614$ km. The right answer is 514 km. Here, the elements of the frame of discernment are obviously exclusive. However, if someone answers 614 km, the error is smaller than if the answer given is 314 km. If we consider imprecise answers as in [15], $\{\omega_3, \omega_4\}$ make sense while $\{\omega_1, \omega_3\}$ does not.

We therefore consider that we should not take into account all the elements of the power set 2^Ω, but only those which have a meaning. Thus disjunctions that do not contain consecutive elements should not be considered. Let us consider an ordinal variable having values in a finite set $\Omega = \{\omega_1, \ldots, \omega_n\}$ of ordered ω_i, $i = 1, \ldots, n$, i being an ordinal number.

Definition. *The ordered power set, noted oPS^Ω, is a subset of the power set composed by the empty set and all the disjunctions of consecutive elements of Ω.*

A disjunction of consecutive elements from endpoint elements is noted by:

$$\{\omega_i, \omega_j\}_o = \{\omega_i, \omega_{i+1}, \ldots, \omega_{j-1}, \omega_j\}, \text{ with } 1 \leq i \leq j \leq n \tag{1}$$

Hence, the ordered power set is given by:

$$oPS^\Omega = \{\emptyset, \{\{\omega_i, \omega_j\}_o\}_{i,j=1,\ldots,n}\} \tag{2}$$

The number of elements of 2^Ω is 2^n, but it is smaller for oPS^Ω.

Proposition. *The number of elements of oPS^Ω, with: $\Omega = \{\omega_1, \ldots, \omega_n\}$ is:*

$$1 + \frac{n(n+1)}{2} \tag{3}$$

Proof. The set oPS^Ω contains the empty set and ordered elements determined by the endpoint elements. The number of these elements is the number of pairs (i, j) with $1 \leq i \leq j \leq n$. This number is $\frac{n!}{2!(n-2)!} = \frac{n(n+1)}{2}$. So $|oPS^\Omega| = 1 + \frac{n(n+1)}{2}$. $\square$

Let $\Omega = \{\omega_1, \ldots, \omega_n\}$ be a frame of discernment of ordered, exclusive and exhaustive elements ω_i, $i = 1, \ldots, n$. A mass function m is the mapping from elements of the ordered power set oPS^Ω onto $[0, 1]$ such that:

$$\sum_{X \in oPS^\Omega} m(X) = 1. \tag{4}$$

A focal element X is an element of oPS^Ω such that $m(X) \neq 0$.

From this definition of mass function, all the definitions of special mass functions are available (simple mass functions, non dogmatic, consonant, etc.). A categorical mass function with $m(X) = 1$ is noted m_X. Definitions of classical belief functions such as credibility, plausibility and pignistic probability are also available on the ordered power set oPS^Ω. The credibility function is given for all $X \in oPS^\Omega$ by:

$$\text{bel}(X) = \sum_{Y \subseteq X, Y \neq \emptyset} m(Y). \tag{5}$$

The plausibility function is given for all $X \in oPS^\Omega$ by:

$$\text{pl}(X) = \sum_{Y \in oPS^\Omega, Y \cap X \neq \emptyset} m(Y). \tag{6}$$

However, it is not possible to compute this function as a dual of the credibility function, because an ordered power set is not invariant by the complement. Indeed, if we consider $\Omega = \{\omega_1, \omega_2, \omega_3\}$, the complementary of ω_2 is $\{\omega_1, \omega_3\} \notin oPS^\Omega$. The pignistic probability [14] can be written for all $\omega \in \Omega$ by:

$$\text{BetP}(\omega) = \sum_{X \in oPS^\Omega, X \neq \emptyset, \omega \in X} \frac{1}{|X|} \frac{m(X)}{1 - m(\emptyset)}. \tag{7}$$

3   Combination of Belief Functions on Ordered Power Set

When considering several mass functions on an ordered power set oPS^Ω from several sources or persons, we need to be able to combine them. The combination operator must therefore produce a mass function in the same ordered power set oPS^Ω. Therefore, an ordered power set must be invariant by the combination operator. There are a large number of combination operators [11] and not all of them verify this property.

Proposition. *An ordered power set is invariant by the conjunctive combinations (normalized, not normalized, Yager [11]).*

Proof. Let two focal elements of an ordered power set oPS^Ω: $\{\omega_{i_1}, \omega_{j_1}\}_o$ and $\{\omega_{i_2}, \omega_{j_2}\}_o$. The intersection of these elements is empty or given by:

$$\{\max(\omega_{i_1}, \omega_{i_2}), \min(\omega_{j_1}, \omega_{j_2})\}_o \in oPS^\Omega$$

Therefore, an ordered power set is invariant by all conjunctive operators. □

Proposition. *An ordered power set is not invariant by the disjunctive combination.*

Proof. Let two categorical mass functions m_{ω_1} and m_{ω_3}, with two non consecutive focal elements such as ω_1 and ω_3. The disjunctive combination of these two mass functions is the categorical mass function with $\{\omega_1, \omega_3\}$ such as unique focal element. As $\{\omega_1, \omega_3\} \notin oPS^\Omega$, the ordered power set is not invariant by the disjunctive combinations. □

Thus, an ordered power set is not invariant by all the mixed combination operators based on the disjunctive operator. Here we propose a new disjunctive combination operator that makes an ordered power set invariant. Let two elements Y_i and $Y_j \in oPS^\Omega$, we note $Y_i = \{\omega_{i_1}, \omega_{i_{n_i}}\}_o$ and $Y_j = \{\omega_{j_1}, \omega_{j_{n_j}}\}_o$. We define the union of these two ordered elements by:

$$Y_i \overset{o}{\cup} Y_j = \{\min(\omega_{i_1}, \omega_{j_1}), \max(\omega_{i_{n_i}}, \omega_{j_{n_j}})\}_o \tag{8}$$

The union of s ordered elements is given by the extension of the previous equation or recursively by $((Y_1 \overset{o}{\cup} Y_2) \overset{o}{\cup} Y_3) \ldots \overset{o}{\cup} Y_s$.

Let two mass functions defined on the ordered power set oPS^Ω, for all $X \in oPS^\Omega$, the disjunctive combination is given by:

$$m_{\text{oDis}}(X) = \sum_{Y_i \overset{o}{\cup} Y_j = X} m_i(Y_i) m_j(Y_j). \tag{9}$$

The disjunction combination of s mass functions on the ordered power set oPS^Ω, is given for all $X \in oPS^\Omega$ by:

$$m_{\text{oDis}}(X) = \sum_{Y_1 \overset{o}{\cup} \ldots \overset{o}{\cup} Y_s = X} \prod_{j=1}^{S} m_j(Y_j). \tag{10}$$

where $Y_j \in oPS^\Omega$ is a focal element of the source S_j, and $m_j(Y_j)$ the associated mass function. We can thus rewrite the mixed combination operators [11], such as the Dubois and Prade one [4] given for all $X \in oPS^\Omega$, $X \neq \emptyset$ by:

$$m_{\text{oDP}}(X) = \sum_{Y_1 \cap \ldots \cap Y_s = X} \prod_{j=1}^{S} m_j(Y_j) + \sum_{\substack{Y_1 \overset{o}{\cup} \ldots \overset{o}{\cup} Y_s = X \\ Y_1 \cap \ldots \cap Y_s = \emptyset}} \prod_{j=1}^{S} m_j(Y_j). \tag{11}$$

Proposition. *An ordered power set is stable by the average combination.*

Proof. The proof is obvious, because the set of focal elements obtained by the combination is the union of the sets focal elements from the mass functions to be combined. □

4   Distances on Belief Functions on Ordered Power Set

In the context of belief functions, we often use a distance between mass functions, in order to measure similarity for clustering, to define some measures or to decide. There are many distances that can be considered [7]. The most commonly used distance between the belief functions is the distance defined in [6]. This distance can obviously be defined for two mass functions m_1 and m_2 on oPS^Ω by:

$$d_J(m_1, m_2) = \sqrt{\frac{1}{2}(m_1 - m_2)^T \underline{D}(m_1 - m_2)}, \tag{12}$$

where $\underline{D}$ is an $1 + \frac{n(n+1)}{2} \times 1 + \frac{n(n+1)}{2}$ matrix based on Jaccard dissimilarity whose elements are:

$$D(A, B) = \begin{cases} 1, \text{ if } A = B = \emptyset, \\ \dfrac{|A \cap B|}{|A \overset{o}{\cup} B|}, \forall A, B \in oPS^\Omega. \end{cases} \tag{13}$$

where $|X|$ is the cardinality of $X \in oPS^\Omega$. Of course we have:

$$\frac{|A \cap B|}{|A \overset{o}{\cup} B|} = \frac{|A \cap B|}{|A \cup B|}$$

because if $A \cap B \neq \emptyset$ then $A \overset{o}{\cup} B = A \cup B$ with $A, B \in oPS^\Omega$.

However, with this distance, we have without any distinction:

$$d_J(m_{\omega_1}, m_{\omega_2}) = d_J(m_{\omega_1}, m_{\omega_3}) = 1.$$

The value 1 is the maximum value of this distance.

4.1 Distance Between Ordered Elements

Let us take the example of the question about the distance between Paris and Lannion. We have noticed that if someone answers 614km, the error is smaller than if the answer given is 314 km. Thus, the distance between elements of an ordered frame of discernment Ω can be considered differently according to their order. On a Likert scale, we have the same argument. We can consider that the minimal distance between two elements of Ω is that between two consecutive elements. Inthe same way, the maximum distance between two elements of Ω is the distance between the first ω_1 and the last ω_n element of Ω.

In order to have a distance with a normality property, we propose the following distance between two elements of Ω:

$$d_o(\omega_i, \omega_j) = \frac{|i - j|}{n - 1} \tag{14}$$

where $|x|$ is the absolute value of $x = 1, \ldots, n$. This distance takes obviously its values in $[0, 1]$.

The distance between an element of Ω and an element $X = \{\omega_{1_x}, \omega_{n_x}\}_o$ of oPS^Ω can be defined by one of the following equations:

$$d_{o_{min}}(\omega_i, X) = \min(d_o(\omega_i, \omega_{1_x}), d_o(\omega_i, \omega_{n_x})) \tag{15}$$

$$d_{o_{max}}(\omega_i, X) = \max(d_o(\omega_i, \omega_{1_x}), d_o(\omega_i, \omega_{n_x})) \qquad (16)$$

$$d_{o_{av}}(\omega_i, X) = \frac{1}{[X]} \sum_{k=1_x}^{n_x} d_o(\omega_i, \omega_k) \qquad (17)$$

The distance between two elements $X = \{\omega_{1_x}, \omega_{n_x}\}_o$ and $Y = \{\omega_{1_y}, \omega_{n_y}\}_o$ of oPS^Ω can be defined by one of the following equations:

$$d_{o_{min}}(X, Y) = \min_{\omega_{y_i} \in Y} d_{o_{min}}(X, \omega_{y_i}) \qquad (18)$$

$$d_{o_{max}}(X, Y) = \max_{\omega_{y_i} \in Y} d_{o_{max}}(X, \omega_{y_i}) \qquad (19)$$

$$d_{o_{av}}(X, Y) = \frac{1}{[XY]} \sum_{k_x=1_x}^{n_x} \sum_{k_y=1_y}^{n_x} d_o(\omega_{k_x}, \omega_{k_y}) \qquad (20)$$

These distances take their values in $[0, 1]$. The use of one of these distances rather than another may depend on the application. For the sake of simplicity, we will note d_o in the following.

4.2   Distance Between Belief Functions

As we have seen, the Jousselme distance does not fit the ordered elements of the frame of discernment. Indeed, with Jaccard dissimilarity, the dissimilarity is zero if the intersection is empty. However, on ordered and exclusive elements, the dissimilarity can be different depending on the order. Therefore, we modify the dissimilarity of Jaccard on empty intersection elements from the distance defined in the previous section. Since the minimum strictly positive value of the Jaccard dissimilarity is $\frac{1}{n}$, the proposed dissimilarity takes its values on $[0, \frac{1}{n}]$. Thus, we define a modified Jaccard dissimilarity for ordered elements by:

$$D_o(A, B) = \begin{cases} 1, \text{ if } A = B = \emptyset, \\ \dfrac{|A \cap B|}{|A \overset{o}{\cup} B|} + (1 - Int(A, B))\dfrac{1 - d_o(A, B)}{n}, \forall A, B \in oPS^\Omega. \end{cases} \qquad (21)$$

where Int is the intersection index defined such as in [7] by $Int(A, B) = 1$ if $A \cap B \neq \emptyset$ and 0 otherwise.

According to [1], D_o define a matrix positive definite. The distance obtained is given for two mass functions m_1 and m_2 on oPS^Ω by:

$$d_{Jo}(m_1, m_2) = \sqrt{\frac{1}{2}(m_1 - m_2)^T \underline{\underline{D_o}}(m_1 - m_2)}, \qquad (22)$$

where $\underline{\underline{D_o}}$ is given by Eq. (21).

If we consider $\Omega = \{\omega_1, \omega_2, \omega_3\}$ and the distance given by Eq. (20), we obtain:

$$\underline{\underline{D_o}} = \begin{pmatrix} 1 & 0 & 0 & 0 & 0 & 0 & 0 \\ 0 & 1 & \frac{1}{6} & \frac{1}{2} & 0 & \frac{1}{2} & \frac{1}{3} \\ 0 & \frac{1}{6} & 1 & \frac{1}{2} & \frac{1}{6} & \frac{1}{2} & \frac{1}{3} \\ 0 & \frac{1}{2} & \frac{1}{2} & 1 & \frac{1}{12} & \frac{1}{3} & \frac{2}{3} \\ 0 & 0 & \frac{1}{6} & \frac{1}{12} & 1 & \frac{1}{2} & \frac{1}{3} \\ 0 & \frac{1}{12} & \frac{1}{2} & \frac{1}{3} & \frac{1}{2} & 1 & \frac{2}{3} \\ 0 & \frac{1}{3} & \frac{1}{3} & \frac{2}{3} & \frac{1}{3} & \frac{2}{3} & 1 \end{pmatrix} \tag{23}$$

Hence:

$$d_{Jo}(m_{\omega_1}, m_{\omega_2}) = \sqrt{\frac{5}{6}} \simeq 0.91 \text{ and } d_{Jo}(m_{\omega_1}, m_{\omega_3}) = 1 \tag{24}$$

In this way, we show that we can take into account the proximity of ordered elements in the distance.

The modification of the Jaccard dissimilarity, could induce an interpretation of the non-exclusivity of the elements. We will discuss this in Sect. 6.

5   Decision and Conflict on Ordered Elements

After combining the mass functions, we generally want to make a decision about the resulting mass function. This mass function with the operators seen in Sect. 3 takes its values on oPS^{Ω}. It is common to make the decision on Ω by maximum of credibility, plausibility or by compromise with the pignistic probability. We have seen in Sect. 2, that we can calculate these belief functions in oPS^{Ω} and if we denote f_d one of these functions the decision is made by:

$$\omega_d = \underset{\omega \in \Omega}{\operatorname{argmax}} \left(f_d(\omega) \right). \tag{25}$$

However, these functions do not consider the difference of proximity of the elements of Ω and do not allow to decide on the disjunctions, i.e. on some elements of oPS^{Ω}. In [5], we introduced another decision process based on a distance given by:

$$A = \underset{X \in \mathcal{D}}{\operatorname{argmax}} \left(d_{Jo}(m, m_X) \right), \tag{26}$$

where m_X is the categorical mass function $m(X) = 1$, m is the mass function coming from the combination rule and d_{Jo} is the distance introduced in Equation (22). The subset $\mathcal{D} \subseteq oPS^{\Omega}$ is the set of elements on which we want to decide.

This last decision process also allows to decide on imprecise elements of the ordered power set oPS^{Ω} and to take into account with the distance the difference of proximity of ordered element of Ω.

Let us take the example of the Paris-Lannion trip, if we get 3 categorical answers from 3 people such as: Person 1 said: 314 km, person 2 said: 514 km, and person 3 said: 614 km. Person 1 disagrees more with person 2 than with person 3. The distance introduced in Sect. 4 can measure this difference on the ordered elements of the frame of discernment. Therefore, we can extend here the conflict measure introduced in [10], between two mass functions m_1 and m_2 by:

$$\text{Conf}(m_1, m_2) = (1 - \delta_{inc}(m_1, m_2))d_{Jo}(m_1, m_2) \tag{27}$$

where d_{Jo} is the distance defined by equation (22) and δ_{inc} is a degree of inclusion measuring how m_1 and m_2 and included each other. For more details on possible degree of inclusion see [10]. To measure the conflict between more than two mass functions, the average of the conflicts two by two can be considered.

6   Belief Functions on Ordered Fuzzy Elements

Now assume that the answers to the question about the distance between Lannion and Paris are:

about 300 km, about 400 km, about 500 km, about 600 km

The possible answers are always ordered, and if they can always be considered exclusive, they are fuzzy answers.

The link between belief functions and fuzzy sets has already been discussed in different ways [2,3]. Here we wish to study the representation and the consideration of ordered and fuzzy elements of the frame of discernment.

As we saw in Sect. 4 by redefining Jaccard's dissimilarity, the exclusivity of the ordered elements can be questioned. The fact of having ordered fuzzy elements does not however call into question the exclusivity of the elements and thus the intersection of the elements. Indeed, there is no semantic sense in having about 400 km and about 500 km at the same time. On the other hand, considering fuzzy elements allows to redefine the cardinality of the intersection of the elements from the membership functions.

In [12], the cardinality of the intersection of ordered elements is defined from a definition of the membership function, which allows them to propose another modified version of Jaccard dissimilarity. Inspired by this work, we define here the membership function of an element of $\omega \in \Omega$ to a set of $X \in oPS^\Omega$ by:

$$\mu_X(\omega) = \begin{cases} 1, \text{ if } \omega \in X, \\ \alpha e^{-\gamma d_{Jo}(\omega, X)}, \text{ otherwise,} \end{cases} \tag{28}$$

with $0 \leq \alpha \leq 1$, $0 \leq \gamma \leq 1$ two thresholds to control the membership function and d_o the distance defined by Eq. (15), (16) or (17). Based on this membership function, we define the cardinality of the intersection of two elements X and $Y \in oPS^\Omega$ by:

$$|X \cap Y|_o = \sum_{\omega \in X \mathring{\cup} Y, \omega \in \Omega} \min(\mu_X(\omega), \mu_Y(\omega)) \tag{29}$$

If $\alpha = 0$ then $|X \cap Y|_o = |X \cap Y|$.

Therefore, we define a modified Jaccard dissimilarity for ordered elements by:

$$D_o(A, B) = \begin{cases} 1, \text{ if } A = B = \emptyset, \\ \dfrac{|A \cap B|_o}{|A \overset{o}{\cup} B|}, \forall A, B \in oPS^{\Omega}. \end{cases} \tag{30}$$

We have $0 \le D_o(A, B) \le 1$ and $D(A, B) \le D_o(A, B)$ [12]. A new distance d_{Jo} between the ordered and fuzzy elements can thus be defined by Eq. (22).

The cardinality of the intersection defined by Eq. (29) allows its use in the mixed rules [8] to regulate the conjunctive/disjunctive behaviour by taking into account the partial combinations according to the cardinality of the elements. The mixed rule is given for m_1 and m_2 for all $X \in 2^{\Omega}$ by:

$$m_{\text{Mix}}(X) = \sum_{Y_1 \overset{o}{\cup} Y_2 = X} \delta_1 m_1(Y_1) m_2(Y_2) \\ + \sum_{Y_1 \cap Y_2 = X} \delta_2 m_1(Y_1) m_2(Y_2). \tag{31}$$

The choice of $\delta_1 = 1 - \delta_2$ can also be made from Jaccard dissimilarity:

$$\delta_2(Y_1, Y_2) = \frac{|Y_1 \cap Y_2|_o}{|Y_1 \overset{o}{\cup} Y_2|}. \tag{32}$$

Thus, if we have a partial conflict between Y_1 and Y_2, $Y_1 \cap Y_2 = \emptyset$, the rule transfers the mass on $Y_1 \overset{o}{\cup} Y_2$ according to the difference in order of the elements of Ω constituting Y_1 and Y_2.

Questionnaires whose responses are in the form of a Likert scale can also be modeled by a frame of discernment of ordered and fuzzy elements and thus use the modified Jaccard dissimilarity defined by Eq. (30).

7   Conclusion

In this paper, we explored the modeling and integration of an ordered element of frame of discernment by belief functions. The fact of considering ordered elements led us to redefine the power space of unions of ordered elements making sense, called ordered power space. Thus, to consider disjunctive combination rules leaving invariant the ordered power space, we defined the union of ordered elements. From a distance between ordered elements we redefined a distance between mass functions that can be used in a conflict measure and for decision making. Without redefining the intersection of ordered elements, which we still consider exclusive, we have redefined, from a membership function, the cardinality of the intersection of ordered elements that can be considered fuzzy.

A large number of applications can be addressed by modeling on ordered sets of elements. We are thinking in particular of the answers to questionnaires which very often use Likert scales. This type of questionnaire can, for example, be used to evaluate the knowledge and skills of students allowing to consider the closest answers in the sense of the order of the right answer, without considering them as totally wrong. Concrete applications will be considered in future work.

References

1. Bouchard, M., Jousseleme, A.-L., Doré, P.-E.: A proof for the positive definiteness of the Jaccard index matrix. Int. J. Approximate Reasoning **4**(5), 615–626 (2013)
2. Denœux, T.: Modeling vague beliefs using fuzzy-valued belief structures. Fuzzy Sets Syst. **116**(2), 167–199 (2000)
3. Denœux, T.: Belief functions induced by random fuzzy sets: a general framework for representing uncertain and fuzzy evidence. Fuzzy Set. Syst. **424**, 63–91 (2021)
4. Dubois, D., Prade, H.: Representation and combination of uncertainty with belief functions and possibility measures. Comput. Intell. **4**, 244–264 (1988)
5. Essaid, A., Martin, A., Smits, G., Ben Yaghlane, B.: A distance-based decision in the credal level. In: Aranda-Corral, G.A., Calmet, J., Martín-Mateos, F.J. (eds.) AISC 2014. LNCS (LNAI), vol. 8884, pp. 147–156. Springer, Cham (2014). https://doi.org/10.1007/978-3-319-13770-4_13
6. Jousselme, A.-L., Grenier, D., Bossé, E.: A new distance between two bodies of evidence. Inf. Fusion **2**, 91–101 (2001)
7. Jousselme, A.-L., Maupin, P.: Distances in evidence theory: comprehensive survey and generalizations. Int. J. Approximate Reasoning **53**(2), 118–145 (2011)
8. Martin, A., Osswald, C.: Toward a combination rule to deal with partial conflict and specificity in belief functions theory. In: International Conference on Information Fusion, Québec, Canada (2007)
9. Martin, A.: Implementing general belief function framework with a practical codification for low complexity, in Advances and Applications of DSmT for Information Fusion, (Collected Works, Vol 3), Smarandache F., Dezert, J. (Eds.) American Research Press Rehoboth, pp. 217–273 (2009)
10. Martin, A.: About conflict in the theory of belief functions. In: Denoeux, T., Masson, M.H. (eds.) Belief Functions: Theory and Applications. Advances in Intelligent and Soft Computing, vol. 164, pp. 161–168. Springer, Berlin, Heidelberg (2012). https://doi.org/10.1007/978-3-642-29461-7_19
11. Martin, A.: Conflict management in information fusion with belief functions. In: Bossé, É., Rogova, G.L. (eds.) Information Quality in Information Fusion and Decision Making. IFDS, pp. 79–97. Springer, Cham (2019). https://doi.org/10.1007/978-3-030-03643-0_4
12. Petković, M., Škrlj, B., Kocev, D., Simidjievski, N.: Fuzzy Jaccard index: a robust comparison of ordered lists. Appl. Soft Comut. **113**(Part A), 107849 (2021)
13. Shafer, G.: A Mathematical Theory of Evidence. Princeton University Press, Princeton (1976)
14. Smets, P.: Constructing the pignistic probability function in a context of uncertainty. Uncertainty Artif. Intell. **5**, 29–39 (1990)
15. Thierry, C., Martin, A., Dubois, J.-C., Le Gall, Y.: Modeling uncertainty and inaccuracy on data from crowdsourcing platforms: MONITOR. In: IEEE 31st International Conference on Tools with Artificial Intelligence, Portland, United States (2019)

On Modelling and Solving the Shortest Path Problem with Evidential Weights

Tuan-Anh Vu[(⊠)], Sohaib Afifi, Éric Lefèvre, and Frédéric Pichon

Univ. Artois, UR 3926, Laboratoire de Genie Informatique et d'Automatique de l'Artois (LGI2A), 62400 Bethune, France
{tanh.vu,sohaib.afifi,eric.lefevre,frederic.pichon}@univ-artois.fr

Abstract. We study the single source single destination shortest path problem in a graph where information about arc weights is modelled by a belief function. We consider three common criteria to compare paths with respect to their weights in this setting: generalized Hurwicz, strong dominance and weak dominance. We show that in the particular case where the focal sets of the belief function are Cartesian products of intervals, finding best, *i.e.*, non-dominated, paths according to these criteria amounts to solving known variants of the deterministic shortest path problem, for which exact resolution algorithms exist.

Keywords: Shortest path · Belief function · Exact method

1 Introduction

The Shortest Path Problem (SPP) is one of the most studied problems in combinatorial optimization with a wide range of applications in, *e.g.*, transportation and telecommunications. In many realistic situations, uncertainty on arc weights is encountered; for instance, the travel times between cities can be affected by external factors such as weather conditions or traffic jams. Many approaches have been proposed to model the uncertainty on arc weights. In particular, robust optimization frameworks have represented uncertainty by discrete scenario sets [3,15] and by intervals [3,10].

In this paper, we investigate the case where the uncertainty on arc weights is *evidential, i.e.*, modelled by a belief function [12]. More specifically, we assume that each focal set of the considered belief function is a Cartesian product of intervals with each interval describing possible values of each arc weight. Such a belief function is a direct and natural generalization of the above-mentioned interval-based uncertainty representation, which arises when considering probabilities that the intervals hold. It can be illustrated as follows: in a network with three cities A, B, and C, under good weather conditions, it may take 20 to 30 min to travel from A to B, and 10 to 20 min to travel from B to C; however under bad weather conditions, the travel times from A to B (resp. B to C) takes 30 to 40 min (resp. 15 to 25 min) and the forecast tells us that the probability of good weather (resp. bad weather) is 0.8 (resp. 0.2).

© The Author(s), under exclusive license to Springer Nature Switzerland AG 2022
S. Le Hégarat-Mascle et al. (Eds.): BELIEF 2022, LNAI 13506, pp. 139–149, 2022.
https://doi.org/10.1007/978-3-031-17801-6_14

In the presence of evidential uncertainty on arc weights, the notion of best, *i.e.*, shortest, paths becomes ill-defined. In a similar vein as [3] and using decision theory under evidential uncertainty [2], best paths are defined in this paper as the non-dominated ones with respect to a preference relation over paths, built from some criterion relying on the notions of upper and lower expected weights of paths. We consider in particular three common criteria, called generalized Hurwicz, strong dominance and weak dominance; the first one induces a complete preference relation while the latter two induce partial relations leading, as will be seen, to more challenging optimisation problems.

Combinatorial optimization problems under evidential uncertainty have received some attention recently. Notably, in [8,14], the authors studied different variants of the Vehicle Routing Problem (VRP) with different uncertainty factors and with similar particular focal sets as in this paper. They proposed approximate resolution methods based on metaheuristics to find non-dominated solutions with respect to a complete relation built from a particular case of the generalized Hurwicz criterion. Guillaume *et al.* [6] studied a general optimization problem in which the coefficients in the objective function are subject to uncertainty. They considered also the generalized Hurwicz criterion and provided results about the complexity of finding a non-dominated solution.

In contrast to [8,14], we provide in this paper *exact* methods to find non-dominated solutions with respect to both complete *and partial* relations, owing to the fact that the SPP is much simpler than the VRP. Furthermore, although Guillaume *et al.* [6] showed that in general it is intractable to find best solutions, our results indicate that it can nonetheless be done when focal sets are of a particular kind. Finally, we may note that the particular optimization problems that we consider allow us to take advantage of specialized (SPP-related) algorithms, in contrast to [11] which also provides means to find best elements according to some criteria, such as strong dominance, but which cannot benefit from such specialized algorithms as it is framed in a more general setting.

The rest of this paper is organized as follows. Section 2 presents necessary background material. Section 3 is devoted to the formalization and resolution of the SPP with evidential weights. The paper ends with a conclusion in Sect. 4.

2   Preliminaries

In this section, we present basic elements necessary for the rest of the paper.

2.1   Deterministic Shortest Path Problem

Let $G = (V, A)$ be a directed graph with set of vertices V, set of arcs A and weight $c_{ij} > 0$ for each arc (i, j) in A. Let s and t be two vertices in V called the source and the destination, respectively. Let $\mathcal{X}$ be the set of all s-t paths in G with the assumption that $\mathcal{X} \neq \emptyset$. If all arc weights c_{ij} are known then finding a s-t shortest path, *i.e.*, a s-t path of lowest weight, can be modelled as the following optimization problem

$$\min \sum_{(i,j)\in A} c_{ij}p_{ij} \qquad (1)$$

$$\sum_{(s,i)\in A} p_{si} - \sum_{(j,s)\in A} p_{js} = 1 \qquad (2)$$

$$\sum_{(t,i)\in A} p_{ti} - \sum_{(j,t)\in A} p_{jt} = -1 \qquad (3)$$

$$\sum_{(k,i)\in A} p_{ki} - \sum_{(j,k)\in A} p_{jk} = 0, \quad \forall k \in V\backslash\{s,t\} \qquad (4)$$

$$p_{ij} \in \{0,1\}, \quad \forall(i,j) \in A \qquad (5)$$

where each path in $\mathcal{X}$ is identified with a set $p = \{p_{ij}|(i,j) \in A\}$ of which element $p_{ij} = 1$ if arc (i,j) is in the path and $p_{ij} = 0$ otherwise.

Example 1. Considering the directed graph depicted in Fig. 1, the set of all s-t paths is $\mathcal{X} = \{s\text{-}a\text{-}t, s\text{-}b\text{-}t, s\text{-}t\}$ and $s\text{-}a\text{-}t$ is the shortest s-t path with weight 2.

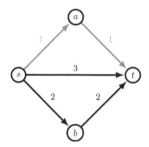

Fig. 1. Shortest path $s\text{-}a\text{-}t$ between vertices s and t.

2.2 Belief Function Theory

Let $\Theta = \{\theta_1,\ldots,\theta_n\}$ be the set, called frame of discernment, of all possible values of a variable θ. In belief function theory [12], partial knowledge about the true (unknown) value of θ is represented by a mapping $m : 2^\Theta \mapsto [0,1]$ called mass function and such that $\sum_{A\subseteq\Theta} m(A) = 1$ and $m(\emptyset) = 0$, where mass $m(A)$ quantifies the amount of belief allocated to the fact of knowing only that $\theta \in A$. A subset $A \subseteq \Theta$ is called a focal set of m if $m(A) > 0$.

Assume θ represents the state of nature and its true value is known in the form of some mass function m. Assume further that a decision maker (DM) needs to choose an act (decision) f from a finite set $\mathcal{Q}$, where each act $f \in \mathcal{Q}$ induces a cost $l(f,\theta_i)$ for each possible state of nature $\theta_i \in \Theta$. In this context, the DM's preference over acts is denoted by $\preceq$, where $f \preceq g$ means that act f is preferred to act g. Relation $\preceq$ is complete if for any two acts f and g, $f \preceq g$ or $g \preceq f$, otherwise, it is partial. Furthermore, f is strictly (resp. equally) preferred to g, which is denoted by $f \prec g$ (resp. $f \sim g$), if $f \preceq g$ but not $g \preceq f$ (resp. if $f \preceq g$ and $g \preceq f$).

Typically, the DM seeks elements in the set Opt of non-dominated acts:

$$Opt = \{f \in \mathcal{Q} : \nexists g \text{ such that } g \prec f\}. \tag{6}$$

If relation $\preceq$ is complete, finding one element in Opt is enough since elements in Opt are preferred equally between each other and strictly preferred to the rest $\mathcal{Q} \backslash Opt$. In this case, elements in Opt are also called optimal acts. On the other hand, if relation $\preceq$ is partial, the DM may need to identify all elements in Opt.

Usually, the DM constructs his preference over acts based on some criterion. We denote by $\preceq_{cr}$ his preference according to some criterion cr and by Opt_{cr} its associated set of non-dominated (or best) acts. In this paper, we consider three common criteria defined as follows for any two acts f and g [2]:

1. Generalized Hurwicz criterion: $f \preceq_{hu} g$ if

$$\alpha \overline{E}_m(f) + (1 - \alpha)\underline{E}_m(f) \leq \alpha \overline{E}_m(g) + (1 - \alpha)\underline{E}_m(g) \tag{7}$$

for some fixed parameter $\alpha \in [0, 1]$, and where $\overline{E}_m(f)$ and $\underline{E}_m(f)$ denote, respectively, the upper and lower expected costs of act f with respect to mass function m defined as

$$\overline{E}_m(f) = \sum_{A \subseteq \Theta} m(A) \max_{\theta_i \in A} l(f, \theta_i), \tag{8}$$

$$\underline{E}_m(f) = \sum_{A \subseteq \Theta} m(A) \min_{\theta_i \in A} l(f, \theta_i). \tag{9}$$

Relation $\preceq_{hu}$ is complete and we have $f \prec_{hu} g$ if (7) is strict.
2. Strong dominance criterion: $f \preceq_{str} g$ if

$$\overline{E}_m(f) \leq \underline{E}_m(g). \tag{10}$$

Relation $\preceq_{str}$ is partial and we have $f \prec_{str} g$ if (10) is strict.
3. Weak dominance criterion: $f \preceq_{weak} g$ if

$$\overline{E}_m(f) \leq \overline{E}_m(g) \text{ and } \underline{E}_m(f) \leq \underline{E}_m(g). \tag{11}$$

Relation $\preceq_{weak}$ is partial and we have $f \prec_{weak} g$ if at least one inequality in (11) is strict.

3   Shortest Path Problem with Evidential Weights

In this section, we formalize what we mean by best paths in a graph with evidential weights and provide methods for finding such paths.

3.1   Modelling

Let us assume that the arc weights c_{ij}, for all $(i,j) \in A$, of the graph introduced in Sect. 2.1 are only partially known. More specifically, we consider the case where information about arc weights is modelled by a mass function. Formally, let Ω_{ij} denote the frame of discernment for the variable c_{ij}, i.e., the set of possible values for the weight c_{ij}. We assume that $\Omega_{ij} \subset \mathbb{N}_{>0}$. Let $\Omega := \times_{(i,j)\in A}\Omega_{ij}$. Any $c \in \Omega$ will be called a scenario: it represents a possible assignment of values for all the weights in the graph. A mass function m on Ω, with set of focals sets denoted by $\mathcal{F} = \{F_1, \ldots, F_K\}$, represents then uncertainty about arc weights.

Example 2. Let c^1 and c^2 be the two scenarios represented by Figs. 2a and 2b, respectively. The mass function m such that $m(F_1) = 0.4$ and $m(F_2) = 0.6$, with $F_1 = \{c^1, c^2\}$ and $F_2 = \{c^1\}$, represents partial knowledge about arc weights.

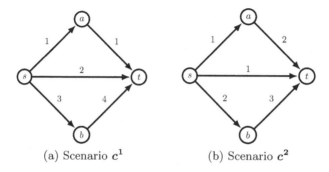

(a) Scenario c^1 (b) Scenario c^2

Fig. 2. Two possible assignments of values, *i.e.*, two scenarios, for the arc weights.

As will be seen, making a particular assumption about the nature of the focal sets of m is useful. This assumption, denoted CI for short, is the following: each focal set of m can be expressed as a Cartesian product of intervals, *i.e.*, $F_r = \times_{(i,j)\in A}[l_{ij}^r, u_{ij}^r]$ for all $r \in \{1, \ldots, K\}$. Such a focal set is illustrated by Example 3.

Example 3. Let F be the Cartesian product of intervals (depicted by Fig. 3):

$$F = [l_{sa}, u_{sa}] \times [l_{sb}, u_{sb}] \times [l_{st}, u_{st}] \times [l_{at}, u_{at}] \times [l_{bt}, u_{bt}]$$
$$= [1, 5] \times [2, 4] \times [2, 4] \times [1, 3] \times [2, 5].$$

F is a subset of Ω: it includes, for instance, the scenario $c = \{c_{sa}, c_{sb}, c_{st}, c_{at}, c_{bt}\}$ with $c_{sa} = 1, c_{sb} = 3, c_{st} = 2, c_{at} = 1$, and $c_{bt} = 3$.

When arc weights are evidential, *i.e.*, there is some uncertainty about them in the form of a mass function m on Ω, the preference over s-t paths with respect to their (uncertain) weights can be established using the decision-making framework recalled in Sect. 2.2. Specifically, the set Ω of scenarios represents the

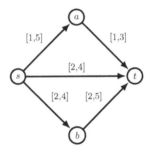

Fig. 3. Focal set as a Cartesian product of intervals.

possible states of nature. The set $\mathcal{X}$ of s-t paths represents the possible acts. The weight $\sum_{(i,j)\in A} c_{ij}p_{ij}$ of path $p = \{p_{ij}|(i,j)\in A\} \in \mathcal{X}$ under scenario $c = \{c_{ij}|(i,j)\in A\} \in \Omega$ represents the cost $l(p,c)$ of path (act) p for the scenario (state of nature) c. The preference over s-t paths, and the associated best s-t paths, can then be defined using any of the three criteria recalled in Sect. 2.2. In the next section, we provide the main results of this paper, which concern best s-t paths according to these three criteria and under assumption CI.

3.2   Solving

In this section, methods for finding best paths according to, in turn, the generalized Hurwicz, strong dominance, and weak dominance criteria, are provided. We can remark that these criteria rely on the notions of upper and lower expected costs of acts, acts being here paths. These costs $\overline{E}_m(p)$ and $\underline{E}_m(p)$ of a path p can be computed easily under assumption CI:

Proposition 1. *Under assumption CI, we have*

$$\overline{E}_m(p) = \sum_{(i,j)\in A} \bar{u}_{ij}p_{ij} \tag{12}$$

$$\underline{E}_m(p) = \sum_{(i,j)\in A} \bar{l}_{ij}p_{ij} \tag{13}$$

with $\bar{u}_{ij} := \sum_{r=1}^{K} m(F_r)u_{ij}^r$ and $\bar{l}_{ij} := \sum_{r=1}^{K} m(F_r)l_{ij}^r$ for all $(i,j) \in A$.

Proof. By definition, the upper and lower expected costs of path p are

$$\overline{E}_m(p) = \sum_{r=1}^{K} m(F_r) \max_{c^r \in F_r} \Big(\sum_{(i,j)\in A} c_{ij}^r p_{ij} \Big), \tag{14}$$

$$\underline{E}_m(p) = \sum_{r=1}^{K} m(F_r) \min_{c^r \in F_r} \Big(\sum_{(i,j)\in A} c_{ij}^r p_{ij} \Big). \tag{15}$$

The inner maximum and minimum in (14) and (15) are obtained when each arc weight c_{ij}^r in scenario c^r equals u_{ij}^r and l_{ij}^r, respectively. By regrouping terms we get the desired result.   □

Proposition 1 is instrumental to uncover exacts methods finding best s-t paths.

Generalized Hurwicz Criterion. Since relation $\preceq_{hu}$ is complete, it is sufficient to find one element of the set Opt_{hu}, as explained in Sect. 2.2. To find one such element, *i.e.*, best path according to the generalized Hurwicz criterion, we need to solve the optimization problem

$$\min \ \alpha \overline{E}_m(p) + (1-\alpha)\underline{E}_m(p) \tag{16}$$

$$p \in \mathcal{X}. \tag{17}$$

The complexity of the problem (16–17), in the case of general focal sets, has been studied in the literature. If $\alpha = 1$, the problem is weakly NP-hard already in the case when mass function m has a single focal set containing two elements [15]. If $\alpha = 0$, the problem is even harder: it is strongly NP-hard and not approximable [6, Theorem 1]. However, under assumption CI, the problem (16–17) becomes much easier to solve:

Proposition 2. *Under assumption CI, solving the problem (16–17) amounts to solving the SPP in graph $G = (V, A)$ with arc weights $c_{ij} = \alpha \bar{u}_{ij} + (1-\alpha)\bar{l}_{ij}$.*

Proof. Using Proposition 1, the problem (16–17) becomes

$$\min \ \sum_{(i,j)\in A} (\alpha \bar{u}_{ij} + (1-\alpha)\bar{l}_{ij})p_{ij} \tag{18}$$

$$p_{ij} \text{ satisfies } (2\text{--}5) \ \forall (i,j) \in A \tag{19}$$

□

According to Proposition 2, to find one element in Opt_{hu}, we can use a fast algorithm for the SPP such as [4].

Strong Dominance Criterion. Since relation $\preceq_{str}$ is partial, it may be necessary to find all elements of the set Opt_{str}, *i.e.*, all best paths according to the strong dominance criterion.

Proposition 3. *Under assumption CI, finding all elements in Opt_{str} amounts to finding all paths, in graph $G = (V, A)$ with arc weights $c_{ij} = \bar{l}_{ij}$, whose weights are lower than or equal to $\bar{d}_\star := \min_{q\in\mathcal{X}} \overline{E}_m(q)$.*

Proof. By definition,

$$p \in Opt_{str} \Leftrightarrow \nexists q \in \mathcal{X} \text{ such that } \overline{E}_m(q) < \underline{E}_m(p) \tag{20}$$

$$\Leftrightarrow \forall q \in \mathcal{X} \text{ then } \overline{E}_m(q) \geq \underline{E}_m(p) \tag{21}$$

$$\Leftrightarrow \min_{q\in\mathcal{X}} \overline{E}_m(q) \geq \underline{E}_m(p) \tag{22}$$

As a special case of Proposition 2, when $\alpha = 1$, $\min_{q \in \mathcal{X}} \overline{E}_m(q)$ is obtained by solving the deterministic SPP in G with arc weights $c_{ij} = \bar{u}_{ij}$. From Proposition 1, we have $\underline{E}_m(p) = \sum_{(i,j) \in A} \bar{l}_{ij} p_{ij}$. Hence, to find p such that $\underline{E}_m(p) \leq \bar{d}_\star$, we set arc weights c_{ij} of G to $\bar{l}_{ij}$ and finding all elements in Opt_{str} amounts to finding all s-t paths in G whose weights are lower or equal than $\bar{d}_\star$.   □

To find all elements in Opt_{str}, we can use efficient algorithms such as the one in [1], where the authors studied a problem of determining near optimal paths; for example, they wished to find all s-t paths in a directed graph whose weights do not exceed more than 10% the lowest weight, which is basically finding all paths whose weights are lower than or equal to a given value.

Weak Dominance Criterion. Similarly as for the strong dominance criterion, all elements of the set Opt_{weak} may need to be found since $\preceq_{weak}$ is partial.

There is a strong connection between the weak dominance criterion and bi-objective optimization. A bi-objective optimization problem can be expressed as

$$\min \ f_1(x) \tag{23}$$
$$\min \ f_2(x) \tag{24}$$
$$x \in X \tag{25}$$

As the objectives (23–24) are typically conflicting, there is usually no solution x that minimizes simultaneously $f_1(x)$ and $f_2(x)$. Instead, we seek to find all so-called efficient solutions of (23–25): a solution x is efficient if there is no feasible solution $y \in X$ such that $f_1(y) \leq f_1(x)$ and $f_2(y) \leq f_2(x)$ where at least one of the inequalities is strict.

The bi-objective SPP is a particular bi-objective optimization problem. Assume that each arc (i,j) in G has two deterministic attributes c_{ij} and t_{ij} that describes, *e.g.*, the distance and the travel time from i to j, respectively. The goal is to find all efficient solutions, *i.e.*, s-t paths of the following problem

$$\min \ \sum_{(i,j) \in A} c_{ij} p_{ij} \tag{26}$$

$$\min \ \sum_{(i,j) \in A} t_{ij} p_{ij} \tag{27}$$

$$p_{ij} \text{ satisfies (2–5) } \forall (i,j) \in A. \tag{28}$$

Proposition 4. *Under assumption CI, finding all elements in Opt_{weak} amounts to finding all efficient solutions of a bi-objective SPP in graph G where each arc $(i,j) \in A$ has two attributes $\bar{u}_{ij}$ and $\bar{l}_{ij}$.*

Proof. Finding all elements in Opt_{weak} is equivalent to finding all efficient solutions $p \in \mathcal{X}$ of a bi-objective optimization problem with objectives $f_1(p) := \overline{E}_m(p)$ and $f_2(p) := \underline{E}_m(p)$, which, using Proposition 1, comes down to a bi-objective SPP in graph G where each arc (i,j) has two attributes $\bar{u}_{ij}, \bar{l}_{ij}$.   □

To find all elements in Opt_{weak}, we can apply fast algorithms developed for the bi-objective SPP such as the one in [5].

We note that any generalized Hurwicz optimal solution for $0 < \alpha < 1$ is also an element in Opt_{weak}, hence finding such solutions for various α will provide an inner approximation of Opt_{weak}. This stems from bi-objective optimization theory, where they are known as the *supported* efficient solutions, which are the solutions of $\min\{\lambda_1 f_1(x) + \lambda_2 f_2(x) : x \in X\}$ for some $\lambda_1, \lambda_2 > 0$.

Example 4. Assume that mass function m has a single focal set, which is the one in Fig. 3. There are three s-t paths with their lower and upper expected costs indicated between parentheses: s-a-t : $(2, 8)$; s-t : $(2, 4)$; s-b-t : $(4, 9)$. If $\alpha = 0$, the two optimal paths according to generalized Hurwicz criterion are s-a-t and s-t with expected cost 2. If $\alpha = 0.5$, s-t is the unique optimal path with expected cost 3. We also have $Opt_{str} = \{s\text{-}a\text{-}t, s\text{-}t, s\text{-}b\text{-}t\}$ and $Opt_{weak} = \{s\text{-}t\}$.

3.3   Sizes of Opt_{weak} and Opt_{str}

It is clear that if $p \prec_{str} q$ then $p \prec_{weak} q$, and that the converse does not hold, hence $Opt_{weak} \subseteq Opt_{str}$. Example 4 showed that $Opt_{weak} \subset Opt_{str}$ in general.

The sets Opt_{weak} and Opt_{str} can be huge so enumerating their elements can be time-consuming. In fact, it is shown in [7, Theorem 1] that in the worst case, the size of the set of efficient paths grows exponentially with $|V|$. Therefore, it is useful to be able to know the size of these sets in advance, without enumerating their elements explicitly. Proposition 5 is a first result in this direction:

Proposition 5. *If $\bar{d}_\star$ and $\bar{l}_{ij}$ in Proposition 3 are rational numbers, $|Opt_{str}|$ (and thus an upper-bound of $|Opt_{weak}|$) can be computed in $O(|V|^2 \times W)$, with $W = \bar{d}_\star \times D$ where D is a common denominator of $\bar{d}_\star$ and of $\bar{l}_{ij}$, for all $(i, j) \in A$.*

Proof. Hereafter, consider graph G with integer arc weights $c_{ij} = \bar{l}_{ij} \times D$. It is easy to show that $|Opt_{str}|$ is equal to the number of s-t paths in G whose weights are lower than or equal to integer value W. Furthermore, let $|V| = n$ and assume, without loss of generality, that vertices are indexed by $0, \ldots, n-1$, with 0 and $n-1$ the source and destination vertices, respectively. Denoting by $N_w(i)$ the number of paths in G from i to $n-1$ whose weights are lower than or equal to w, then we need to calculate $N_W(0)$ since it is equal to $|Opt_{str}|$. We have clearly, for all $i \in \{0, \ldots, n-2\}$ and all $w \in \{1, \ldots, W\}$,

$$N_w(i) = \sum_{j \text{ such that } (i,j) \in A \text{ and } c_{ij} \leq w} N_{w-c_{ij}}(j), \tag{29}$$

with $N_0(i) = 0$ for all $i \in \{0, \ldots, n-2\}$ and $N_w(n-1) = 1$ for all $w \in \{0, \ldots, W\}$.

Consider a $(W+1) \times n$ 2-dimensional array M with each cell $M[w][i]$, $w \in \{0, \ldots, W\}$, $i \in \{0, \ldots, n-1\}$, storing $N_w(i)$; by filling this array row-wise starting with row $w = 0$, computing each row costs $O(|V|^2)$. This leads to the desired complexity. $\qquad\square$

We note that given [9, Theorem 1] and the above proof, computing $|Opt_{str}|$ is actually NP-hard. Nonetheless, in practice, W in Proposition 5 may not be too big, so that computing $|Opt_{str}|$ may be quite fast.

4   Conclusion

In this paper, we have considered the case where uncertainty about arc weights in a graph is represented by a mass function. We have proposed extensions of the notion of shortest path to this context, as the sets of non-dominated paths according to the generalized Hurwicz, strong dominance, and weak dominance criteria. We have shown that if the focal elements of the mass function are Cartesian products of intervals, these sets can be found by applying algorithms developed for variants of the deterministic SPP. Future works include considering other criteria, such as maximality [2] or minimization of expected costs according to Shenoy's expectation operator [13].

References

1. Byers, T.H., Waterman, M.S.: Determining all optimal and near-optimal solutions when solving shortest path problems by dynamic programming. Oper. Res. **32**(6), 1381–1384 (1984)
2. Denoeux, T.: Decision-making with belief functions: a review. Int. J. Approx. Reason. **109**, 87–110 (2019)
3. Dias, L.C., Clímaco, J.N.: Shortest path problems with partial information: models and algorithms for detecting dominance. Eur. J. Oper. Res. **121**(1), 16–31 (2000)
4. Dijkstra, E.W.: A note on two problems in connexion with graphs. Numer. Math. **1**(1), 269–271 (1959)
5. Duque, D., Lozano, L., Medaglia, A.L.: An exact method for the biobjective shortest path problem for large-scale road networks. Eur. J. Oper. Res. **242**(3), 788–797 (2015)
6. Guillaume, R., Kasperski, A., Zieliński, P.: Robust optimization with scenarios using random fuzzy sets. In: Proceedings of FUZZ-IEEE 2021, pp. 1–6. IEEE (2021)
7. Hansen, P.: Bicriterion path problems. In: Fandel, G., Gal, T. (eds.) Multiple Criteria Decision Making Theory and Application. Lecture Notes in Economics and Mathematical Systems, vol. 177, pp. 109–127. Springer, Berlin, Heidelberg (1980). https://doi.org/10.1007/978-3-642-48782-8_9
8. Helal, N., Pichon, F., Porumbel, D., Mercier, D., Lefèvre, E.: The capacitated vehicle routing problem with evidential demands. Int. J. Approx. Reason. **95**, 124–151 (2018)
9. Mihalák, M., Srámek, R., Widmayer, P.: Approximately counting approximately-shortest paths in directed acyclic graphs. Theory Comput. Syst. **58**(1), 45–59 (2016)
10. Montemanni, R., Gambardella, L.M.: An exact algorithm for the robust shortest path problem with interval data. Comput. Oper. Res. **31**(10), 1667–1680 (2004)
11. Nakharutai, N., Troffaes, M.C.M., Caiado, C.C.S.: Improving and benchmarking of algorithms for γ-maximin, γ-maximax and interval dominance. Int. J. Approx. Reason. **133**, 95–115 (2021)
12. Shafer, G.: A Mathematical Theory of Evidence. Princeton University Press, Princeton (1976)
13. Shenoy, P.P.: An expectation operator for belief functions in the Dempster-Shafer theory. Int. J. Gen Syst **49**(1), 112–141 (2020)

14. Tedjini, T., Afifi, S., Pichon, F., Lefèvre, E.: The vehicle routing problem with time windows and evidential service and travel times: a recourse model. In: Vejnarová, J., Wilson, N. (eds.) ECSQARU 2021. LNCS (LNAI), vol. 12897, pp. 381–395. Springer, Cham (2021). https://doi.org/10.1007/978-3-030-86772-0_28

15. Yu, G., Yang, J.: On the robust shortest path problem. Comput. Oper. Res. **25**(6), 457–468 (1998)

Data and Information Fusion

Heterogeneous Image Fusion for Target Recognition Based on Evidence Reasoning

Shuyue Wang$^{(\boxtimes)}$, Zhunga Liu, Zuowei Zhang, and Yang Li

School of Automation, Northwestern Polytechnical University, Xi'an, China
`wangshuyue@mail.nwpu.edu.cn`

Abstract. Multi-source fusion is an efficient strategy in complex image target recognition since it can exploit the complementary knowledge in different sources to improve the classification performance. In this paper, we propose a new end-to-end framework for heterogeneous (i.e. visible & infrared) image fusion target recognition (HIFTR). Firstly, two networks are built for the visible and infrared images respectively and jointly trained based on mutual learning. It aims to transfer heterogeneous information mutually and improve the generalization performance of the networks. Secondly, a weighted decision-level fusion method based on evidence reasoning is developed to combine the classification results of visible and infrared images for the final target recognition. In the training process, the weight of each image is automatically optimized in the networks. Finally, the performance of the proposed HIFTR has been evaluated by comparing with other related methods, and the experimental results show that the HIFTR method can efficiently improve the classification accuracy.

Keywords: Heterogeneous image fusion · Evidence reasoning · Mutual learning · Target recognition

1 Introduction

Multi-source fusion plays an important role in the complex image target recognition [4–6] since the given images from different sensors can provide more or less complementary information thereby improving the classification performance. Visible and infrared images are two kinds of important sources, and we usually consider that visible detectors are consistent with the human visual system, which can provide high spatial resolution and clear texture details. However, they are susceptible to interference from environmental factors such as illumination and weather. In contrast, infrared detectors can distinguish the target from the background based on the difference in radiation and can work well in all-weather and all-day/night conditions [9], while infrared images have low resolution and a relatively blurred. The fusion of these two kinds of heterogeneous images tends to complement each other's strengths and thus improve the performance of target recognition.

© The Author(s), under exclusive license to Springer Nature Switzerland AG 2022
S. Le Hégarat-Mascle et al. (Eds.): BELIEF 2022, LNAI 13506, pp. 153–162, 2022.
https://doi.org/10.1007/978-3-031-17801-6_15

Deep learning techniques are widely used in image target recognition due to the ability to extract features automatically [7,10,14]. In training, the supervision information usually comes from the true class of the target (i.e. hard labels), while the information contained in hard labels is often insufficient. Interestingly, the deep mutual learning (DML) strategy [15] provides an alternative since it enables a cohort of networks to be trained simultaneously. In this case, each network in the queue is supervised by the true labels and constrained by the posterior probabilities provided by its peer networks. Posterior probability can be regarded as a kind of soft label, which reflects the intra-class and inter-class relationship of the images. Therefore, mutual learning using visible and infrared heterogeneous images can improve the performance of the networks in case of the insufficient prior information.

In this paper, we propose an end-to-end DML framework for heterogeneous (i.e. visible & infrared) image fusion target recognition (HIFTR) based on evidence reasoning [11] due to its advantage in dealing with uncertain information in the decision-level fusion. The proposed HIFTR consists of two networks for recognizing visible and infrared images. During training, pairs of visible and infrared images are input to the two networks, and the DML strategy is exploited to realize the mutual supervision between the networks. The classification results output by different networks are combined by a weighted fusion method based on evidence reasoning to make the final decision, in particular, where the optimization of weights is performed automatically in the training process.

The rest of this paper is organized as follows. A brief recall of evidence reasoning is given in Sect. 2. Section 3 describes the details of our proposed method. Section 4 presents a detailed experimental evaluation and discusses the performance of the proposed method. Section 5 makes some conclusions about this paper.

2   Brief Recall of Evidence Reasoning

Evidence reasoning is established under the discernment framework represented by $\Omega = \{\omega_1, \ldots, \omega_n\}$. The power set of Ω is denoted as 2^Ω, which contains all the subsets of Ω . The basic belief assignment (BBA) of the discernment framework, also known as the mass function, is a mapping function from 2^Ω to [0,1]. It satisfies the conditions expressed by

$$\begin{cases} \sum_{A \in 2^\Omega} m(A) = 1 \\ m(\emptyset) = 0 \end{cases} \tag{1}$$

where A represents a focal element in 2^Ω when $m(A) > 0$. In classification tasks, if all focal elements of BBA are singleton classes, $m(A)$ represents the probability that the target belongs to A.

In image target recognition problem, each classification result can be considered as a piece of evidence denoted as a BBA. Dempster-Shafer (DS) rule [2] is widely exploited to combine multiple BBA's due to its commutative and associative properties, which makes it relatively easy for implementation [8]. Assuming that there are two BBA's denoted as $\mathbf{m}_1$ and $\mathbf{m}_2$ ($\mathbf{m}_i = m_i(.)$), and $B, C \in 2^\Omega$, the combination of $\mathbf{m}_1$ and $\mathbf{m}_2$ by DS rule is defined by

$$
\begin{cases}
m(A) = \mathbf{m}_1 \oplus \mathbf{m}_2 \\
\quad = \dfrac{\sum_{B \cap C = A} m_1(B) m_2(C)}{1 - \sum_{B \cap C = \emptyset} m_1(B) m_2(C)}, A \neq \emptyset \\
m(\emptyset) = 0
\end{cases}
\tag{2}
$$

where $\oplus$ represents DS rule. It is can be seen that DS rule is applicable only when the conflicting mass $\sum_{B \cap C = \emptyset} m_1(B) m_2(C) < 1$.

When pieces of evidence have different reliabilities, the Shafer's discounting method [11] is used to reduce their weights in the fusion process. For a BBA with the reliability denoted as α, the discounted BBA is given by

$$
\begin{cases}
{}^\alpha m(A) = \alpha \cdot m(A), A \neq \Omega \\
{}^\alpha m(\Omega) = 1 - \alpha + \alpha \cdot m(\Omega).
\end{cases}
\tag{3}
$$

In Eq. (3), the mass values of all focal elements are proportionally redistributed to the ignorant element Ω. By doing this, the influence of each evidence in the fusion can be well controlled.

3 Heterogeneous Image Fusion for Target Recognition

In this section, we design an end-to-end framework for visible and infrared image fusion target recognition. The architecture of our framework is shown in Fig. 1. The whole model is divided into two streams, which are the visible stream for recognizing visible images and the infrared stream for recognizing infrared images. For convenience, the images input to the two streams are represented by $X_{VIS} = \{x_{VIS}^1, \ldots, x_{VIS}^N\}$ and $X_{IR} = \{x_{IR}^1, \ldots, x_{IR}^N\}$, respectively. The patterns in X_{VIS} and X_{IR} are paired, and N represents the number of patterns. The label set is denoted as $L = \{\omega_1, \ldots, \omega_K\}$, where K represents the number of classes. The networks of the visible and infrared streams are denoted as θ_{VIS} and θ_{IR}, respectively. Their predicted labels for X_{VIS} and X_{IR} are written as $Y_{VIS} = \{y_{VIS}^1, \ldots, y_{VIS}^N\}$ and $Y_{IR} = \{y_{IR}^1, \ldots, y_{IR}^N\}$. $\boldsymbol{p}_{VIS}$ and $\boldsymbol{p}_{IR}$ are the probabilistic predictions for the target by θ_{VIS} and θ_{IR}. Moreover, the probability that the pattern x belongs to class ω_k is denoted as $p^k(x)$, where $k = 1, \ldots, K$. The networks θ_{VIS} and θ_{IR} are jointly trained, where $\boldsymbol{p}_{VIS}$ and $\boldsymbol{p}_{IR}$ are used as the additional information to supervise θ_{IR} and $\theta_{VIS.}$. Then, two classification results (i.e. probabilistic predictions) for the target can be obtained by θ_{VIS} and θ_{IR}. Finally, the discounted results with proper weights are combined by DS rule to get the final classification result. The weights for visible and infrared images are obtained by minimizing an error criterion.

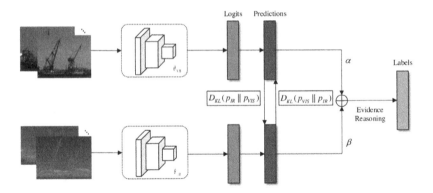

Fig. 1. The architecture of the proposed framework.

3.1 Mutual Learning of the Networks for Heterogeneous Images

The information in heterogeneous images can be transferred to each other through the mutual learning of the networks. This strategy is beneficial to get more robust classification results. When training the networks with the deep mutual learning strategy, the loss of each network consists of two parts. One is the gap between the predicted values and the true values, and the other is the gap between the predicted values from different sources. The following will describe how to calculate the loss using the visible stream as an example.

In multi-classification tasks, the cross-entropy loss is usually utilized as the objective function to evaluate the difference between the predicted values and the true values, so it is applied as the traditional supervised loss of the networks in this work. The cross-entropy loss of θ_{VIS} denoted as $L_{C_{VIS}}$ is calculated by

$$L_{C_{VIS}} = -\sum_{n=1}^{N}\sum_{k=1}^{K} I(y_{VIS}^n, \omega_k) log(p^k(x_{VIS}^n)) \tag{4}$$

where x_{VIS}^n represents the nth pattern in X_{VIS} and $p^k(x_{VIS}^n)$ denotes the probability that the x_{VIS}^n belongs to class ω_k. y_{VIS}^n represents the predicted label for x_{VIS}^n by θ_{VIS}. The indicator function I in Eq. (4) is defined by

$$I(y_{VIS}^n, \omega_k) = \begin{cases} 1, & y_{VIS}^n = \omega_k \\ 0, & y_{VIS}^n \neq \omega_k. \end{cases} \tag{5}$$

The purpose of the mutual learning between θ_{VIS} and θ_{IR} is to exploit the posterior probabilities of the peer networks for extra supervision information. It aims to make the distributions of $\boldsymbol{p}_{VIS}$ and $\boldsymbol{p}_{IR}$ as similar as possible. Kullback-Leible Divergence (KLD) can evaluate the difference between two distributions, so it is used here to measure the gap between the $\boldsymbol{p}_{VIS}$ and $\boldsymbol{p}_{IR}$. For the visible stream, the KLD between $\boldsymbol{p}_{VIS}$ and $\boldsymbol{p}_{IR}$ is represented as $D_{KL}(\boldsymbol{p}_{IR} \parallel \boldsymbol{p}_{VIS})$, which is given by

$$D_{KL}(\boldsymbol{p}_{IR} \parallel \boldsymbol{p}_{VIS}) = \sum_{n=1}^{N} \sum_{k=1}^{K} p^k(x_{IR}^n) log \frac{p^k(x_{VIS}^n)}{p^k(x_{IR}^n)} \qquad (6)$$

where x_{IR}^n represents the nth pattern in X_{IR} and $p^k(x_{IR}^n)$ denotes the probability that the x_{IR}^n belongs to class ω_k. Here, x_{VIS}^n and x_{IR}^n are paired. $D_{KL}(\boldsymbol{p}_{IR} \parallel \boldsymbol{p}_{VIS})$ will be close to zero when $\boldsymbol{p}_{VIS}$ and $\boldsymbol{p}_{IR}$ are very similar.

The loss of the visible stream is the sum of the $L_{C_{VIS}}$ and $D_{KL}(\boldsymbol{p}_{IR} \parallel \boldsymbol{p}_{VIS})$, which can be expressed by

$$L_{\theta_{VIS}} = L_{C_{VIS}} + D_{KL}(\boldsymbol{p}_{VIS} \parallel \boldsymbol{p}_{IR}). \qquad (7)$$

Similarly, the loss of the infrared stream can be written as

$$L_{\theta_{IR}} = L_{C_{IR}} + D_{KL}(\boldsymbol{p}_{IR} \parallel \boldsymbol{p}_{VIS}). \qquad (8)$$

Two classification results about one target can be obtained with θ_{VIS} and θ_{IR}. The next step to consider is how to combine them efficiently to improve the classification performance, and it is represented in the next subsection.

3.2   Weighted Fusion of Multiple Classification Results

In this subsection, evidence reasoning is used to combine the classification results of θ_{VIS} and θ_{IR}. Moreover, each result will be weighted according to its reliability. The classification results of θ_{VIS} and θ_{IR} can be regarded as BBAs, and DS rule is used to combine them. Because the reliability of different sources varies, it is necessary to assign a weight to each classification result for a good fusion result. The weights for visible and infrared images can be represented by a vector (α, β). The process of fusion is denoted as

$$\boldsymbol{p}^n = {}^{\alpha}\boldsymbol{p}_{VIS}^n \oplus {}^{\beta}\boldsymbol{p}_{IR}^n \qquad (9)$$

where $\boldsymbol{p}^n$ represents the fusion result obtained by combining the classification results of the nth patterns in X_{VIS} and X_{IR}. The class with the highest probability in $\boldsymbol{p}^n$ is the label predicted by the model. $\boldsymbol{p}_{VIS}^n$ and $\boldsymbol{p}_{IR}^n$ are the probabilistic outputs of the peer networks θ_{VIS} and θ_{IR} for the nth training patterns in X_{VIS} and X_{IR}.

The optimal weights should make the fusion results as close as possible to the true values, so (α, β) can be estimated by minimizing the distances between the fusion results and the ground truth of the training patterns, as shown in

$$\{\alpha, \beta\} = \arg \min_{\alpha, \beta} \sum_{n=1}^{N} \| {}^{\alpha}\boldsymbol{p}_{VIS}^n \oplus {}^{\beta}\boldsymbol{p}_{IR}^n - \boldsymbol{t}_n \|_2 \qquad (10)$$

where $\|.\|_2$ represents the Euclidean distance. Note that the images in X_{VIS} and X_{IR} are paired, so $\boldsymbol{p}_{VIS}^n$ and $\boldsymbol{p}_{IR}^n$ are predictions for the same target. $\boldsymbol{t}_n$ denotes the ground-truth vector of the nth target, written as a one-hot vector

$t_n = (t_{n1}, t_{n2}, \ldots, t_{nk})$. If the true label of the target is ω_i, the value of the ith element in $t_n = (t_{n1}, t_{n2}, \ldots, t_{nk})$ is equal to 1, and the values of the remaining elements are 0. The optimal α and β can be obtained when the distance is minimized. Moreover, the values of the weights satisfy $\alpha \in [0, 1]$ and $\beta \in [0, 1]$. The optimization process shown in Eq. (10) is a classical constrained least squares problem. Here, the sequential least squares programming (SLSQP) optimization algorithm is used to optimize the weights.

The ultimate goal of the proposed method is to make the final classification result as accurate as possible, so the gap between the fusion results and the ground truth of the patterns should be taken into account. To this end, a fusion loss L_F is additionally applied to train the model. The cross-entropy loss is also utilized here to calculate L_F, and it is given by

$$L_F = -\sum_{n=1}^{N}\sum_{k=1}^{K} I(y^n, \omega_k) log(p^k(x^n)) \qquad (11)$$

where x^n represents the target corresponding to the nth pair of visible and infrared images. y^n is the predicted label obtained by p^n in Eq. (9). $p^k(x^n)$ denotes the probability that the target belongs to class ω_k. $I(y_n, \omega_k)$ is an indicator function, which has the similar form as Eq. (5).

The total losses of the visible and infrared streams are denoted as $L_{\theta_{VIS}}$ and $L_{\theta_{IR}}$, calculated by

$$\begin{cases} L_{\theta_{VIS}} = L_{C_{VIS}} + D_{KL}(p_{VIS} \parallel p_{IR}) + L_F \\ L_{\theta_{IR}} = L_{C_{IR}} + D_{KL}(p_{IR} \parallel p_{VIS}) + L_F. \end{cases} \qquad (12)$$

4   Experiment

4.1   Datasets and Preprocessing

We apply our method on ship image classification and scene image classification to valid the efficiency and performance. Both the VAIS ship image dataset [13] and the RGB-NIR scene image dataset [1] contain paired visible and infrared images. The specific division of two kinds of images is shown in Table 1 and Table 2. The nighttime test set in Table 1 contains only infrared images. We use bilateral interpolation to resize the images to 224×224 pixels before feeding them into the model. Because the infrared image is a single-channel image, we replicate the single channel three times to obtain a pseudo RGB image.

Table 1. The division of each class in the VAIS ship dataset

No.	Class	Train	Test	
			Daytime	Nighttime
0	Cargo	83	63	34
1	Medium-other	62	76	14
2	Passenger	58	59	12
3	Sailing	148	136	15
4	Small	158	195	30
5	Tug	30	20	49

Table 2. The division of each class in the RGB-NIR scene dataset

No	Class	Train	Test
0	Country	41	11
1	Field	40	11
2	Forest	42	11
3	Indoor	45	11
4	Mountain	44	11
5	Old building	40	11
6	Street	39	11
7	Urban	47	11
8	Water	40	11

4.2 Experimental Environment and Parameter Settings

The hardware environment includes Intel(R) Xeon(R) Silver4210R CPU@ 2.40GHz processor and NVIDIA GeForce GTX 3080Ti GPU. All the experiments were performed on Pytorch framework. Stochastic Gradient Descent(SGD) is used as the optimization method. The learning rate, batch size and momentum parameter are set to 0.001, 16 and 0.92. For the VAIS ship dataset and the RGB-NIR scene dataset, the epoch is set to 70 and 30. Pre-trained VGG16 [12] and ResNet50 [3] are utilized as the backbone of the model, respectively. We repeat the experiment 5 times and the average accuracy is reported.

4.3 Effectiveness of the Mutual Learning of Heterogeneous Images

To demonstrate the information can be transferred using the mutual learning strategy, we train the model with paired visible and infrared images in the VAIS ship dataset and use the images in the test set to verify the performance. The fusion loss L_F is not added to the training loss in this part of the experiments.

Table 3. Comparison of classification accuracy of infrared and visible images before and after using the mutual learning strategy (in %)

Method	Infrared images			Visible images
	Daytime	Nighttime	All-day	
VGG16	69.47 ± 1.13	66.62 ± 1.18	68.43 ± 0.26	87.47 ± 0.47
VGG16 (ML)	$\mathbf{70.27 \pm 2.73}$	$\mathbf{68.05 \pm 2.27}$	$\mathbf{69.79 \pm 2.53}$	$\mathbf{87.76 \pm 1.65}$
ResNet50	72.93 ± 1.18	53.90 ± 4.00	68.76 ± 1.40	85.28 ± 1.25
ResNet50 (ML)	72.93 ± 1.08	$\mathbf{57.14 \pm 2.72}$	$\mathbf{69.47 \pm 0.76}$	$\mathbf{86.01 \pm 0.69}$

In Table 3, VGG16 and ResNet50 represent fine-tuning of models pre-trained on ImageNet. We use VGG16 (ML) and ResNet50 (ML) to denote models based on VGG16 and ResNet50 using mutual learning strategy, respectively. VGG16 (ML) and ResNet50 (ML) also involve fine-tuning of the VGG16 and ResNet50. We can see from the table that the classification accuracy of VGG16 (ML) and ResNet50 (ML) is higher than that of VGG16 and ResNet50. Moreover, it is worth noting that there are no nighttime images in the training patterns, but the classification accuracy of nighttime infrared images is still improved. The experimental results show that the mutual learning of heterogeneous images plays a positive role in the training process.

4.4   Results and Analysis

For the VAIS ship dataset, we choose 7 methods for comparison. VGG16-IR, VGG16-VIS, ResNet50-IR and ResNet50-VIS represent the classification results of infrared and visible images obtained by fine-tuned VGG16 and ResNet50. Pixel-level average method fuses the two images before they are fed into the network, and it only needs to optimize a single network. We utilize VGG16 and ResNet50 to classify fused images, respectively. CNN+Gnostic Fields is the baseline of this dataset and is a decision-level fusion method. As shown in Table 4, it is difficult to obtain relatively good classification results with a single source. The poor quality of infrared images leads to low classification accuracy. Fusion of heterogeneous images is a good way to improve accuracy. The HIFTR method based on VGG16 achieves highest classification accuracy, indicating that the new method can effectively improve the classification performance.

For the RGB-NIR scene dataset, we select 7 methods for comparison. VGG16-IR, VGG16-VIS, ResNet50-IR, ResNet50-VIS, Pixel-level average and the HIFTR method all utilize deep neural networks to automatically learn features. MSIFT (PCA) [1] learns features manually and it is the baseline of this dataset. As shown in Table 4, methods utilizing deep neural networks outperform the methods that extract features manually, which indicates that deep neural networks have strong feature extraction ability to produce good results. Moreover, the HIFTR method based on ResNet50 outperforms the comparative methods, reflecting its effectiveness and superiority.

Table 4. Classification accuracy of VAIS ship dataset and RGB-NIR scene dataset using different methods (in %)

Method	VAIS ship dataset	RGB-NIR scene dataset
ResNet50-IR	72.93	84.65
ResNet50-VIS	85.28	89.50
VGG16-IR	69.47	84.44
VGG16-VIS	87.47	88.49
Pixel-level average (ResNet50)	85.65	88.69
Pixel-level average (VGG16)	87.36	88.08
CNN+Gnostic Fields	87.4	–
MSIFT (PCA)	–	73.1
HIFTR (ResNet50)	89.22	**90.51**
HIFTR (VGG16)	**90.09**	89.09

5 Conclusion

In this paper, an end-to-end framework for heterogeneous (i.e.visible & infrared) image fusion target recognition is proposed. The heterogeneous information is transferred between the networks using the mutual learning strategy, so that the generalization performance of the networks is enhanced. The classification results of visible and infrared images are combined by DS rule using the optimal weights for a more accurate result. The experimental results show that our HIFTR method is able to improve classification accuracy comparing with other related method. In the future work, we will deal with a more severe case where patterns of some classes in different sources are missing.

Acknowledgement. This work was supported in part by the National Natural Science Foundation of China under Grants U20B2067, 61790552 and 61790554, and in part by the Aeronautical Science Foundation of China under Grant 201920007001.

References

1. Brown, M., Süsstrunk, S.: Multi-spectral sift for scene category recognition. In: CVPR 2011, pp. 177–184. IEEE (2011)
2. Dempster, A.P.: Upper and lower probabilities induced by a multivalued mapping. In: Yager, R.R., Liu, L. (eds.) Classic works of the Dempster-Shafer theory of belief functions, vol. 219, pp. 57–72. Springer, Heidelberg (2008). https://doi.org/10.1007/978-3-540-44792-4_3
3. He, K., Zhang, X., Ren, S., Sun, J.: Deep residual learning for image recognition. In: Proceedings of the IEEE Conference on Computer Vision and Pattern Recognition, pp. 770–778 (2016)
4. Jia, S., et al.: Multiple feature-based superpixel-level decision fusion for hyperspectral and lidar data classification. IEEE Trans. Geosci. Remote Sens. **59**(2), 1437–1452 (2020)

5. Kulkarni, S.C., Rege, P.P.: Pixel level fusion techniques for SAR and optical images: a review. Inf. Fusion **59**, 13–29 (2020)
6. Li, H., Wu, X.J.: Densefuse: a fusion approach to infrared and visible images. IEEE Trans. Image Process. **28**(5), 2614–2623 (2019). https://doi.org/10.1109/TIP.2018.2887342
7. Liu, X., Jiao, L., Li, L., Tang, X., Guo, Y.: Deep multi-level fusion network for multi-source image pixel-wise classification. Knowl.-Based Syst. **221**, 106921 (2021)
8. Liu, Z., Pan, Q., Dezert, J., Han, J.W., He, Y.: Classifier fusion with contextual reliability evaluation. IEEE Trans. Cybern. **48**(5), 1605–1618 (2018). https://doi.org/10.1109/TCYB.2017.2710205
9. Ma, J., Ma, Y., Li, C.: Infrared and visible image fusion methods and applications: a survey. Inf. Fusion **45**, 153–178 (2019)
10. Qiu, X., Li, M., Zhang, L., Zhao, R., et al.: Deep convolutional feature fusion model for multispectral maritime imagery ship recognition. J. Comput. Commun. **8**(11), 23 (2020)
11. Shafer, G.: A mathematical theory of evidence turns 40. Int. J. Approximate Reasoning **79**, 7–25 (2016)
12. Simonyan, K., Zisserman, A.: Very deep convolutional networks for large-scale image recognition. arXiv preprint arXiv:1409.1556 (2014)
13. Zhang, M.M., Choi, J., Daniilidis, K., Wolf, M.T., Kanan, C.: Vais: A dataset for recognizing maritime imagery in the visible and infrared spectrums. In: Proceedings of the IEEE Conference on Computer Vision and Pattern Recognition Workshops, pp. 10–16 (2015)
14. Zhang, T., et al.: Hog-shipclsnet: a novel deep learning network with hog feature fusion for SAR ship classification. IEEE Trans. Geosci. Remote Sens. **60**, 1–22 (2022). https://doi.org/10.1109/TGRS.2021.3082759
15. Zhang, Y., Xiang, T., Hospedales, T.M., Lu, H.: Deep mutual learning. In: Proceedings of the IEEE Conference on Computer Vision and Pattern Recognition, pp. 4320–4328 (2018)

Cluster Decomposition of the Body of Evidence

Alexander Lepskiy[✉][iD]

Higher School of Economics, 20 Myasnitskaya Ulitsa, Moscow 101000, Russia
alepskiy@hse.ru
https://www.hse.ru/en/org/persons/10586209

Abstract. Two algorithms for the body of evidence clustering are developed and studied in this paper. The first algorithm is based on the use of the distribution density function of conflicting focal elements of the body of evidence. The second algorithm is similar to the k-means algorithm, but it uses the external conflict measure instead of the metric. It is shown that cluster decomposition can be used to evaluate the internal conflict of the body of evidence.

Keywords: Body of evidence · Measure of conflict · Decomposition of evidence

1 Introduction

The body of evidence may have a complex structure in applied problems of the theory of belief functions. For example, it may consist of many focal elements with a complex intersection structure. Such evidence is difficult to interpret. In addition, since many of the operations of the theory of belief functions (for example, combined rules) are computationally difficult, applying these operations to evidence bodies with many focal elements also becomes computationally difficult.

Therefore, the following tasks are relevant: 1) analysis of the structure of the set of focal elements $\mathcal{A}$ of the body of evidence $F = (\mathcal{A}, m)$ (m is the mass function); 2) finding an enlarged (simplified) structure of the set of focal elements $\widetilde{\mathcal{A}}$; 3) redistribution of masses of focal elements of the set $\mathcal{A}$ to focal elements from $\widetilde{\mathcal{A}}$. As a result, we obtain a new mass function $\widetilde{m}$, etc.

The paper proposes to solve these problems based on the clustering of a set of focal elements. We suggest that the complex structure of the body of evidence may be the result of aggregation of heterogeneous information. This information, which is obtained from various sources, may be contradictory (conflict). Therefore, the general approach to clustering the body of evidence can be as follows.

The financial support from the Government of the Russian Federation within the framework of the implementation of the 5–100 Programme Roadmap of the National Research University Higher School of Economics is acknowledged.

The inconsistency should be minimal within clusters and maximal between clusters in the resulting partition into clusters of the original set of focal elements. This approach is similar to the compactness principle in cluster analysis. Distances should be minimal between elements of the same cluster and maximal between clusters. But in this article, we will use the measure of external conflict (contradiction) between evidence bodies [8] as a proximity functional between clusters instead of a metric, and/or the measure of internal conflict of evidence bodies [9] as the proximity functional of focal elements within a cluster.

The idea of approximating the body of evidence F by a simpler body of evidence $\widetilde{F}$ using hierarchical clustering of focal elements was proposed in [6,10] and developed in [5]. Clustering was carried out by union and intersection of 'close' focal elements and summing their masses. A clustering algorithm based on the concept of conflict density was proposed in [2].

In the general case, we get a partition (or coverage) $\{\mathcal{A}_1, \ldots, \mathcal{A}_l\}$ of the set of focal elements $\mathcal{A}$ as a result of its clustering, which can be associated with the set of bodies of evidence $\{F_1, \ldots, F_l\}$, where $F_i = (\mathcal{A}_i, m_i)$, $i = 1, \ldots, l$.

In addition to revealing the structure and simplifying the body of evidence, clustering can be used to evaluate the internal conflict $Con_{in}(F)$ of the original body of evidence $F = (\mathcal{A}, m)$ according to the formula $Con_{in}(F) = Con(F_1, \ldots, F_l)$, where Con is a measure of external conflict.

Two algorithms for clustering the body of evidence are proposed in the article. The first algorithm is based on the use of the conflict distribution density function. The second algorithm is analogous to the k-means algorithm, but instead of a metric, a measure of external conflict is used.

2   Basic Concepts of the Evidence Theory

Let us briefly recall the basic concepts of evidence theory [4,12]. Let X be a finite set, 2^X be the set of all subsets on X. A body of evidence on the set X is a pair $F = (\mathcal{A}, m)$, where $\mathcal{A}$ is a set of non-empty subsets (focal elements) from the X, $m : 2^X \to [0, 1]$ is a mass function that satisfies the conditions: $m(A) > 0 \Leftrightarrow A \in \mathcal{A}$, $\sum_A m(A) = 1$. Let $\mathcal{F}(X)$ be the set of all bodies of evidence on the X.

Special cases of bodies of evidence are categorical evidence $F_A = (\{A\}, 1)$, vacuous evidence $F_X = (\{X\}, 1)$, simple evidence $F_A^\alpha = \alpha F_A + (1 - \alpha)F_X$, $\alpha \in [0, 1]$. An arbitrary body of evidence $F = (\mathcal{A}, m)$ can be represented in the form $F = \sum_{A \in \mathcal{A}} m(A)F_A$.

Some set functions are associated with the body of evidence $F = (\mathcal{A}, m)$: belief function $Bel(A) = \sum_{B \subseteq A} m(B)$, plausibility function $Pl(A) = 1 - Bel(\neg A) = \sum_{A \cap B \neq \emptyset} m(B)$, etc. These functions uniquely define the entire body of evidence.

Inconsistency (conflict) is an important joint characteristic of two or more bodies of evidence. The external conflict of two evidence bodies $F_1 = (\mathcal{A}_1, m_1)$ and $F_2 = (\mathcal{A}_2, m_2)$ is some measure $Con(F_1, F_2) : \mathcal{F}(X) \times \mathcal{F}(X) \to [0, 1]$, which takes on a greater value in the case of the existence of a large number of pairs (A, B) with large masses of non-overlapping focal elements of two bodies of evidence: $A \in \mathcal{A}_1$, $B \in \mathcal{A}_2$, $A \cap B = \emptyset$. An overview of articles on external conflict measures can be found in [8]. Below we will use the canonical measure of external conflict associated with the Dempster rule: $Con(F_1, F_2) = \sum_{A \cap B = \emptyset} m_1(A) m_2(B)$.

Along with the measure of external conflict between several bodies of evidence, the inconsistency of information provided by one body of evidence is also considered. An evidence body that results from combining multiple bodies of conflicting evidence bodies can have a large internal conflict. The inconsistency of the body of evidence $F = (\mathcal{A}, m)$ is evaluated using some internal conflict measure $Con_{in}(F) : \mathcal{F}(X) \rightarrow [0, 1]$ [9].

3 Evidence Clustering

3.1 Restriction and Extension of the Mass Function

Let $F = (\mathcal{A}, m)$ be the body of evidence, where $\mathcal{A}$ is the set of focal elements of this evidence. Let's consider some subset $\mathcal{A}' \subseteq \mathcal{A}$. The set function $m' : 2^X \rightarrow [0, 1]$, $m'(A) = m(A) \ \forall \ A \in \mathcal{A}'$ and $m'(A) = 0 \ \forall \ A \notin \mathcal{A}'$ is called the restriction of the mass function m to the set $\mathcal{A}' \subseteq \mathcal{A}$.

In the general case, the mass function m' does not satisfy the normalization condition $\sum_A m'(A) = 1$. Therefore, the pair $(\mathcal{A}', m')$ does not define any body of evidence. It is necessary to extend the set function m' to the mass function $\tilde{m}'$ so that this extension reflects the distribution of the m'.

This can be done in many ways. Examples of some extensions:

1) proportional extension: $\tilde{m}'(A) = m'(A) / \sum_{B \in \mathcal{A}'} m'(B) \ \forall A \in \mathcal{A}'$.
2) vacuous extension: $\tilde{m}'(A) = m'(A)$, $\tilde{m}'(X) = m'(X) + 1 - \sum_{B \in \mathcal{A}'} m'(B)$.

Note that various extensions of the set function to the mass function of some body of evidence are used in the combination rules. For example, proportional extension is used in Dempster's rule [4], and inconsistent continuation is used in Yager's rule [13].

If a certain rule for the extension of the mass function is fixed, then the new body of evidence $F' = (\mathcal{A}', \tilde{m}')$ will be uniquely determined by the original body of evidence $F = (\mathcal{A}, m)$ by the set $\mathcal{A}' \subseteq \mathcal{A}$. Therefore, such a body of evidence will be denoted as $F(\mathcal{A}') = (\mathcal{A}', \tilde{m}')$.

In particular, if the vacuous extension is used, then the body of evidence $F(\{A\}) = F_A^{m(A)} = m(A) F_A + (1 - m(A)) F_X$ will be simple for any set $A \in \mathcal{A}$.

3.2 Statement of the Problem of Clustering the Body of Evidence Based on Conflict Optimization

Various formulations of the clustering problem are possible. Let's note some of them. Suppose we have a body of evidence $F = (\mathcal{A}, m)$. It is required to find a partition of the set of all focal elements $\mathcal{A}$ into subsets $\{\mathcal{A}_1, \ldots, \mathcal{A}_l\}$ such that:

1) maximize external conflict between bodies of evidence (clusters):
   $Con(F(\mathcal{A}_1), \ldots, F(\mathcal{A}_l)) \to$ max, where Con is a measure of external conflict;
2) minimize total internal conflict within clusters
   $\sum_{i=1}^{l} Con_{in}(F(\mathcal{A}_i)) \to$ min, where Con_{in} is a measure of internal conflict;
3) minimize the overall conflict between the centers of clusters and the bodies
   of evidence formed by the focal elements of these clusters
   $\sum_{i=1}^{l} \sum_{B \in \mathcal{A}_i} Con(F(\{B\}), C_i) \to$ min, where C_i is the reference body of
   evidence corresponding to the i-th cluster, $i = 1, \ldots, l$.

In a more general setting, it is required to find a covering of the set of focal elements $\mathcal{A}$ instead of a partition.

3.3   Cluster Decomposition of Evidence Based on the Conflict Density Function

Density Function. The concept of conflict density was introduced in [2]. Let $F = (\mathcal{A}, m)$ be the body of evidence. A mapping $\psi_F : 2^X \to [0,1]$ is called a conflict density function of the body of evidence F if it satisfies the following conditions:

1) $\psi_F(A) = 0$, if $B \cap A \neq \emptyset \; \forall B \in \mathcal{A}$;
2) $\psi_F(A) = 1$, if $B \cap A = \emptyset \; \forall B \in \mathcal{A}$;
3) $\psi_{\alpha F_1 + \beta F_2} = \alpha \psi_{F_1} + \beta \psi_{F_2} \; \forall F_1, F_2 \in \mathcal{F}(X)$, where $\alpha + \beta = 1$, $\alpha \geq 0$, $\beta \geq 0$.

It is easy to show [2] that a set function satisfying conditions 1)–3) is equal to $\psi_F(A) = \sum_{B: A \cap B = \emptyset} m(B) = 1 - Pl(A)$. Note that the function was considered in [11] and was called the inconsistency function.

The main idea of the clustering algorithm for a set of focal elements $\mathcal{A}$ based on the conflict density function is that the 'centers' of the clusters should have a large value of the conflict density function.

We will use the function $\varphi_F(A) = m(A)\psi_F(A)$, $A \in \mathcal{A}$ instead of the density function ψ_F itself. The function φ_F will take on large values for those focal elements that have not only a high density, but also a large mass.

Algorithm for Cluster Decomposition of Evidence Based on Conflict Density Functions. This algorithm will consist of the following steps.

Algorithm 1

1. Let's calculate the values of the set function $\varphi_F(A)$, $A \in \mathcal{A}$. If we have $\varphi_F(A) = 0$ for all $A \in \mathcal{A}$, then we stop the algorithm. In this case, we have non-conflict body of evidence $\mathcal{A}$: $B \cap A \neq \emptyset \; \forall A, B \in \mathcal{A}$. Therefore, there will be no clustering.
2. If there are $A \in \mathcal{A}$ for which $\varphi_F(A) > 0$, then we arrange such focal elements in descending order of function values φ_F: $\varphi_F(A_1) \geq \varphi_F(A_2) \geq \ldots$. We choose the number of clusters l by analyzing the rate of decrease of the sequence $\{\varphi_F(A_i)\}$. Selected focal elements will be initial clusters: $\mathcal{A}_i^{(0)} = \{A_i\}$, $i = 1, \ldots, l$.

3. The remaining focal elements are redistributed among clusters $\mathcal{A}_1^{(0)}, ..., \mathcal{A}_l^{(0)}$ according to the principle of maximizing the conflict between evidence clusters. We will assign a focal element $B \in \mathcal{A} \backslash \left\{ \mathcal{A}_1^{(0)}, ..., \mathcal{A}_l^{(0)} \right\}$ to the cluster $\mathcal{A}_i^{(0)}$ for which the maximum conflict measure is reached:

$$\mathcal{A}_i^{(0)} = \underset{j:B \in \mathcal{A}_j^{(0)}}{\arg \max} Con \left(F\left(\mathcal{A}_1^{(0)}\right), ..., F\left(\mathcal{A}_j^{(0)} \cup \{B\}\right), ..., F\left(\mathcal{A}_l^{(0)}\right) \right).$$

If equal maximum values of the conflict are obtained by assigning B to several clusters $\mathcal{A}_j^{(0)}$, $j \in J$, then we include B in all these clusters, and the mass value $m(B)$ is evenly distributed among the updated clusters. In this case, B will be included in each cluster with weight $m(B)/|J|$. As a result, we obtain a coverage $\{\mathcal{A}_1, ..., \mathcal{A}_l\}$ of the set of all focal elements $\mathcal{A}$. The values of the mass function m_i on the $\mathcal{A}_i$, $i = 1, ..., l$ are calculated using the given restriction and extension procedures.

Example 1. Let we have $X = \{1, 2, 3\}$ and the body of evidence $F = 0.3F_{\{1\}} + 0.2F_{\{2\}} + 0.3F_{\{1,3\}} + 0.2F_{\{2,3\}}$ is given on X, i.e. $\mathcal{A} = \{\{1\}, \{2\}, \{1,3\}, \{2,3\}\}$.

Step 1. Find the values of the function φ_F: $\varphi_F(\{1\}) = \varphi_F(\{2\}) = 0.12$, $\varphi_F(\{1,3\}, \{2,3\}) = 0.06$.

Step 2. Let us assign the number of clusters $l = 2$ and $\mathcal{A}_1^{(0)} = \{\{1\}\}$, $\mathcal{A}_2^{(0)} = \{\{2\}\}$.

Step 3. Let's distribute the remaining two focal elements among clusters. We have for $B = \{1,3\}$. If $B \in \mathcal{A}_1$, then $F\left(\{B\} \cup \mathcal{A}_1^{(0)}\right) = 0.3F_{\{1\}} + 0.3F_{\{1,3\}} + 0.4F_X$, $F\left(\mathcal{A}_2^{(0)}\right) = 0.2F_{\{2\}} + 0.8F_X$. Consequently $Con\left(F\left(\{B\} \cup \mathcal{A}_1^{(0)}\right), F\left(\mathcal{A}_2^{(0)}\right)\right) = 0.12$. But if $B \in \mathcal{A}_2$, then $F\left(\mathcal{A}_1^{(0)}\right) = 0.3F_{\{1\}} + 0.7F_X$, $F\left(\{B\} \cup \mathcal{A}_2^{(0)}\right) = 0.2F_{\{2\}} + 0.3F_{\{1,3\}} + 0.5F_X$ and $Con\left(F\left(\mathcal{A}_1^{(0)}\right), F\left(\{B\} \cup \mathcal{A}_2^{(0)}\right)\right) = 0.06$. Thus, we assign $B = \{1,3\}$ to the cluster $\mathcal{A}_1$.

We have for focal element $B = \{2,3\}$. If $B \in \mathcal{A}_1$, then $F\left(\{B\} \cup \mathcal{A}_1^{(0)}\right) = 0.3F_{\{1\}} + 0.2F_{\{2,3\}} + 0.5F_X$, $F\left(\mathcal{A}_2^{(0)}\right) = 0.2F_{\{2\}} + 0.8F_X$ and $Con\left(F\left(\{B\} \cup \mathcal{A}_1^{(0)}\right), F\left(\mathcal{A}_2^{(0)}\right)\right) = 0.06$. But if $B \in \mathcal{A}_2$, then $F\left(\mathcal{A}_1^{(0)}\right) = 0.3F_{\{1\}} + 0.7F_X$, $F\left(\{B\} \cup \mathcal{A}_2^{(0)}\right) = 0.2F_{\{2\}} + 0.2F_{\{2,3\}} + 0.6F_X$ and $Con\left(F\left(\mathcal{A}_1^{(0)}\right), F\left(\{B\} \cup \mathcal{A}_2^{(0)}\right)\right) = 0.12$. Thus, we assign $B = \{2,3\}$ to the cluster $\mathcal{A}_2$ and we get the final focal element clustering $\mathcal{A}_1 = \{\{1\}, \{1,3\}\}$, $\mathcal{A}_2 = \{\{2\}, \{2,3\}\}$.

Remark 1. The distance between the selected focal elements can also be considered at step 2 of the algorithm in addition to calculating the conflict density (function φ_F), as was done in [2]. In this case, the focal elements are selected in descending order of the function φ_F. If $\mathcal{A}'$ is a set of already selected focal elements, then the next element A_k is added to this set, provided that $\min_{A\in\mathcal{A}'} d(F(\{A\}), F(\{A_k\})) > h$, where d is some metric on the set of evidence bodies [7], h is the threshold value.

3.4 The k-Means Algorithm for the Body of Evidence

Let $F = (\mathcal{A}, m)$ be the body of evidence. It is required to find such a coverage of the set of all focal elements $\mathcal{A}$ by subsets (clusters) $\mathcal{C} = \{\mathcal{A}_1, \ldots, \mathcal{A}_l\}$ that would minimize intracluster conflict. We will use the concept of center of a set (cluster) of focal elements by analogy with the classical k-means algorithm. By the center of the i-th cluster $\mathcal{A}_i$, we mean some body of evidence C_i constructed from the pair $(\mathcal{A}_i, m_i)$, where m_i is the restriction of the mass function to $\mathcal{A}_i \subseteq \mathcal{A}$, $i = 1, \ldots, l$. We will consider the total conflict between the centers of clusters and the bodies of evidence generated by the focal elements of these clusters as a minimized functional by analogy with the k-means algorithm:

$$\Phi = \sum_{i=1}^{l} \sum_{B\in\mathcal{A}_i} Con\left(F(\{B\}), C_i\right), \tag{1}$$

where $F(\{B\})$ is the evidence generated from the set $\{B\}$ using the restriction and extension procedures (see Subsect. 3.1). In particular, if the vacuous extension is chosen, then $F(\{B\}) = m(B)F_B + (1 - m(B))F_X$.

In this algorithm, the number of evidence bodies l into which the evidence body $F = (\mathcal{A}, m)$ is decomposed will be considered predetermined (it is determined from some other heuristic considerations). Also, the method of extension the mass function will be considered predetermined.

Let us assume that the covering $\mathcal{C} = \{\mathcal{A}_1, \ldots, \mathcal{A}_l\}$ is fixed and the center of the i-th cluster has the form

$$C_i = \sum_{A\in\mathcal{A}_i} \alpha_i(A)F_A, \tag{2}$$

where $\alpha_i = (\alpha_i(A))_{A\in\mathcal{A}_i} \in S_{|\mathcal{A}_i|}$, $S_k = \left\{(t_1, \ldots, t_k) : t_i \geq 0, i = 1, \ldots, k, \sum_{i=1}^{k} t_i = 1\right\}$ is the k-dimensional simplex. Then we have for the vacuous extension

$$\Phi = \sum_{i=1}^{l} \sum_{B\in\mathcal{A}_i} Con\left(F(\{B\}), C_i\right) = \sum_{i=1}^{l} \sum_{B\in\mathcal{A}_i} m(B) \sum_{\substack{A\in\mathcal{A}_i: \\ A\cap B=\emptyset}} \alpha_i(A)$$

$$= \sum_{i=1}^{l} \sum_{B\in\mathcal{A}_i} m(B) \left(1 - \sum_{\substack{A\in\mathcal{A}_i: \\ A\cap B\neq\emptyset}} \alpha_i(A)\right) = k_{\mathcal{C}} - \sum_{i=1}^{l} Q_i(\alpha_i),$$

where $k_{\mathcal{C}} = \sum_{i=1}^{l} \sum_{B\in\mathcal{A}_i} m(B) \geq 1$ ($k_{\mathcal{C}} = 1 \Leftrightarrow \mathcal{C} = \{\mathcal{A}_1, \ldots, \mathcal{A}_l\}$ is the partition of the set of focal elements), $Q_i(\alpha_i) = \sum_{A\in\mathcal{A}_i} \alpha_i(A)Pl_{\mathcal{A}_i}(A)$ and $Pl_{\mathcal{A}_i}(A) =$

$\sum_{\substack{B \in \mathcal{A}_i: \\ A \cap B \neq \emptyset}} m(B)$ is the restriction of the plausibility function to the set $\mathcal{A}_i$. The minimum of the functional Φ for a fixed coverage $\mathcal{C} = \{\mathcal{A}_1, \ldots, \mathcal{A}_l\}$ will be achieved when the linear functions $Q_i(\alpha_i)$ reach maxima on the simplices $S_{|\mathcal{A}_i|}$, $i = 1, \ldots, l$. But

$$\max_{\alpha \in S_{|\mathcal{A}_i|}} Q_i(\alpha) = \max_{A \in \mathcal{A}_i} Pl_{\mathcal{A}_i}(A), \quad i = 1, \ldots, l.$$

Let $\overline{\mathcal{A}}_i = \left\{ A \in \mathcal{A}_i : A = \arg\max_{A \in \mathcal{A}_i} Pl_{\mathcal{A}_i}(A) \right\}$. If

$$C_i = \sum_{A \in \overline{\mathcal{A}}_i} \alpha_i(A) F_A, \quad \alpha_i = (\alpha_i(A))_{A \in \overline{\mathcal{A}}_i} \in S_{|\overline{\mathcal{A}}_i|}, \quad i = 1, \ldots, l, \qquad (3)$$

then the functional Φ will reach a minimum in the case of a fixed coverage $\mathcal{C} = \{\mathcal{A}_1, \ldots, \mathcal{A}_l\}$ with such a choice of cluster centers. This minimum will be

$$\min \Phi = k_{\mathcal{C}} - \sum_{i=1}^{l} \max_{A \in \mathcal{A}_i} Pl_{\mathcal{A}_i}(A) \qquad (4)$$

and does not depend on the choice of parameters $\alpha_i = (\alpha_i(A))_{A \in \overline{\mathcal{A}}_i} \in S_{|\overline{\mathcal{A}}_i|}$, $i = 1, \ldots, l$.

Then the evidence body clustering algorithm, by analogy with the classical k-means algorithm, will be as follows.

Algorithm 2

1. Let's choose and fix the number of clusters l. Let's assign some evidence bodies as initial cluster centers $C_i^{(0)}$, $i = 1, \ldots, l$. We fix the threshold of maximum conflict within clusters $Con_{\max} \in [0, 1]$. We install $s = 0$.
2. We redistribute focal elements among clusters according to the principle of minimizing the conflict between evidence clusters and cluster centers. The focal element $B \in \mathcal{A}$ is assigned to the cluster $\mathcal{A}_i^{(s)} = \arg\min_j Con\left(F(\{B\}), C_j^{(s)}\right)$ and $\min_i Con\left(F(\{B\}), C_i^{(s)}\right) \leq Con_{\max}$. If $\min_i Con\left(F(\{B\}), C_i^{(s)}\right) > Con_{\max}$, then the focal element B is assigned as the center of the new cluster. As a result, we get clusters $\mathcal{A}_i^{(s)}$, $i = 1, \ldots, l$.
3. Let us calculate new cluster centers using the formula. We increase the counter $s \leftarrow s + 1$.
4. Steps 2 and 3 are repeated until the clusters (or their centers) stabilize.

Proposition 1. *Algorithm 2 converges in a finite number of steps.*

The proof follows from the fact that the functional Φ does not increase at the 2nd and 3rd steps of the algorithm and we have a finite number of possible configurations.

Remark 2. Evidence bodies $C_i^{(0)} = F_{A_i}$, $i = 1, ..., l$ can be chosen as the initial centers of clusters at the 1st step of the algorithm, where focal elements A_i, $i = 1, ..., l$ are chosen arbitrarily or, for example, using the density maximization algorithm.

Remark 3. Cluster centers may depend on parameters $\alpha = (\alpha(A))_{A \in \overline{\mathcal{A}}_i} \in S_{|\overline{\mathcal{A}}_i|}$ (see formula (3)). In this case, it is necessary to use additional procedures for choosing parameters at the 2nd or 3rd steps of the algorithm. The selection criteria can be considered, for example:

1) coverage minimization, i.e., we choose the parameters at the 2nd step of the algorithm so that the coverage $\mathcal{C} = \{\mathcal{A}_1, ..., \mathcal{A}_l\}$ is 'closer' to the partition. For example, $\sum_{i=1}^{l} |\mathcal{A}_i| \to \min$.
2) minimizing the uncertainty of evidence-centers of clusters C_i, $i = 1, ..., l$. This procedure is applied at the 3rd step of the algorithm. The uncertainty of evidence can be assessed using one of the imprecision indices [1]. For example, it can be the generalized Hartley measure $H(C_i) = \sum_{A \in \overline{\mathcal{A}}_i} \alpha_i(A) \ln |A|$.
3) minimizing the distance between cluster centers and the original evidence body with respect to some metric between evidence bodies [7]: $d(C_i, F) \to \min$, $i = 1, ..., l$;
4) maximizing distance between cluster centers $d(C_i, C_j) \to \max$ or maximizing conflict $Con(C_i, C_j) \to \max$, $i, j = 1, ..., l$ $(i \neq j)$ etc.

Remark 4. One way to evaluate the internal conflict [9] of a body of evidence F on the X is based on finding the maximum of the contour function [3]: $Con(F) = 1 - \max_{x \in X} Pl(x)$. Then formula (4) can be interpreted as a total intra-cluster internal conflict.

Remark 5. It is possible to search for cluster centers C_i, $i = 1, ..., l$ in the form (2), minimizing the functional (1) with a fixed coverage $\mathcal{C} = \{\mathcal{A}_1, ..., \mathcal{A}_l\}$ and under the condition that uncertainties of cluster centers C_i are bounded:

$$H(C_i) = \sum_{A \in \mathcal{A}_i} \alpha_i(A) \ln |A| \leq u_i, \quad i = 1, ..., l,$$

where u_i, $i = 1, ..., l$ are some threshold values. Then the problem of minimizing the functional (1) for a fixed coverage $\mathcal{C} = \{\mathcal{A}_1, ..., \mathcal{A}_l\}$ is reduced to solving l linear programming problems:

$$\sum_{A \in \mathcal{A}_i} \alpha_i(A) Pl_{\mathcal{A}_i}(A) \to \max$$

subject to $\alpha_i = (\alpha_i(A))_{A \in \mathcal{A}_i} \in S_{|\mathcal{A}_i|}$, $\sum_{A \in \mathcal{A}_i} \alpha_i(A) \ln |A| \leq u_i$, $i = 1, ..., l$.

Example 2. Algorithm 2 will give the following result of clustering the body of evidence from Example 1 ($X = \{1, 2, 3\}$, $F = 0.3F_{\{1\}} + 0.2F_{\{2\}} + 0.3F_{\{1,3\}} + 0.2F_{\{2,3\}}$ into two clusters and the vacuous extension.

Step 1. We have $l = 2$. Let the initial centers of the clusters be equal to $C_1^{(0)} = F_{\{1\}}$, $C_2^{(0)} = F_{\{2\}}$; $s = 0$.

Step 2. We have $Con\left(F(\{1\}), C_1^{(0)}\right) = Con\left(F(\{1,3\}), C_1^{(0)}\right) = 0$,

$Con\left(F(\{2\}), C_1^{(0)}\right) = Con\left(F(\{2,3\}), C_1^{(0)}\right) = 0.2$,

$Con\left(F(\{1\}), C_2^{(0)}\right) = Con\left(F(\{1,3\}), C_2^{(0)}\right) = 0.3$, $Con\left(F(\{2\}), C_2^{(0)}\right) =$
$Con\left(F(\{2,3\}), C_2^{(0)}\right) = 0$ (for example, $Con\left(F(\{1,3\}), C_2^{(0)}\right) =$
$Con\left(0.3F_{\{1,3\}} + 0.7F_X, F_{\{2\}}\right) = 0.3$).

Therefore, according to the principle of minimizing the conflict between evidence clusters and cluster centers, the initial clustering will have the form $\mathcal{A}_1^{(0)} = \{\{1\}, \{1,3\}\}$, $\mathcal{A}_2^{(0)} = \{\{2\}, \{2,3\}\}$.

Step 3. Let us calculate the new cluster centers using the formula (3):
$Pl_{\mathcal{A}_1^{(0)}}(\{1\}) = 0.3 + 0.3 = 0.6$, $Pl_{\mathcal{A}_1^{(0)}}(\{1,3\}) = 0.3 + 0.3 = 0.6$, $Pl_{\mathcal{A}_2^{(0)}}(\{2\}) = 0.2 + 0.2 = 0.4$, $Pl_{\mathcal{A}_2^{(0)}}(\{2,3\}) = 0.2 + 0.2 = 0.4$.

Therefore $C_1^{(1)} = \alpha F_{\{1\}} + (1-\alpha)F_{\{1,3\}}$ and $C_2^{(1)} = \beta F_{\{2\}} + (1-\beta)F_{\{2,3\}}$, $\alpha, \beta \in [0,1]$.

If we require the minimization of the generalized Hartley measure (see Remark 2), we get $C_1^{(1)} = \underset{0 \le \alpha \le 1}{\arg\min} H\left(\alpha F_{\{1\}} + (1-\alpha)F_{\{1,3\}}\right) = F_{\{1\}}$, $C_2^{(1)} =$
$\underset{0 \le \beta \le 1}{\arg\min} H\left(\beta F_{\{2\}} + (1-\beta)F_{\{2,3\}}\right) = F_{\{2\}}$ and the algorithm will stop its work, since the centers of the clusters have not changed. If, however, we apply the coverage minimization rule (see Remark 3), we move on to the next step.

Step 4. We redistribute focal elements according to the criterion of least conflict with new centers:
$Con\left(F(\{1\}), C_1^{(1)}\right) = Con\left(F(\{1,3\}), C_1^{(1)}\right) = 0$, $Con\left(F(\{2\}), C_1^{(1)}\right) = 0.2$,

$Con\left(F(\{2,3\}), C_1^{(1)}\right) = 0.2\alpha$, $Con\left(F(\{1\}), C_2^{(1)}\right) = 0.3$,

$Con\left(F(\{1,3\}), C_2^{(1)}\right) = 0.3\beta$, $Con\left(F(\{2\}), C_2^{(1)}\right) = Con\left(F(\{2,3\}), C_2^{(1)}\right) = 0$.

We will get clusters $\mathcal{A}_1^{(1)} = \{\{1\}, \{1,3\}\}$, $\mathcal{A}_2^{(1)} = \{\{2\}, \{2,3\}\}$ after applying the coverage minimization rule (see Remark 3). Clusters have stabilized. Stop of the algorithm.

As a result, we get, in fact, a new body of evidence defined on the base set $\mathcal{A}$, the found coverage sets (clusters) $\mathcal{C} = \{\mathcal{A}_1, \ldots, \mathcal{A}_l\}$ will be focal elements, the mass function will be equal to $m(\mathcal{A}_i) = \sum_{B \in \mathcal{A}_i} m(B)/n(B)$, where $n(B) = |\{\mathcal{A}_i : B \in \mathcal{A}_i\}|$ (the number of clusters containing the set B). Such a body of evidence can be considered second-order evidence, which reflects the enlarged structure of the original evidence.

4   Evaluation of the Internal Conflict of the Body of Evidence Based on Its Clustering

Let us assume that in one way or another, a cluster coverage $\mathcal{C} = \{\mathcal{A}_1, \ldots, \mathcal{A}_l\}$ (in a particular case, partitioning) of the body of evidence $F = (\mathcal{A}, m)$ is obtained. Then we can offer the following ways to evaluate the internal conflict of this body of evidence using some measure of external conflict Con:

1) $Con_1(F) = Con(F(\mathcal{A}_1), \ldots, F(\mathcal{A}_l))$;
2) $Con_2(F) = Con(C_1, \ldots, C_l)$, where $C_1, \ldots, C_l$ are centers of clusters $\mathcal{A}_1, \ldots, \mathcal{A}_l$ respectively.

For example, we will obtain the following estimates of the internal conflict for the body of evidence considered in Example 2, the vacuous extension, and the canonical measure of the external conflict. We have $\mathcal{A}_1 = \{\{1\}, \{1, 3\}\}$, $\mathcal{A}_2 = \{\{2\}, \{2, 3\}\}$ and $Con_1(F) = Con(F(\mathcal{A}_1), F(\mathcal{A}_2)) = 0.18$, $Con_2(F) = Con(C_1, C_2) = \alpha + (1 - \alpha)\beta$, $\alpha, \beta \in [0, 1]$.

Proposition 2. *The following equality is true*

$$Con_1(F) = Con(F(\mathcal{A}_1), \ldots, F(\mathcal{A}_l)) = \sum_{A_1 \in \mathcal{A}_1, \ldots, A_l \in \mathcal{A}_l} Con(F(\{A_1\}), \ldots, F(\{A_l\})).$$

5   Conclusion

Two methods of evidence body clustering are discussed in this paper. Each of these methods assumes that weakly conflicting focal elements should belong to one cluster, and strongly conflicting focal elements should belong to different clusters. This requirement is similar to the basic principle of compactness in cluster analysis: the distances between elements of one cluster should be minimal, and between clusters should be maximum.

The first algorithm is based on the use of the distribution density function of conflicting focal elements. The second algorithm implements the idea of the k-means method. In this case, the cluster centers are formed in some optimal way. Further, focal elements are redistributed according to the principle of minimizing conflict with cluster centers.

It shows how clustering can be used to evaluate the internal conflict of a body of evidence.

References

1. Bronevich, A., Lepskiy, A.: Imprecision indices: axiomatic, properties and applications. Int. J. Gen Syst **44**(7–8), 812–832 (2015)
2. Bronevich, A., Lepskiy, A.: Measures of conflict, basic axioms and their application to the clusterization of a body of evidence. Fuzzy Sets Syst. **446**, 277–300 (2021). https://doi.org/10.1016/j.fss.2021.04.016
3. Daniel, M.: Properties of plausibility conflict of belief functions. In: Rutkowski, L., Korytkowski, M., Scherer, R., Tadeusiewicz, R., Zadeh, L.A., Zurada, J.M. (eds.) ICAISC 2013. LNCS (LNAI), vol. 7894, pp. 235–246. Springer, Heidelberg (2013). https://doi.org/10.1007/978-3-642-38658-9_22
4. Dempster, A.P.: Upper and lower probabilities induced by multivalued mapping. Ann. Math. Stat. **38**, 325–339 (1967)
5. Denœux, T.: Inner and outer approximation of belief structures using a hierarchical clustering approach. Int. J. Uncertainty Fuzziness Knowl.-Based Syst. **9**(4), 437–460 (2001)
6. Harmanec, D.: Faithful approximations of belief functions. In: Laskey, K.B., Prade, H. (eds.) Uncertainty in Artificial Intelligence 15 (UAI 1999), Stockholm, Sweden (1999)
7. Jousselme, A.-L., Maupin, P.: Distances in evidence theory: comprehensive survey and generalizations. Int. J. Approx. Reason. **53**, 118–145 (2012)
8. Lepskiy, A.: Analysis of information inconsistency in belief function theory. Part I: external conflict. Control Sci. **5**, 2–16 (2021)
9. Lepskiy, A.: Analysis of information inconsistency in belief function theory. Part II: internal conflict. Control Sci. **6**, 2–12 (2021)
10. Petit-Renaud, S., Denœux, T.: Handling different forms of uncertainty in regression analysis: a fuzzy belief structure approach. In: Hunter, A., Parsons, S. (eds.) ECSQARU 1999. LNCS (LNAI), vol. 1638, pp. 340–351. Springer, Heidelberg (1999). https://doi.org/10.1007/3-540-48747-6_31
11. Pichon, F., Jousselme, A.L., Abdallah, N.B.: Several shades of conflict. Fuzzy Sets Syst. **366**, 63–84 (2019)
12. Shafer, G.: A Mathematical Theory of Evidence. Princeton University Press, Princeton (1976)
13. Yager, R.: On the Dempster-Shafer framework and new combination rules. Inf. Sci. **41**, 93–137 (1987)

Evidential Trustworthiness Estimation for Cooperative Perception

Antoine Lima[(✉)][iD], Véronique Cherfaoui[iD], and Philippe Bonnifait[iD]

Université de Technologie de Compiègne, CNRS UMR 7253, Heudiasyc,
Compiègne, France
{antoine.lima,veronique.cherfaoui,philippe.bonnifait}@hds.utc.fr

Abstract. Intelligent Vehicles can exchange their perception information using wireless technology in a cooperative and decentralized manner. This has the potential to extend the range of perception and thus improve anticipation for complex driving maneuvers and decision making. However, information received from other peers can be erroneous and has to be used carefully. In this paper, we present a method that allows each peer to assign a trust in the information received from other peers based on comparisons with its current knowledge of the world. We describe how this process is managed using the Dempster-Shafer theory. We also present how positive and negative evidence cues can be developed in this problem, in particular by using detectability grids. An experimental evaluation, carried out with real vehicles, is reported to show that this formalism behaves correctly.

Keywords: Cooperative perception · Trust · Multi-robot system · Belief functions

1 Introduction

In order to navigate safely, intelligent vehicles need to perceive their environment. Their on-board sensors, such as cameras or LiDARs, are generally sufficient for local navigation tasks but not for more complex maneuvers because of the limited range of the sensors and because there are occlusions in their Field Of Views (FOVs). For example, in Fig. 1, v_1 cannot see if a vehicle is coming from behind the building on its right and will thus have to either be cautious and engage slowly or break strongly once the vehicle becomes visible. Cooperative Perception (CP) aims at improving the navigation performance in such situations by taking advantage of perceptual information captured by others. Indeed, using upcoming wireless technologies, it is possible for vehicles and the infrastructure to exchange perceptual information with each other. By integrating this information to its own, one's knowledge of its surroundings can be extended further and behind obstructions. For example, in Fig. 1, v_3 could warn v_1 of its presence and v_2 could warn v_1 that a vehicle is present in front of it. However, although the authenticity of peers can be cryptographically guaranteed using public-private key pairs [5],

S. Le Hégarat-Mascle et al. (Eds.): BELIEF 2022, LNAI 13506, pp. 174–183, 2022.
https://doi.org/10.1007/978-3-031-17801-6_17

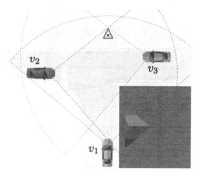

Fig. 1. Three cooperative vehicles at an intersection. v_1 cannot see v_3 because it is hidden by a building but v_2 and v_3 can see each other.

security vulnerabilities or perception malfunctions (e.g. sensors failures) can still generate erroneous information that should not be incorporated to one's own.

To prevent this, we propose an information processing and data fusion system that confronts the information received and the information from the embedded system to estimate the trustworthiness of the peers. This information can then be used in a cooperative tracker to attenuate or ignore information from untrustworthy peers. This process is done locally by each peer, without communicating its trust, by verifying that the received information matches with its knowledge of the world. For example, detecting objects at the same location creates trustworthiness while mismatching or illogical information creates untrustworthiness. In Fig. 1, because v_1 partially shares objects and FOV with v_2 and v_3, it will trust them and thus anticipate v_3 earlier.

After a review of related works in Sect. 2, we will introduce the problem at hand using a dense representation of the detectable or undetectable space from the point of view of different peers in Sect. 3. In Sect. 4, trustworthiness estimation will be formulated. Finally in Sect. 5, a simulation study and experimental results based on real data will be given and analyzed.

2   Related Works

The field of Cooperative Perception began with [7] demonstrating its potential for safety in intelligent transportation systems. Since then, the European Telecommunications Standards Institute (ETSI) standardized the Cooperative Perception Message (CPM) [6], composed of the sender position, sensor descriptions and a list of objects. It is used in many cooperative approaches, as studied in [4].

A part of the research effort is focused on preventing attacks as CP works on a public network. The most common form of attack prevention is misbehavior detection, as reviewed in [8]. In this paper, the authors list and classify numerous approaches as being standalone or shared, distributed or centralized

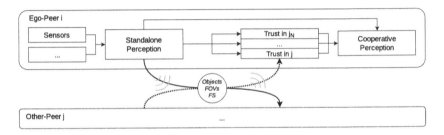

Fig. 2. Decentralized cooperative perception with trustworthiness estimation.

and node or data centric. For example, in [12], errors are detected by comparing received positions with detections from embedded sensors. In [3], four levels of checks ranging from simple bound check to object comparison are used to emit reports on misbehaving peers. When enough reports are received about a faulty peer, their certificate is revoked, excluding them from communication. In [2], a probability of trustworthiness is estimated for each pair of peers, by checking the consistency of their object lists and their respective detection probabilities across space. More recently, [11] compared occupied or free space in the form of grids and verified that detected objects matched with these grids. In [1], sensors estimate a probability of existence for each object. When fusing object from multiple sensors, they switch to an evidential representation and use a persistence probability to model the field of view of fused sensors. A trust parameter representing the sensor's information reliability is fixed for each sensors.

Our method can be seen as an unification of [2,3,11] where fault detection generates untrustworthiness and confirmation creates trustworthiness.

3   Problem Statement with Object Detectability

Consider a driving situation composed of N vehicles $v_1, v_2 \dots$. Every cooperative vehicle perceives surrounding objects $o \in O$ and Free Space FS. Objects can be any kind of road user (e.g.: pedestrians, cooperative or non-cooperative vehicles) or static features (e.g.: traffic signs) whereas FS are areas explicitly characterized as being free. Objects and FS are broadcast to peers in the communication range, supposed to be further than the perception range. In addition, the Fields Of View (FOV) of the sensors are shared as per the CPM such that all exchanged information is vectorial in order to reduce the amount of data exchanged. Upon reception, each peer assesses the trustworthiness of the sender, as illustrated in Fig. 2.

As the problem at hand combines multiple point of views, a method to define which area was seen by the cooperating peers is needed. For this, we extend the detection probability of [2] with the capacity to state that there cannot be objects in the measured FS by using an evidential representation. In this representation, the ground plane is divided into cells of fixed size that contain a mass function ${}^i m^{\mathcal{D}}_{x,y}$ defined on $2^{\Omega^{\mathcal{D}}}$ where $\Omega^{\mathcal{D}} = \{D_i, \mathcal{D}\}$. D_i represents that peer i can detect

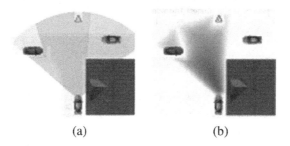

(a)   (b)

Fig. 3. Illustration of a detectability grid from the point of view of v_1. (a) Hidden area in grey, FS in red and detectable in green. (b) Resulting grid: non-detectability in red, detectability in blue and unknown in grey. (Color figure online)

an object in the cell at position $[x, y]^T$ and $\mathcal{D}$ that an object cannot be detected at that point. Conversely, the global grid is noted $m^\mathcal{D}$ and the detectability of an object is $m_o^\mathcal{D} = m_{x_o, y_o}^\mathcal{D}$.

This grid (called detectability grid) is built with the process illustrated in Fig. 3. Outside the FOVs and behind buildings, the state is unknown. The FS characterized by local sensors (e.g. LiDAR points that hit the ground) is used to express the impossibility to detect objects within it. In space neither free nor unknown, an object is likely to be detected.

Object detectability is used in two different ways when receiving information. First, the sender's detectability grid is reconstructed to assess its objects coherency (e.g. there is no object in the FS or out of the FOV). Then, the receiver's detectability grid is constructed to verify the coherency of received objects and to only compare objects that are detectable by it and the sender.

4   Evidential Trustworthiness Estimation

Trustworthiness in the information sent by other peers is estimated by every peer individually as a mass function noted m_j^T about peer j. It is designed to be used in a subsequent cooperative fusion to ignore or discount objects originating from untrustworthy peers. As such, it is defined on 2^{Ω^T} with $\Omega^T = \{T, \mathcal{T}\}$ to express that information from j is trustworthy and can be integrated without hesitation or conversely not integrated at all. In the rest of this paper, trust mass functions are normalized and will be given in the following order: $T, \mathcal{T}, \Omega^T$.

Mass functions are particularly adapted to the problem at hand for several reasons. Firstly, similar to humans, trust evolves over time and can be forgotten when peers are not interacting anymore, which can be managed with discounting. Secondly, as new peers have an unknown degree of trustworthiness, choosing a wrong prior could lead to ignoring good information or including misleading one during transient phases. Finally, this provides more information for a subsequent cooperative tracker to make more or less cautious decisions when peers are only partly trustworthy (because their information contains both valid and invalid values).

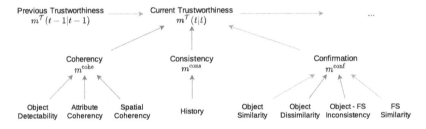

Fig. 4. Evidential network for trustworthiness computation at time t. Red arrows only convey untrustworthiness, green trustworthiness, and orange both. (Color figure online)

Trustworthiness is sequentially estimated using an evidential network. Similar to state filtering, the current estimate at time t $m_j^T(t|t)$ is derived as the combination of the previous estimate $m_j^T(t-1|t-1)$ and new evidence about "Coherency", "Consistency" and "Confirmation" as illustrated in Fig. 4. They are respectively denoted m_j^{cohe}, m_j^{cons} and m_j^{conf}, defined on 2^{Ω^T} and group leaves of the network as described later on. Those leaves express simple and non-dogmatic constraints on either trustworthiness or non-trustworthiness based on different aspects of the received information.

In the combination process, every term is discounted by an associated factor that is not be explicitly noted here for the sake of clarity. Therefore, leaves always express some degree of belief on $m(\Omega^T)$. Dempster's rule $\oplus$ is well adapted in this case and is used for combination:

$$m_j^T(t|t) = {}_{\Lambda_{\Delta t}} m_j^T(t-1|t-1) \oplus m_j^{\text{cohe}} \oplus m_j^{\text{cons}} \oplus m_j^{\text{conf}} \qquad (1)$$

where $\Lambda_{\Delta t}$ is a discounting factor that depends on the elapsed time Δt, moving an $\Lambda_{\Delta t}$-proportion of every focal set to the unknown [9].

4.1 Coherency

m^{cohe} models that the information contained in a message has to be coherent within itself. Multiple constraints are combined using Dempster's rule, three of them are given here as an example:

$$m_j^{\text{cohe}} = m_j^{\text{obd}} \oplus m_j^{\text{atc}} \oplus m_j^{\text{spc}} \qquad (2)$$

m_j^{obd} expresses that objects cannot exist inside the FS or outside the perception range of the peer. For this, the detectability measure $^j m^{\mathcal{D}}$ of the sending peer j is used. For example, an object that is in the FS is by definition undetectable and its detectability will be low. Similarly objects outside the FOV are unknown and will have a low detectability. We use a constant $D^{\min}$ threshold to assign a mass on the untrustworthiness parametrized with a constant β^{pen} for such objects:

$$m_j^{\text{obd}} = \bigoplus_{\substack{o \in O_j \\ {}^j m_o^{\mathcal{D}}(D) < D^{\min}}} \begin{bmatrix} 0 & \beta^{\text{pen}} & 1 - \beta^{\text{pen}} \end{bmatrix} \qquad (3)$$

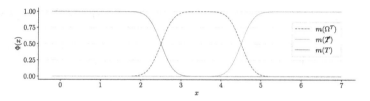

Fig. 5. Sigmoid function $\Phi(x)$ with parameters $\mu = 2$, $\sigma = 0.25$ and $\delta = 2$ producing an arbitrary mass function m.

m_j^{atc} expresses that object attributes have to be likely. For example, the speed v_o of the object o has to be coherent with a normal behaviour which is handled by a scalar sigmoid function Φ illustrated in Fig. 5 with parameters μ^v, σ^v and δ^v chosen to reflect speed limits:

$$m_j^{\text{atc}} = \bigoplus_{o \in O_j} \Phi(v_o, \mu^v, \sigma^v, \delta^v) \qquad (4)$$

with

$$\Phi(x, \mu, \sigma, \delta) = \left[m_1 = \frac{1 - \text{CDF}\left(\frac{x - \mu - 2\sigma}{\sigma\sqrt{2}}\right)}{2} \quad m_2 = \frac{1 + \text{CDF}\left(\frac{x - \mu - 2\sigma - \delta}{\sigma\sqrt{2}}\right)}{2} \quad 1 - m_1 - m_2 \right] \qquad (5)$$

m_j^{spc} expresses that objects have to be spatially coherent. For example, cars should be close to the road network. Again a scalar metric is computed and used as input to a sigmoid function Φ. For instance, let C_j be the subset of O_j classified as cars and d_o the distance between a car o and the road. With adapted parameters μ^d, σ^d and δ^d, the mass is:

$$m_j^{\text{spc}} = \bigoplus_{o \in C_j} \Phi(d_o, \mu^d, \sigma^d, \delta^d) \qquad (6)$$

Please note that other constraints can be added following the same formalism, such as modeling that object sizes or covariances should be of reasonable values.

4.2 Consistency

m^{cons} models that objects must follow coherent trajectories in time and not change their dynamics in an unpredictable way (by making improbable position jumps between two messages for instance). For this, previously received objects are predicted $O_j(t|t-1)$ and associated with newly received objects O_j using an assignment function noted A. The similarity function described in [13] is used to compare objects and is noted $m_{a,b}^{\mathcal{S}}$. It compares different characteristics of the objects and yields a mass function defined on $\Omega^{\mathcal{S}} = \{S, \mathcal{S}\}$ to express that a and b can correspond to the same physical object or to two different ones:

$$m_{Oj}^{\mathcal{S}} = \bigoplus_{o(t-1), o \in A(O_j(t|t-1), O_j)} m_{o(t-1), o}^{\mathcal{S}}$$
$$m_j^{\text{cons}} = \left[0 \quad m_{Oj}^{\mathcal{S}}(\{\mathcal{S}\}) \quad 1 - m_{Oj}^{\mathcal{S}}(\{\mathcal{S}\}) \right] \qquad (7)$$

Thus m_j^{cons} expresses untrustworthiness when objects mismatch with their past but is vacuous otherwise.

As we have seen, the coherency and consistency constraints (m_j^{cohe} and m_j^{cons}) can only express untrustworthiness. Therefore, other criteria have to be used to allow trustworthiness to increase.

4.3   Confirmation Through Free Space and Objects

m^{conf} models that received objects and FS should match with the current knowledge of the world of the receiving peer. For this, the detectability $m^{\mathcal{D}}$ of the receiving and sending peers are used to represent that peers have different FOVs and that comparisons cannot be made on non-overlapping areas.

$$m_j^{\text{conf}} = m_j^{\text{osi}} \oplus m_j^{\text{odi}} \oplus m_j^{\text{ofi}} \oplus m_j^{\text{fsi}} \tag{8}$$

m_j^{osi} and m_j^{fsi} model that trustworthy information should match with the local one. The received FS is compared using the method of [11] in m_j^{fsi}. Received objects O_j are matched using an assignment function noted A and compared using the similarity function $m^{\mathcal{S}}$ defined in [13]. The local object detectability grid $m^{\mathcal{D}}$ is used as a discounting factor to only compare objects that are locally detectable:

$$m_{Oj}^{\text{osi}} = \bigoplus_{o,o_j \in \mathcal{A}(O,O_j)} \left(1 - m_{o_j}^{\mathcal{D}}(D)\right) m_{o,o_j}^{\mathcal{S}}$$
$$m_j^{\text{osi}} = \begin{bmatrix} m_{Oj}^{\text{osi}}(S) & 0 & 1 - m_{Oj}^{\text{osi}}(S) \end{bmatrix} \tag{9}$$

Conversely, m_j^{odi} models that received objects must not mis-match local ones O. For this, the j-detectability of objects not matched with the assignment function A is used:

$$m_{Oj}^{\text{odi}} = \bigoplus_{o \in \mathcal{A}(O,O_j)} \left(1 - m_o^{\mathcal{D}}(D)\right)^j m_o^{\mathcal{D}}$$
$$m_j^{\text{odi}} = \begin{bmatrix} 0 & m_{Oj}^{\text{odi}}(D) & 1 - m_{Oj}^{\text{odi}}(D) \end{bmatrix} \tag{10}$$

Similarly m_j^{ofi} models that the received objects O_j must not be inconsistent with the free space FS estimated locally:

$$m_j^{\text{ofi}} = \bigoplus_{o_j \in O_j} \begin{bmatrix} 0 & m_{oj}^{\mathcal{D}}(\bar{D}) & 1 - m_{oj}^{\mathcal{D}}(\bar{D}) \end{bmatrix} \tag{11}$$

5   Results

In order to illustrate and validate our approach, we implemented the equations detailed in Sect. 4, first in a simple situation in Sect. 5.1 then on real data in Sect. 5.2.

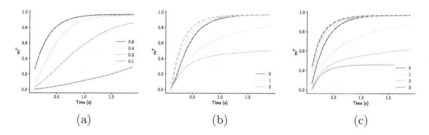

Fig. 6. Simulation results: Trustworthiness between two vehicles with different levels of object detectability in (a), number of ghost objects in (b) and number of objects with incoherent sizes in (c). Continuous lines are $m(T)$ and dashed lines are $m(T) + m(\mathcal{T})$.

5.1 Simulation Study

In this section, trustworthiness estimates are obtained by running simulations implementing Fig. 2 on the situation of Fig. 1 for 2 seconds with varying parameters. Curves of Fig. 6 correspond to the trustworthiness a vehicle attributed to another one under different conditions. Simulation results about a particular parameter are plotted on top of each other to compare its impact, while others remain unchanged and optimal.

One can see in Fig. 6a that the object detectability value plays a major role. When it is low, trustworthiness converges more slowly, which is a desired behavior. In Fig. 6b, the presence of objects that do not exist creates untrustworthiness while matched objects creates trustworthiness. The same can be seen in Fig. 6c, where erroneous sizes generate untrustworthiness.

5.2 Experimental Results

To validate our approach, it has been applied to real-world data using the same dataset as in [10]. In it, three vehicles v_1, v_2 and v_3 were driven in an busy roundabout with v_1 stopped at one of the roundabout entrance while v_2 and v_3 followed each other inside of it. In post-processing, LiDAR point clouds and RTK GNSS receivers have been processed to generate object lists and FSs. The different parameters (e.g. discounting factors) have been tuned on some preliminaries tests to get smooth trust variations. Figure 7 shows the trust estimated by the three vehicles in each other over the course of 22 seconds. At the beginning, trustworthiness in the others is completely unknown. Exchanged information is faithful up to time $t = 12$ s when v_3 starts sending erroneous information (incoherent sizes, omitted and ghosts objects) then stops at time $t = 16$ s. In this case, v_3 is voluntary lying to v_1 and v_2 but its internal information remains correct.

Note that v_2 and v_3 always share perceived areas, but only do with v_1 from $t = 6$ to $t = 12$ s. As a result, v_1 is uncertain in the trustworthiness of v_2 and v_3 and reciprocally when they do not share objects, while v_2 and v_3 trust each other rapidly. A transient phase can be observed when v_1 starts perceiving common areas with v_2 and v_3. At first, only untrustworthiness is expressed because small

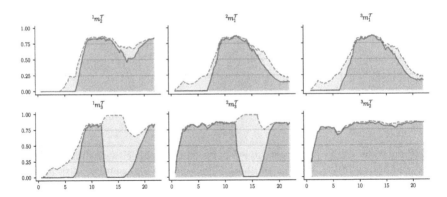

Fig. 7. Real data: Estimated trustworthiness of v_1, v_2 and v_3 in each other, where $^i m_j^T$ denotes the trust of i in j. The green curve is $m(T)$ and the red curve $m(T) + m(\mathcal{T})$.

inconsistencies between objects at the boundary of their FOVs accumulate without enough positive information to counteract. Once enough FS and objects are shared, trustworthiness can increase.

When v_3 sends erroneous information, one can see on $^1 m_3^T$ and $^2 m_3^T$ that it is detected by v_1 and v_2, with untrustworthiness being rapidly estimated and maintained. Once v_3 stops sending erroneous information at $t = 16$ s, v_1 and v_2 increase their trustworthiness in it.

These results illustrate well that to have trustworthiness in another peer, overlapping and coherent objects and free space are necessary. In our opinion, the laws of physics of the real world are the most reliable way to induce trust. Finally, by comparing $^2 m_1^T$ with $^3 m_1^T$ and $^1 m_3^T$ with $^2 m_3^T$, one can see that the trustworthiness estimated about a particular peer differs from the other due to different points of view. This is another consequence of our choice to manage trust in a decentralized way.

In terms of computation performance, the trustworthiness estimation can take up to 500 ms per iteration with our current implementation in Python/C++. As such, it cannot run in real time as communications are at 10 Hz. However, this is not necessarily an issue as this process can be run asynchronously at lower frequencies.

6 Conclusion

In this paper, we have proposed a method to estimate trustworthiness in other peers in the context of decentralized cooperative perception. This formulation combines misbehavior detection techniques and positive confirmation thanks to mass functions to express trustworthiness, untrustworthiness or lack of information about another peer. The trust information that peers create in others is personal and never shared. Thus, two peers will have different trusts in a third peer. The convergence of the method has been illustrated in a simple simulation, then confirmed on a real-world situation. It has shown to react quickly to erroneous information to prevent its propagation. In future work, the interaction trust and object estimation will be studied. In addition, a cooperative dataset with a ground truth on the existence of objects will be acquired to illustrate the effectiveness of this formulation and the impact of trustworthiness estimation on the non-propagation of erroneous information.

Acknowledgments. This work was carried out within SIVALab, a shared laboratory between Renault and Heudiasyc (UTC/CNRS).

References

1. Aeberhard, M., Paul, S., Kaempchen, N., Bertram, T.: Object existence probability fusion using dempster-shafer theory in a high-level sensor data fusion architecture. In: IEEE Intelligent Vehicles Symposium (2011)
2. Allig, C., Leinmüller, T., Mittal, P., Wanielik, G.: Trustworthiness estimation of entities within collective perception. In: IEEE Vehicular Networking Conference (2019)
3. Ambrosin, M., Yang, L.L., Liu, X., Sastry, M.R., Alvarez, I.J.: Design of a misbehavior detection system for objects based shared perception V2X applications. In: IEEE Intelligent Transportation Systems Conference (2019)
4. Caillot, A., Ouerghi, S., Vasseur, P., Boutteau, R., Dupuis, Y.: Survey on cooperative perception in an automotive context. In: IEEE Transactions on Intelligent Transportation Systems (2022)
5. Chowdhury, M., Gawande, A., Wang, L.: Secure information sharing among autonomous vehicles in NDN. In: International Conference on IoT Design and Implementation (2017)
6. ETSI: TR 103 562: Analysis of the Collective Perception Service (2019)
7. Günther, H.J., Mennenga, B., Trauer, O., Riebl, R., Wolf, L.: Realizing collective perception in a vehicle. In: IEEE Vehicular Networking Conference (2016)
8. van der Heijden, R.W., Dietzel, S., Leinmüller, T., Kargl, F.: Survey on misbehavior detection in cooperative intelligent transportation systems. In: IEEE Communications Surveys Tutorials (2019)
9. Kurdej, M., Cherfaoui, V.: Conservative, proportional and optimistic contextual discounting in the belief functions theory. In: International Conference on Information Fusion (2013)
10. Lima, A., Bonnifait, P., Cherfaoui, V., Hage, J.A.: Data fusion with split covariance intersection for cooperative perception. In: IEEE International Intelligent Transportation Systems Conference (2021)
11. Liu, X., et al.: MISO- V: misbehavior detection for collective perception services in vehicular communications. In: IEEE Intelligent Vehicles Symposium (2021)
12. Obst, M., Hobert, L., Reisdorf, P.: Multi-sensor data fusion for checking plausibility of V2V communications by vision-based multiple-object tracking. In: IEEE Vehicular Networking Conference (2014)
13. Zoghby, N., Berge-Cherfaoui, V., Denœux, T.: Optimal object association from pairwise evidential mass functions. In: International Conference on Information Fusion (2013)

An Intelligent System for Managing Uncertain Temporal Flood Events

Manel Chehibi[1]([✉])([iD]), Ahlem Ferchichi[1,2], and Imed Riadh Farah[1]

[1] RIADI/ENSI, Campus Universitaire de la Manouba, 2010 Manouba, Tunisia
webmaster@ensi.rnu.tn
[2] University of Hail, Ha'il, Saudi Arabia
info@uoh.edu.sa
https://ensi.rnu.tn/, http://www.uoh.edu.sa/en/pages/default.aspx

Abstract. During the analysis and management of flood events, the issue of time is crucial. In fact, such events require accurate knowledge of the time of their occurrence and their temporal relationships. However, in most cases, temporal information about flood events are uncertain. In this paper, we propose an intelligent system for managing the temporal uncertainty of these events. The modeling of this uncertainty is based on the belief functions theory and Allen's Interval Algebra. In fact, an uncertain flood event is represented using temporal intervals. Each interval is associated with a belief mass. If there is different uncertain information provided from different sources, a reasoning phase is carried out on these data in order to obtain more reliable information.

Keywords: Flood · Uncertainty · Temporal representation and reasoning · Belief functions theory · Allen's interval algebra

1 Introduction

The concept of time is extremely significant in our daily lives. Indeed, we must always be aware of the time, duration, and temporal relationships of events that occur around us. Flood events are among those events for which one must be particularly aware of their date of occurrence. This allows us to analyze these events and thus avoid many risks and damages and prepare well for this type of natural disaster.

However temporal information about flood events are often uncertain. In fact, this type of information is typically obtained through remote sensing or provided by experts. In the first case, coverage during periods of significant flooding, poor sun light, temporal resolution and mixed pixel and image quality are all factors that influence information. In the second case, information are subjective. Both cases explain and provide insight into the causes of information uncertainty. This uncertainty is problematic in terms of time representation and reasoning.

In the literature, several works have been proposed for modeling, reasoning about and managing uncertain temporal information. Dyreson and Snodgrass

[1] proposed a probabilistic approach to representing the indeterminacy of an event's occurrence time. The probability that an indeterminate instant i occurs during chronon a is expressed by a probability mass function. In [2], Dubois and Prade suggest a method for modeling uncertain times based on possibility theory [3,4]. The possibility distribution in this work represents the degree of likelihood that the date a is equals to the instant t. Nagypal and Motik [5] propose a fuzzy set theory-based approach [6] to deal with the subjectivity and uncertainty of historical information. This uncertainty arises from the fact that such data are often graduel. In [7], Dutta uses fuzzy sets to represent uncertain events with the membership function defining the possibility of an event occurring in a given time interval.

The purpose of this article is to manage the uncertainty of temporal information on the one hand and user requests on the other hand. In fact, in this work, we propose a question/answer system that allows the administrator to represent uncertain beliefs about the date of the flood using the theory of belief functions and the users to express their temporal queries in a non-rigorous way using Allen's temporal relations.

This article is organized as follows: Section 2 reviews the basic concepts needed to understand the following sections, Sect. 3 presents our method of modeling and reasoning about uncertain temporal information, Sect. 4 introduces the proposed system for handling uncertain temporal information and user queries before concluding in Sect. 5.

2   Preliminaries

In this section, we give a brief recall on the theory of belief functions (This section is mainly taken from the article [8]), and on the Allen's interval algebra.

2.1   Theory of Belief Functions

The theory of belief functions, also called Dempster-Shafer theory, was first introduced by Dempster [9] and mathematically formalized by Shafer [10]. This theory models imprecise, uncertain and missing data.

In the theory of belief functions, a *frame of discernment*, noted $\Theta = \{H_1, ..., H_N\}$, is a set of N exhaustive and mutually exclusive hypotheses $H_i, 1 \leq i \leq N$ where only one of them is likely to be true.

The *power set*, $2^\Theta = \{A/A \subseteq \Theta\} = \{\emptyset, H_1, ..., H_N, H_1 \cup H_2, ..., \Theta\}$, enumerates 2^N sub-assemblies of Θ. It includes not only hypotheses of Θ, but also, disjunctions of these hypotheses.

The true hypothesis in Θ is unknown; thus, a degree of belief is assessed to subsets of 2^Θ reflecting our degree of faith on the truth of each subset of 2^Θ.

A *basic belief assignment (bba)*, also called *mass function*, is noted m^{Θ} and defined such that:

$$m^{\Theta} : 2^{\Theta} \rightarrow [0,1]$$
$$m^{\Theta}(\emptyset) = 0 \qquad\qquad (1)$$
$$\sum_{A \subseteq \Theta} m(A) = 1$$

The mass $m^{\Theta}(A)$ represents the degree of belief on the truth of $A \in 2^{\Theta}$. When $m^{\Theta}(A) > 0$, A is called *focal element*.

2.2 Allen's Interval Algebra

Allen's interval algebra [12] is a well-known formalism often used in temporal reasoning. This algebra consists of thirteen primitive and mutually exclusive relationships that can be applied between two time intervals $A = [a, a^{'}]$ and $B = [b, b^{'}]$. These relationships are: before (b), meets (m), overlaps (o), starts (s), during (d), finishes (f), after (bi), met by (mi), overlapped by (oi), started by (si), contains (di), finished by (fi) and equals (e). Each of these relations corresponds to a specific order of the four bounds of the two intervals. For example, the statement A overlaps B $(A \text{ o } B)$ corresponds to $(a < b) \wedge (b < a^{'}) \wedge (a^{'} < b^{'})$ (Table 1).

3 Temporal Representation and Reasoning Under Uncertainty

3.1 Modeling Uncertain Temporal Flood Events

In this work, temporal uncertainty means that the flood date is uncertain, which means that various dates are possible. Let X the frame of discernment over which temporal intervals are defined. An interval-based approach is used to represent uncertain flood events. This representation consists of a less certain exterior interval T_{LC} of the flood date and a more certain interior interval T_{MC}. The modeling of the temporal uncertainty is done by means of a mass function m which represents the degree of certainty on each of these two intervals where:

$T_U = \{T_{LC} = [T_{LC1}, T_{LC2}], m(T_{LC}); T_{MC} = [T_{MC1}, T_{MC2}], m(T_{MC})\}$
with $T_{MC} \subseteq T_{LC}$ and $m(T_{LC}) = \bar{m}(T_{MC}) = 1 - m(T_{MC})$

3.2 Temporal Reasoning Under Uncertainty

Similarity of Flood Events. In many applications, in particular those intended for the management of flood risks, the comparison of the similarity of temporal information can help in judgment or decision making. In addition, if two information are similar, merging them will provide more reliable information. Here, we present our approach to determine the similarity of flood events.

Table 1. Allen's temporal relations

Relation	Inverse	Signification	Relations between bounds
A b B	B bi A		b> a'
A m B	B mi A		a'=b
A o B	B oi A		b> a ∧ a'>b ∧ b'>a'
A d B	B di A		a>b ∧ b'>a'
A s B	B si A		a=b ∧ b'>a'
A f B	B fi A		a>b ∧ b'=a'
A e B	B e A		a=b ∧ a'=b'

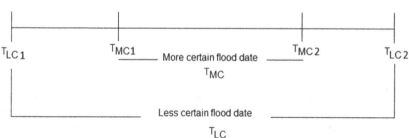

Fig. 1. Temporal uncertainty.

This approach is based on the relationship between the interior or exterior time intervals which each represent temporal information on the flood event.

Let $I1$ and $I2$ be two intervals, and any intersection between them is the interval INT. We base the similarity measure on the relationship between the length of INT and the length of $I1$ and $I2$. We therefore have for $Sim(I1, I2)$

$Sim(I1, I2) = (|INT|/|I1| + |INT|/|I2|)/2$

Since the interior intervals of flood events are more certain, we can rely on them to determine the degree of similarity between events. The exterior intervals can also be taken into account to determine the similarity of events, but as a secondary factor since they are less certain. It should be noted that if we have a strong belief in interior intervals, we can overlook the exterior intervals' similarity and say that E1 and E2 are thought to be quite comparable events.

Aggregation of Events. If the events' similarity is assessed and they appear to be considerably similar, then a combination of events could be considered. As a result, we obtain the aggregated event $T12_U$. The merge operation is performed by applying the operator max on the upper limits of the interior or exterior intervals of the two events and min on the lower limits. The mass of a combined interval is equal to the mass of the first interval multiplied by the mass of the second interval. To satisfy the condition of sum of belief masses, it is necessary to normalize them to be equal to 1. Let $I1 = [I1a, I1b]$ and $I2 = [I2a, I2b]$ be two time intervals, we therefore have for $T12_U$

$T12_U = [min[I1a, I2a],\ max[I1b, I2b]];\ m(T12_I) = norm(\ m(T1_I) * m(T2_I))$

When there is more than one source of information, the similarity measurements are performed for each pair of sources. After that, all the similar sources are grouped together in order to combine the information from these sources. This method is used to avoid quickly obtaining a very large uncertainty after several aggregation steps.

4 Intelligent Query-Answering System

In this section, we propose a system that allows its users to view answers to their flood-related queries.

User queries are often intended to display floods events that have occurred over a certain period of time or rather events such as their flood start dates are included in a given time interval. In general, the system allows the user to visualize flood events according to the nature of the relationship he has chosen with the time interval he has entered. On the other hand, the system allowing its administrators to store and manage flood events which are often of an uncertain nature. Uncertain temporal information on floods events can be provided by several sources.

4.1 System Architecture

The architecture of the proposed system is presented in this section. Figure 2 shows an overview of the system architecture and the different modules needed for the management of uncertain flood events and the processing of user requests It consists of two major parts, each of which contains a set of modules. Each module's functions are summarized below:

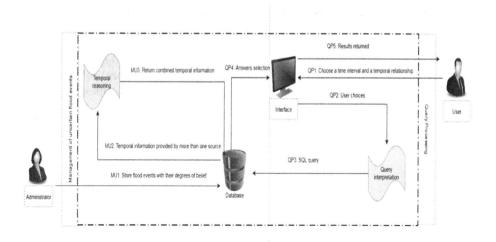

Fig. 2. System architecture.

Query Processing: In the "Interface" module, the user chooses a temporal interval and a temporal relation using a graphical interface. Then, the system transmits the request to the "Query interpretation" module which transforms the request into an SQL query. This request is then sent to the database management system in order to select the records meeting the criteria of the temporal request. The returned results are attached with degrees of belief and then displayed on the user interface.

Management of uncertain flood events: In the "temporal representation" module, the system administrator stores each flood event in the temporal database with their degree of belief as follows: exterior flood start date, exterior flood end date, exterior interval mass, interior flood start date, interior flood end date, interior interval mass. If there is more than one source of information, which is expected, a reasoning phase is necessary, the system calls the "Reasoning" module. This module uses the proposed reasoning mechanisms and calculates the degree of similarity. Depending on this degree a combination operation will be performed or not.

4.2 Illustrative Examples

In our work, we are interested in the events of the floods that occurred in Chad. In fact, floods are a common occurrence in this country. Many areas, particularly those on the shores of Lake Chad or traversed by the Logone or Chari rivers, are subject to flooding, causing displacement, loss of life and destruction of property. Flood risk management requires the availability of temporal information about the flood. It is critical to estimate the date of the flood in order to provide information and recommendations on the best flood-response strategies. This estimate is based on an examination of previous floods in the same area. In addition to the fact that the dates of floods, i.e. temporal information in general, are often uncertain in nature, these estimates are always subject to a certain degree of uncertainty. In this section, we show how our proposed system is very useful for modeling, reasoning and visualizing uncertain temporal information. To make our proposition more clear, we have included some examples below.

Example 1: Let X the frame of discernment over which temporal intervals are defined. Suppose an expert has an uncertain judgment on the date of a flood event E1 that occurred in Chad in 2014. Table 2 presents this uncertain temporal information.

Table 2. The temporal information provided by the source S1

Record	Source	Flood event	More uncertain flood time interval I_1	Mass of I_1	Less uncertain flood time interval I_2	Mass of I_2
1	S1	E1	[29/07/2014– 05/09/2014]	0.41	[14/08/2014– 21/08/2014]	0.59

If a user asks to display flood events that occurred in Chad in 2014, the system will automatically display the record E1.

Example 2: Suppose another source of information provides a different and uncertain judgment on the date of the flood that occurred in Chad in 2014. Table 3 presents this uncertain temporal information.

If a user asks to display flood events that occurred in Chad in 2014, the system must go through a reasoning phase before displaying the response to the user. This passage is mandatory since we have different information from more than one source of information on the same flood event. As indicated in Sect. 3, the temporal reasoning phase consists of two steps: measuring the similarity and combining the information.

In this example, the similarity between the information provided by sources s1 and s2, based on the most certain time intervals is: $Sim(I1, I2) = (|INT|/|I1| + |INT|/|I2|)/2 = (7/8 + 7/9)/2 = (0,88 + 0.78)/2 = 0.83$.

We notice that there is a very great similarity between the two pieces of information. It is then very interesting to carry out a combination operation on these uncertain information. The combined information is:

$T12_U = [min[I1a, I2a], \ max[I1b, I2b]]; \ m(T12_I) = norm(\ m(T1_I) * m(T2_I))$

$T12_U = [min[14/08, 12/08], \ max[21/08, 20/08]]; \ m(T12_I) = norm(0.59 * 0.55) =$

$0.3245 + 0, 2455 = 0.57$

$T12_U = [12/08, \ 20/08]; \ m(T12_I) = 0.57$

Then, the system displays this aggregate information to the user which is more reliable than the two other information provided separately. In fact, this

Table 3. The temporal information provided by the source S2

Record	Source	Flood event	More uncertain flood time interval I_1	Mass of I_1	Less uncertain flood time interval I_2	Mass of I_2
3	S3	E1	[29/07/2014–05/09/2014]	0.45	[12/08/2014–20/08/2014]	0.55

combined information is the closest to reality (see Table 4). Table 4 presents a row of a real database of floods that occurred in Chad between 2012 and 2015.

Table 4. Flood event database

Dates of flood events	Flood end dates	Duration of floods	Extent of floods (km)
13/08/2014	19/08/2014	6	312

This information can be very useful to estimate the future date of the flood and then adopt the best response strategies to possible future floods.

Example 3: Suppose that the second source S2 provides the information presented in Table 5 and not that presented in Table 3.

As in the second example, if a user asks to display flood events that occurred in Chad in 2014, the system must go through a reasoning phase before displaying the response to the user since we have two different information about the same flood event from two different information sources. Except that in this example the system will perform the combination operation on the two pieces of information without going through the similarity measurement step.

Then combined information is:

$T12_U = [min[I1a, I2a], max[I1b, I2b]]; m(T12_I) = norm(m(T1_I) * m(T2_I))$

$T12_I = [min[14/08, 30/07], max[21/08, 02/09]]; m(T12_I) = norm(0.59 * 0.7) = 0,413 + 0.2056 = 0.62$

$T12_I = [30/07, 02/09]; m(T12_I) = 0.62$

We note here that the resulting information is far from reality. We then deduce that by ignoring the step of calculating the similarity, we can have false results. Then, it is very important to go through the measurement of the similarity before proceeding to the combination of information. It is even possible to set the similarity threshold from which this information can be combined.

Table 6 summarizes all these examples about the date of the flood event E1 that occurred in Chad in 2014 and shows that the combination process provides information that is more reliable and closer to reality provided there is a similarity between information.

Table 5. The supposed temporal information provided by the source S2

Record	Source	Flood event	More uncertain flood time interval I_1	Mass of I_1	Less uncertain flood time interval I_2	Mass of I_2
2	S2	E1	[20/07/2014– 10/09/2014]	0.3	[30/07/2014– 2/09/2014]	0.7

Table 6. Summary of illustrative examples results

Uncertain tempoal information $T1_U$	Combination of $T1_U$ with a non-similar information $T2_U$	Combination of $T1_U$ with another similar information $T2_U$	Real date of flood event
$T1_{MC} = [14/08, 21/08]$; $m(T1_{MC}) = 0.59$	$T12_I = [30/07, 02/09]$; $m(T12_I) = 0.62$	$T12_U = [12/08, 20/08]$; $m(T12_I) = 0.57$	$[13/08, 19/08]$

5   Conclusions and Future Work

In this article we propose a system for managing uncertain temporal information. The proposed idea can be applied in several fields such as medical diagnosis, archeology, and so on. In this work, we are interested in the field of natural disasters more specifically, floods. Our method makes it possible to express uncertain temporal knowledge about flood events, store and manage these uncertain information and display it to the user who can then analyze it and draw inferences and predictions about possible flood events. In fact, uncertain information about the date of flood events are represented using belief function theory. In the case of several pieces of information coming from several sources, an operation to measure the similarity of these information provided is carried out. If these information are similar then a combination operation is performed in order to obtain more reliable information. As future work, we plan to apply our approach on real data.

References

1. Dyreson, C., Snodgrass, R.: Supporting valid-time indeterminacy. ACM Trans. Database Syst. **23**(1), 1–57 (1998)
2. Dubois, D., Prade, H.: Processing fuzzy temporal knowledge. IEEE Trans. Syst. Man Cybern. **19**(4), 729–744 (1989)
3. Zadeh, L.: Fuzzy sets as a basis for a theory of possibility. Fuzzy Sets Syst. **1**, 3–28 (1978)
4. Dubois, D., Prade, H.: Théorie des possibilités: applications à la représentation des connaissances en informatique, Collection Méthode+ Programmes, Masson, Paris. 1 MAGE EVALUATION TEST TARGET
5. Nagypál, G., Motik, B.: A fuzzy model for representing uncertain, subjective, and vague temporal knowledge in ontologies. In: Meersman, R., Tari, Z., Schmidt, D.C. (eds.) OTM 2003. LNCS, vol. 2888, pp. 906–923. Springer, Heidelberg (2003). https://doi.org/10.1007/978-3-540-39964-3_57

6. Zadeh, L.A.: Fuzzy sets. Inf. Control **8**, 338–353 (1965)
7. Dutta, S.: An event based fuzzy temporal logic. In: Proceedings of The 18th International Symposium Multi-Valued Logic, Fukuoka, Japan, pp. 545–551 (1998)
8. Chehibi, M., Chebbah, M., Martin, A.: Independence of sources in social networks. In: Medina, J., et al. (eds.) IPMU 2018, Part I. CCIS, vol. 853, pp. 418–428. Springer, Cham (2018). https://doi.org/10.1007/978-3-319-91473-2_36
9. Dempster, A.P.: Upper and lower probabilities induced by a multiple valued mapping. Ann. Math. Stat. **38**, 325–339 (1967)
10. Shafer, G.: A Mathematical Theory of Evidence. Princeton University Press, Princeton (1976)
11. Smets, P.: Decision making in the TBM: the necessity of the pignistic transformation. Int. J. Approx. Reason. **38**, 133–147 (2005)
12. Allen, J.F.: Maintaining knowledge about temporal intervals. Commun. ACM **26**(11), 832–843 (1983)

Statistical Inference - Graphical Models

A Practical Strategy for Valid Partial Prior-Dependent Possibilistic Inference

Dominik Hose[1], Michael Hanss[1], and Ryan Martin[2(✉)]

[1] Institute of Engineering and Computational Mechanics, University of Stuttgart,
Pfaffenwaldring 9, 70569 Stuttgart, Germany
{dominik.hose,michael.hanss}@itm.uni-stuttgart.de
[2] Department of Statistics, North Carolina State University, Raleigh, NC 27695, USA
rgmarti3@ncsu.edu

Abstract. This paper considers statistical inference in contexts where only incomplete prior information is available. We develop a practical construction of a suitably valid inferential model (IM) that (a) takes the form of a possibility measure, and (b) depends mainly on the likelihood and partial prior. We also propose a general computational algorithm through which the proposed IM can be evaluated in applications.

Keywords: Bayesian · Choquet integral · Credal set · Fisher · Frequentist · Inferential model · Likelihood · Possibility theory

1 Introduction

Many—including Fisher, Popper, etc.—would argue that *inductive inference* is not a domain of application suited for ordinary/precise probability. In fact, none of the "standard" statistical methods used day-to-day by practitioners making inductive inferences involve direct probability statements about the unknowns. For example, on tests of significance, Fisher (1973, p. 63) writes

> [probability and p-values] *should none the less be distinguished, for valid tests of significance at all levels may exist without the possibility of deducing by an accurate argument, a probability distribution for the unknown parameter.*

If it is not the familiar probabilistic reasoning behind those classical statistical methods, then what is it? Martin (2021) gave a possibility-theoretic characterization, showing that behind every statistical method having exact frequentist error rate control lies a so-called *inferential model* (IM) that is *valid* (see below) and takes the mathematical form of a possibility measure. Therefore, the reasoning behind those classical statistical methods must be possibilistic.

D. Hose—Partially supported by the Deutsche Forschungsgemeinschaft (DFG) project no. 319924547 (HA2798/9-2).
M. Hanss—Partially supported by the U.S. National Science Foundation, SES–205122.

S. Le Hégarat-Mascle et al. (Eds.): BELIEF 2022, LNAI 13506, pp. 197–206, 2022.
https://doi.org/10.1007/978-3-031-17801-6_19

Towards advancing this understanding that possibility theory plays a fundamental role in statistical inference, there is one key observation. The classical/frequentist setting should not be interpreted as one in which there is "no prior". Instead, the vacuous prior information state means that no prior distribution can be ruled out. In other words, the frequentist's credal set consists of *every prior*. If the frequentist extreme is "every prior" and the Bayesian extreme is "one prior", then this suggests a spectrum of different scenarios indexed by the extent of *partial prior information* available to the data analyst. This concrete understanding of frequentist and Bayesian as opposite ends of a spectrum provides an opportunity for meaningful unification between the two schools as well as for the development of new and improved methods. Moreover, the fact that this spectrum is indexed by partial beliefs highlights how statistical inference fits squarely in the world of imprecise probability.

Martin (2022) formulates this notion of partial priors using credal sets, $\mathscr{Q}$, i.e., sets of prior probability distributions compatible with available prior information, and develops a notion of validity relative to $\mathscr{Q}$ which is crucial to the Fisherian logic of statistical inference. He also discusses several approaches to constructing a valid IM in this context. In Sect. 4 we build on these developments in a different direction. We propose a general construction that employs an *imprecise probability-to-possibility transform* (Hose and Hanss 2021) that combines the likelihood function and partial prior information—encoded as a possibility measure—to get an IM that is a possibility measure and provably valid. We also present a general, Monte Carlo strategy to implement this new partial-prior IM framework in practical applications. One illustration is presented in Sect. 5 and some concluding remarks are given in Sect. 6.

2   Background

Here we will assume that the reader is familiar with the basics of possibility theory, e.g., that possibility measures are consonant plausibility functions and their numerical values assigned to general assertions is determined by maximizing an associated possibility contour function over a relevant set.

Suppose that observable data Y is available from a statistical model $\mathsf{P}_{Y|\theta}$ indexed by an unknown parameter $\theta \in \mathbb{T}$. The classical situation, in which there is insufficient information available to rule out any specific prior distribution, has been treated recently by the authors—see, e.g., Martin and Liu (2015), Martin (2019) and Hose (2022)—in the context of possibility theory. In particular, for inference on θ in light of an observation $Y = y$ from the posited statistical model, they construct a so-called *inferential model* (IM), which is a y-dependent lower and upper probability pair, denoted by $(\underline{\varPi}_y, \overline{\varPi}_y)$. The constructions vary, e.g., Martin and Liu (2015) employ nested random sets to get it while Liu and Martin (2021) directly use possibility contours, but the end result is an IM whose mathematical form is that of a possibility measure determined by a contour function π_y. That is, the IM's upper probability, which is a possibility measure in this case, are determined by π_y and the relationship

$$\overline{\varPi}_y(A) = \sup_{\vartheta \in \mathbb{T}} \pi_y(\vartheta), \quad A \subseteq \mathbb{T}.$$

The lower probability, a necessity measure, is given by $\underline{\varPi}_y(A) = 1 - \overline{\varPi}_y(A^c)$.

There is nothing inherent about the use of possibility measures or imprecise probabilities more generally that makes the IM output meaningful; its meaningfulness is determined by the statistical properties it satisfies. Indeed, through the specifics of its construction, the IM satisfies a so-called *validity property* (Martin and Liu 2013, 2015) which states that

$$\sup_{\theta \in A} \mathsf{P}_{Y|\theta}\{\overline{\varPi}_Y(A) \le \alpha\} \le \alpha, \quad \text{for all } A \subseteq \mathbb{T} \text{ and all } \alpha \in [0,1]. \tag{1}$$

Intuitively, an IM would not be useful for quantification of uncertainty if its upper probability tended to be relatively small for assertions about θ that are true. The validity property in (1) rules this out, and justifies interpreting $\underline{\varPi}_y(A)$ and $\overline{\varPi}_y(A)$ as measures of the degrees of belief and plausibility, respectively, in the truthfulness of the assertion "$\theta \in A$" based on the observation $Y = y$:

> *The force with which such a conclusion is supported is logically that of the simple disjunction: Either an exceptionally rare chance has occurred, or* [the assertion] *is not true* (Fisher 1973, p. 42).

There is an equivalent and more economical definition of validity ideally suited to IMs that take the form of possibility measures. An IM determined by the possibility contour π_y is *valid* if

$$\sup_{\theta} \mathsf{P}_{Y|\theta}\{\pi_Y(\theta) \le \alpha\} \le \alpha, \quad \text{for all } \alpha \in [0,1]. \tag{2}$$

An immediate and practically relevant conclusion is that, for any $\alpha \in [0,1]$, the IM's possibility contour can be used to construct a $100(1-\alpha)\%$ plausibility region, $C_\alpha(y) = \{\vartheta : \pi_y(\vartheta) > \alpha\}$, and that this is a nominal $100(1-\alpha)\%$ confidence set in the classical/frequentist sense, i.e.,

$$\sup_{\theta} \mathsf{P}_{Y|\theta}\{C_\alpha(Y) \not\ni \theta\} \le \alpha, \tag{3}$$

or, in words, the coverage probability of $C_\alpha(Y)$ is at least $1 - \alpha$.

3 Valid Inference Under Partial Priors

3.1 Partial Priors

Now consider a more general situation where, instead of prior information being vacuous, suppose we do know something. Here, the "something" is what we refer to as *partial prior information*. For example, perhaps a reliable expert says "I'm 50% confident that the unknown parameter is in the interval $[a, b]$". How should such information be used? Frequentists nor Bayesians have a natural way to treat this kind of partial information—they will typically ignore it or elaborate on it to the extent that it becomes unbelievable. However, thanks to our broader view on probability, this is relatively straightforward for us here.

Martin (2022) proposed to encode this partial prior information via a credal set $\mathscr{Q}$ of probability distribution supported on $\mathbb{T}$. Generic members of the credal set, denoted by Q, can be interpreted as Bayesian priors for the unknown parameter. Since we will have occasion to think of θ as having a prior distribution, we will use the notation Θ when it is to be understood as a random variable. The credal set provides partial prior information about θ in the sense that, for each assertion $A \subseteq \mathbb{T}$, there is a (possibly trivial) range of probabilities, from $\underline{Q}(A) = \inf\{Q(A) : Q \in \mathscr{Q}\}$ to $\overline{Q}(A) = \sup\{Q(A) : Q \in \mathscr{Q}\}$.

In Sect. 4, we will focus on cases where $\overline{Q}$ is a possibility measure on $\mathbb{T}$. Then it is completely determined by its contour function $q : \mathbb{T} \to [0, 1]$, via

$$\overline{Q}(A) = \sup_{\vartheta \in A} q(\vartheta), \quad A \subseteq \mathbb{T}.$$

This covers the vacuous prior case, with $q(\vartheta) \equiv 1$, along with many others; in fact, possibility measures are sufficiently general to cover all the statistical applications we have considered so far. There are credal sets whose upper envelopes are not possibility measures, but we will discuss these elsewhere.

3.2 Validity and Its Consequences

Let $\overline{P}_{\mathscr{Q}}$ denote the upper envelope for the collection of joint distributions for (Y, Θ) induced by the credal set $\mathscr{Q}$ and the statistical model $P_{Y|\theta}$. That is,

$$\overline{P}_{\mathscr{Q}}(Y \in B, \Theta \in A) = \sup_{Q \in \mathscr{Q}} P_{Y,\Theta|Q}(Y \in B, \Theta \in A)$$

$$= \sup_{Q \in \mathscr{Q}} \int_A P_{Y|\theta}(B)\, Q(d\theta), \quad A \subseteq \mathbb{T}, \ B \subseteq \mathbb{Y}.$$

For a partial prior-dependent IM with upper probability $\overline{\Pi}_{y,\mathscr{Q}}$, Martin defined *validity with respect to $\mathscr{Q}$* to mean that

$$\overline{P}_{\mathscr{Q}}\{\overline{\Pi}_Y(A) \le \alpha, \Theta \in A\} \le \alpha, \quad \text{all } \alpha \in [0, 1] \text{ and all } A \subseteq \mathbb{T}. \tag{4}$$

This generalizes (1) and the rationale—Fisher's "disjunction"—is the same here as above. There is an analogue to (2) when, as we consider here, the partial-prior IM's output is a possibility measure with contour function $\pi_{y,\mathscr{Q}}$. Indeed, Martin (2022) defines the IM to be *strongly valid with respect to $\mathscr{Q}$* if

$$\overline{P}_{\mathscr{Q}}\{\pi_{Y,\mathscr{Q}}(\Theta) \le \alpha\} \le \alpha, \quad \text{for all } \alpha \in [0, 1]. \tag{5}$$

While (1) and (2) are equivalent in the vacuous-prior case, strong validity in (5) is genuinely stronger than validity in (4). This is not immediate from the formulas, but see Definition 3 and the subsequent lemma in Martin (2022).

Martin goes on to establish several relevant consequences of (strong) validity. On the statistical side, Martin's main result is that, if the partial-prior IM is strongly valid, then the set $C_\alpha(y; \mathscr{Q}) = \{\vartheta : \pi_{y,\mathscr{Q}}(\vartheta) > \alpha\}$ is a $100(1 - \alpha)\%$ confidence set in the sense that $\overline{\mathsf{P}}_\mathscr{Q}\{C_\alpha(Y; \mathscr{Q}) \not\ni \Theta\} \leq \alpha$, for $\alpha \in [0,1]$. This generalizes (3) to coverage probability guarantees under cases where the prior information is not necessarily vacuous.

3.3   How to Achieve (Strong) Validity

Martin (2022) presents a few different ways partial prior validity can be achieved. For brevity, here we will focus on just one that is directly relevant for what we propose in Sect. 4 below; this is the strategy Martin referred to as *validifica-tion*. The basic idea is to define a suitable "plausibility ordering", i.e., a data-dependent ranking of the candidate parameter values, and then adjust this in a suitable way, if necessary, so that strong validity is achieved. More concretely, let $(y, \vartheta) \mapsto h_y(\vartheta)$ be a map from $\mathbb{Y} \times \mathbb{T}$ to, say, $[0,1]$, such that for each y, there exists θ_y such that $h_y(\theta_y) = 1$. The interpretation is that, if $h_y(\vartheta_1) < h_y(\vartheta_2)$, then ϑ_2 is more compatible with observation y than ϑ_1. Of course, whether this ranking is real-world meaningful depends on the choice of h, and we present our recommendation—and justification—in Sect. 4.1 below.

Once h is specified, the validification strategy is conceptually straightforward. Indeed, inspired by the *imprecise-probability-to-possibility transform* (Hose 2022; Hose and Hanss 2021), we define a data-dependent possibility contour as

$$\pi_{y,\mathscr{Q}}(\vartheta) = \overline{\mathsf{P}}_\mathscr{Q}\{h_Y(\Theta) \leq h_y(\vartheta)\}. \tag{6}$$

We claim that the interpretation of the partial-prior IM determined by $\overline{\mathit{\Pi}}_{y,\mathscr{Q}}$ is the same as that determined by $\overline{\mathit{\Pi}}_y$ in Sect. 2. For this interpretation to hold, however, in the sense of Fisher's logic of a simple disjunction, we need calibration. It turns out that this partial-prior IM is provably strongly valid, thereby supporting both its statistical and behavioral interpretation.

Theorem 1. *The partial-prior IM with contour function* (6) *is strongly valid relative to $\mathscr{Q}$ in the sense of* (5).

The proof in Martin (2022) is straightforward, but we will not reproduce it here. Instead, we say a few words about what is actually happening. The h function determines a plausibility ordering, but that particular possibility contour may not determine a strongly valid partial-prior IM because the level sets are not of the right size. Then the validification transformation in (6) simply modifies the level sets of h to make them the correct size to achieve strong validity.

4   Practical IM Construction

4.1   Likelihood-Based Contour

As indicated above, we will focus here on cases where the prior credal set's upper probability, $\overline{\mathsf{Q}}$, is a possibility measure determined by its corresponding possibility contour function q. Next, let L_y denote the likelihood function for θ determined by the data $Y = y$ and the posited statistical model. Recall that $\vartheta \mapsto L_y(\vartheta)$ is a minimal sufficient statistic for the posited model so, in a certain sense, this is the most economical summary of the data. Our proposal here is to combine the prior contour q and the likelihood function L_y—in a straightforward and almost-familiar way—to create an h-contour function as in Sect. 3.3 for which strong validity relative to $\mathscr{Q}$ can be achieved according to Theorem 1.

Our specific proposal is to define the h-contour as

$$h_y(\vartheta) = \frac{L_y(\vartheta)\,q(\vartheta)}{\sup_{t \in \mathbb{T}}\{L_y(t)\,q(t)\}}, \quad t \in \mathbb{T}. \tag{7}$$

This has a strong similarity to the familiar Bayesian "likelihood times prior" updating/combination operation, the key difference is in the normalization—the h above has unit maximum rather than unit integral. For another perspective, write $r_y(\vartheta) = L_y(\vartheta)/\sup_t L_y(t)$ for the *relative likelihood* (e.g., Denoeux 2014). If it is not trivially 0, then the h function in (7) is numerically equivalent to that with L_y replaced by r_y. After making that replacement, (7) resembles one of those combination rules commonly used in the possibility theory literature, e.g., Dubois and Prade (1988), motivated by fuzzy set intersection. While the interpretation of h is important, it is the statistical properties of the IM in (6), i.e., strong validity, that make it useful for statistical inference.

To us, what motivates this likelihood-driven contour h in (7) is that it is simple, it incorporates all the information in the data via the likelihood, and that it allows the prior contour to perform its main function: regularization. Consequently, the partial-prior IM contour $\pi_{y,\mathscr{Q}}$ in (6) depends on the two relevant pieces of evidence about θ: the (data, model) pair and the (partial) prior. We also have some ideas for why this choice is "best" in a certain sense, or at least nearly so, but those details will be reported elsewhere. In any case, while the choice of h is important, it is the strong validity property we get for (6) in Theorem 1 that makes it useful for basing statistical inference on.

4.2   Computation

Computation of the expression in (6) is non-trivial. At first glance, it appears to require optimization over $\mathscr{Q}$, which could easily be intractable. When the upper envelope $\overline{\mathsf{Q}}$ is a possibility measure, however, some crucial simplifications can be made, allowing for efficient numerical evaluation. This section describes these simplifications and puts forth a general computational strategy.

From a broader perspective, (6) is a Choquet integral. But since $\mathscr{Q}$ is determined by a possibility measure, this Choquet integral can be greatly simplified. Indeed, if we write $\mathcal{I}_\eta(t) = \mathsf{P}_{Y|t}\{h_Y(t) \leq \eta\}$ for the distribution function of the random variable $h_Y(t)$ under the model $Y \sim \mathsf{P}_{Y|t}$ for $t \in \mathbb{T}$, then it follows from Proposition 7.14 in Troffaes and de Cooman (2014) that the Choquet integral becomes an ordinary Riemann integral of a supremum:

$$\pi_{y,\mathscr{Q}}(\vartheta) = \int_0^1 \sup_{t \in \mathbb{T}\,:\,q(t) > \alpha} \mathcal{I}_{h_y(\vartheta)}(t)\,\mathrm{d}\alpha. \tag{8}$$

Again, this simplified expression of the Choquet integral holds only for $\mathscr{Q}$ determined by a possibility measure with contour function q.

The inner expression $\mathcal{I}_\eta(t)$ is a simple probability integral which, depending on the statistical model, can be evaluated in closed-form or, more realistically, with efficient Monte Carlo algorithms. In general, there is a vanilla Monte-Carlo quadrature rule (Hose 2022, Ch. 5) that can be followed, which gives

$$\mathcal{I}_\eta(t) \approx J^{-1} \sum_{j=1}^{J} 1\{h_{Y^{(j)}}(t) \leq \eta\},$$

where $1\{\cdot\}$ denotes the indicator function and $\{Y^{(j)} : j = 1, \ldots, J\}$ are independent and identically distributed samples drawn from $\mathsf{P}_{Y|t}$. Since h only depends on data through a minimal sufficient statistic, the $Y^{(j)}$'s can be replaced by draws of that minimal sufficient statistic if those are readily available.

Evaluating the outer expression in (8) is slightly more involved but can be achieved numerically by replacing the (typically continuous) prior possibility contour q by a discretized version $\tilde{q}$. This approximation is non-zero only at a finite number of parameter candidates $\vartheta_1, \ldots, \vartheta_M \in \mathbb{T}$, where it coincides with the values $q_m = q(\vartheta_m)$, $m = 1, \ldots, M$, respectively, and is zero everywhere else.

One might think of these tuples $(\vartheta_m, q_m)_{m=1}^{M}$ as samples of Θ with imprecise weights. If we had a precise prior Q, then one of the most basic ways to numerically represent it would be via the samples $\Theta^{(1)}, \ldots, \Theta^{(M)} \sim \mathsf{Q}$, or, more generally, via a set of weighted samples $(\vartheta_m, w_m)_{m=1}^{M}$, such that the prior probability of $\Theta \in A$ for $A \subseteq \mathbb{T}$ is approximated via $\mathsf{Q}(A) \approx \sum_{m=1}^{M} w_m 1\{\vartheta_m \in A\}$. The advantage of the latter is that $\vartheta_1, \ldots, \vartheta_m \in \mathbb{T}$ can be drawn according to some other sampling procedure, such as Latin hypercube sampling, and need not be sampled from Q. The challenge is in finding the appropriate probability weights $(w_1, \ldots, w_M)$ in the simplex Δ^M. In our imprecise prior setting, the prior Q is to be contained in the credal set of the possibility contour q, more precisely in that of its approximation $\tilde{q}$. A necessary and sufficient criterion for this credal set membership is given by the system of affine constraints

$$\alpha \geq \mathsf{Q}\{\tilde{q}(\Theta) \leq \alpha\} = \sum_{m=1}^{M} w_m 1\{q_m \leq \alpha\}, \quad \text{for all } \alpha \in \{q_1, \ldots, q_M\} \qquad (9)$$

on these probability weights. That is, we approximate the credal set $\mathscr{Q}$ of q by the credal set $\tilde{\mathscr{Q}}$ of $\tilde{q}$, which is simply a set of weighted samples

$$\tilde{\mathscr{Q}} \equiv \{(\vartheta_m, w_m)_{m=1}^{M} : (w_1, \ldots, w_M) \in \mathcal{W}\},$$

where $\mathcal{W} = \{(w_1, \ldots, w_M) \in \Delta^M : (w_1, \ldots, w_M) \text{ satisfies (9)}\}$ is a convex set. Therefore, we call $(\vartheta_m, q_m)_{m=1}^{M}$ imprecisely weighted samples.

Without loss of generality, we now assume that $\vartheta_1, \ldots, \vartheta_M$ are ordered such that their prior possibilities are strictly increasing, $q_1 \leq \ldots \leq q_M = 1$, and we formally define $q_0 = 0$. Replacing q by $\tilde{q}$ in the right-hand side of (8) produces

$$\int_0^1 \sup_{t \in \mathbb{T}: q(t) > \alpha} \mathcal{I}_{h_y(\vartheta)}(t) \, d\alpha \approx \sum_{m=1}^M (q_m - q_{m-1}) \cdot \max_{k=m,\dots,M} \mathcal{I}_{h_y(\vartheta)}(\vartheta_k). \qquad (10)$$

The latter expression is straightforward to evaluate numerically by following the steps outlined below.

1. Sample $\vartheta_1, \dots, \vartheta_{M-1}$ on $\mathbb{T}$, e.g., via uniform or Latin hypercube sampling.
2. Find $\vartheta_M = \arg \sup_{\vartheta \in \mathbb{T}} q(\vartheta)$, such that $q(\vartheta_M) = 1$.
3. For all $m = 1, \dots, M$,
   (a) compute $q_m = q(\vartheta_m)$ and rearrange the indices such that $q_1 \le \dots \le q_M$,
   (b) draw J iid samples $Y^{(m,1)}, \dots, Y^{(m,J)} \overset{\text{iid}}{\sim} \mathsf{P}_{Y|\vartheta_m}$, and
   (c) evaluate $\mathcal{I}_m = J^{-1} \sum_{j=1}^J 1\{h_{Y^{(m,j)}}(\vartheta_m) \le h_y(\vartheta)\}$.
4. Approximate the valid IM's possibility contour by

$$\pi_{y,\mathscr{Q}}(\vartheta) \approx \sum_{m=1}^M (q_m - q_{m-1}) \cdot \max_{k=m,\dots,M} \mathcal{I}_k,$$

where $q_0 = 0$. This will, typically, underestimate $\pi_{y,\mathscr{Q}}(\vartheta)$ slightly.

5   Illustration

A practically important problem is that of comparing the means of two normal populations. Suppose that $X_{j,1}, \dots, X_{j,n_j}$ are iid $\mathsf{N}(\mu_j, \sigma_j^2)$ for $j = 1, 2$. To keep things relatively simple at this point, we assume that the variances, σ_1^2 and σ_2^2, are known; the only unknown parameters are the means, μ_1 and μ_2. The quantity of interest is $\theta = \mu_2 - \mu_1$, the difference in the populations means. This would be relevant if, for example, μ_j represents the mean weight loss of patients on diet j, so that $\theta = \mu_2 - \mu_1$ represents the *diet* or, more generally, the *treatment effect*. Our goal is to construct a suitably valid IM for inference on the treatment effect θ. Of course, under the stated assumptions, the model can be simplified considerably. Indeed, if $\bar{X}_j$ denote the sample mean from group j, then $Y = \bar{X}_2 - \bar{X}_2$ is a minimal sufficient statistic and, itself, has a normal distribution with mean θ and variance $\tau^2 = n_1^{-1}\sigma_1^2 + n_2^{-1}\sigma_2^2$.

From here, the classical/frequentist solutions to the inference problem are standard. On the Bayesian side, with incomplete prior information, one is likely to choose a non- or weakly-informative prior distribution, which amounts to taking the prior for θ to have a density that is relatively flat over the entire real line. Here we consider an alternative approach based on different interpretation and processing of the kind of partial prior information that is often available. Suppose that Treatment 2 is newly developed and Treatment 1 is the standard. Since progress tends to be incremental, researchers may not expect the new treatment to be drastically different—better or worse—than the standard treatment. In particular, suppose the researchers *expect* $|\theta| \le 1$. Following Dubois et al. (2004), this incomplete/partial prior knowledge can be modeled via a so-called *Markov prior* which has possibility contour

$$q(\vartheta) = \min(1, |\vartheta|^{-1}), \quad \vartheta \in \mathbb{T} = \mathbb{R}.$$

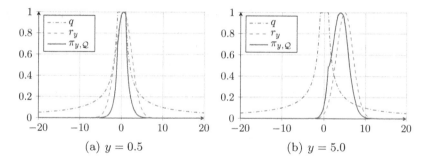

Fig. 1. Inference on the treatment effect, $\theta = \mu_2 - \mu_1$.

The corresponding credal set $\mathscr{Q}$ consists of all priors Q for which $\int |\theta| \, \mathsf{Q}(d\theta) \leq 1$. It is in this sense that q is compatible with the available prior knowledge. Specifically, this partial prior finds all increments or decrements smaller than unity to be entirely plausible, but the plausibility decreases steeply for anything larger than that; but it does not entirely rule out cases where the new treatment is substantially better or worse than the standard treatment.

Following our proposal, we consider the likelihood-driven contour h as in (7). In this case, h can be evaluated in closed-form so it remains to evaluate the IM's possibility contour using the strategy presented in Sect. 4.2. Below we use the proposed Monte-Carlo approximation with $J = 10^3$ samples and $M = 400$ Latin hypercube samples on $\mathbb{T} = [-20, 20]$. Figure 1 shows the posterior possibility contours obtained for different observations $Y = y$, where $\tau^2 = 2$, compared to the relative likelihood $r_y(\vartheta) = L_y(\vartheta)/\sup_t L_y(t)$. Panel (a) shows a case where the data $y = 0.5$ is consistent with the partial prior information and there we see an efficiency gain relative to standard methods that ignore this partial prior information. Indeed, the textbook 95% confidence interval is $\{\vartheta : r_y(\vartheta) > c\}$, with $c = \exp\{-\frac{1}{2}\chi_1^2(0.95)\} \approx 0.147$, which is noticeably wider than the valid, partial-prior IM's 95% plausibility interval $\{\vartheta : \pi_{y,\mathscr{Q}}(\vartheta) > 0.05\}$. On the other hand, Panel (b) shows a case where the data, $y = 5.0$, borderline disagrees with the partial prior information. In this case, as expected, we see some shrinkage of the IM's contour towards the prior center, but little or no reduction in the contour's spread compared to that of the relative likelihood. Further illustrations and comparisons will be presented elsewhere.

6   Conclusion

This paper builds on recent work in Martin (2022), developing a practical and general framework for valid statistical inference when partial prior information is available. Importantly, this framework falls squarely in the scope of possibility theory, highlighting the fundamental importance of possibilistic reasoning to statistical/scientific inference. It remains to consider applications where θ is high-dimensional, where the common/necessary low-dimensional structural assumptions (e.g., sparsity) can be naturally encoded as partial prior information. The validity properties discussed here are not dimension-dependent, so the question boils down to computation: can algorithms like the one proposed here scale up efficiently to handle modern high-dimensional problems?

Details aside, inference under partial priors is territory that neither Bayesians nor frequentists are equipped to handle. Of course, there are still open questions, but we are very excited about the "possibilities" (pun intended)!

References

Denoeux, T.: Likelihood-based belief function: justification and some extensions to low-quality data. Int. J. Approx. Reason. **55**(7), 1535–1547 (2014)

Dubois, D., Foulloy, L., Mauris, G., Prade, H.: Probability-possibility transformations, triangular fuzzy sets, and probabilistic inequalities. Reliable Comput. **10**(4), 273–297 (2004)

Dubois, D., Prade, H.: Representation and combination of uncertainty with belief functions and possibility measures. Comput. Intell. **4**(3), 244–264 (1988)

Fisher, R.A.: Statistical Methods and Scientific Inference, 3rd edn. Hafner Press, New York (1973)

Hose, D.: Possibilistic reasoning with imprecise probabilities: statistical inference and dynamic filtering. Ph.D. thesis, University of Stuttgart (2022)

Hose, D., Hanss, M.: A universal approach to imprecise probabilities in possibility theory. Int. J. Approx. Reason. **133**, 133–158 (2021)

Liu, C., Martin, R.: Inferential models and possibility measures. Handbook of Bayesian, Fiducial, and Frequentist Inference (2021, to appear). arXiv:2008.06874

Martin, R.: False confidence, non-additive beliefs, and valid statistical inference. Int. J. Approx. Reason. **113**, 39–73 (2019)

Martin, R.: An imprecise-probabilistic characterization of frequentist statistical inference (2021). https://researchers.one/articles/21.01.00002

Martin, R.: Valid and efficient imprecise-probabilistic inference across a spectrum of partial prior information (2022). https://researchers.one/articles/21.05.00001

Martin, R., Liu, C.: Inferential models: a framework for prior-free posterior probabilistic inference. J. Amer. Statist. Assoc. **108**(501), 301–313 (2013)

Martin, R., Liu, C.: Inferential Models: Reasoning with Uncertainty. Monographs on Statistics and Applied Probability, vol. 147. CRC Press, Boca Raton (2015)

Troffaes, M.C.M., de Cooman, G.: Lower Previsions. Wiley Series in Probability and Statistics, Wiley, Chichester (2014)

On Conditional Belief Functions
in the Dempster-Shafer Theory

Radim Jiroušek[1,2] , Václav Kratochvíl[1,2] , and Prakash P. Shenoy[3(✉)]

[1] Faculty of Management, University of Economics, Jindřichův Hradec, Czechia
[2] Institute of Information Theory and Automation, Czech Academy of Sciences,
Prague, Czechia
{radim,velorex}@utia.cas.cz
[3] School of Business, University of Kansas, Lawrence, KS 66045, USA
pshenoy@ku.edu
https://pshenoy.ku.edu

Abstract. The primary goal is to define conditional belief functions in the Dempster-Shafer theory. We do so similar to the notion of conditional probability tables in probability theory. Conditional belief functions are necessary for constructing directed graphical belief function models in the same sense as conditional probability tables for constructing Bayesian networks. Besides defining conditional belief functions, we state and prove a few basic properties of conditionals. We provide several examples of conditional belief functions, including those obtained by Smets' conditional embedding.

Keywords: Dempster-Shafer belief function theory · Conditional belief functions · Smets' conditional embedding

1 Introduction

The main goal of this article is to review the concept of conditional belief functions in the Dempster-Shafer (D-S) theory of belief functions [4,13], provide a formal definition, state some basic properties, and provide some examples.

Several theories of belief functions use the representation of belief functions but differ in the combination rules and corresponding semantics. The D-S theory uses Dempster's combination rule. [5] proposes an alternative combination rule interpreting belief functions as credal sets [7]. These two theories of belief functions are different. A comparison of these two theories is outside the scope of this paper. Here, we are concerned exclusively with the D-S theory.

One of the earliest to define conditional belief functions for the D-S theory is Smets [18]. Other contributions on conditional belief functions are (in chronological order) Shafer [14,15], Cano *et al.* [3], Shenoy [16], Almond [1], and Xu and Smets [19].

Shafer [14] is concerned about parametric models. There is a discrete parameter variable Θ and a data variable X. We have a prior basic probability assignment (BPA) m_Θ for Θ. We have a conditional model for the data, BPA $m_{X|\theta}$

© The Author(s), under exclusive license to Springer Nature Switzerland AG 2022
S. Le Hégarat-Mascle et al. (Eds.): BELIEF 2022, LNAI 13506, pp. 207–218, 2022.
https://doi.org/10.1007/978-3-031-17801-6_20

for X given $\theta \in \Omega_\Theta$. Based on a dataset of n independent observations of X, the task is to compute the posterior belief function for Θ. The BPAs $m_{X|\theta}$ for X given $\theta \in \Theta$ are converted to a conditional BPA $m_{\theta,X}$ for (Θ, X) using Smets' conditional embedding. The marginal of $m_{\theta,X}$ for Θ is vacuous. For all $\theta \in \Omega_\Theta$, the conditionals BPA $m_{\theta,X}$ are then combined using Dempster's rule resulting in the conditional $m_{X|\Theta}$. This assumes that the BPAs $m_{\theta,X}$ are distinct, which may be reasonable if the number of elements of Ω_Θ is small. Shafer also looks at the case where BPAs $m_{\theta,X}$ are not independent, and some known distributions describe the dependency.

Shafer [15] discusses conditionals abstractly as potentials that extend the domain of a potential. He calls conditionals 'continuers.' Thus, ψ is a continuer of σ from a to $a \cup b$ if and only if $\sigma^{\downarrow a} \oplus \psi = \sigma^{\downarrow a \cup b}$. Here, $\sigma^{\downarrow a}$ denotes the marginal of σ for a, $\oplus$ denotes Dempster's combination operator, and a and b are disjoint subsets of variables. The paper's focus is on the computation of marginals, but there are some interesting properties of continuers stated.

Cano et al. [3] define conditionals abstractly in the framework of valuation-based systems, but they do require that the marginal $m(b|a)^{\downarrow a}$ of conditional $m(b|a)$ is a vacuous valuation for a. The focus is on finding marginals by propagating conditional valuations in a directed acyclic graph.

Shenoy [16] describes conditional valuations using the removal operator, which is an inverse of the combination operator. For the D-S theory, the removal operator corresponds to pointwise division of commonality functions followed by normalization. If σ is a BPA for subset s of variables, and a and b are disjoint subsets of s, then conditional belief function $\sigma(b|a)$ is defined as $\sigma^{\downarrow a \cup b} \ominus \sigma^{\downarrow a}$. A consequence of this definition is that the marginal of $\sigma(b|a)$ for a is vacuous for a. One disadvantage of this definition is that conditionals are defined starting from the joint. This is not helpful in constructing joint belief functions. We say $\sigma^{\downarrow a}$ is *included* in $\sigma^{\downarrow a \cup b}$ if $\sigma^{\downarrow a \cup b} = \sigma^{\downarrow a} \oplus \sigma(b|a)$. Another disadvantage is that if $\sigma^{\downarrow a}$ is not included in $\sigma^{\downarrow a \cup b}$, $\sigma(b|a)$ may result in a BPA with negative masses. Such BPAs are called quasi-BPAs[1].

Almond [1] defines conditional belief functions as those obtained from a joint BPA by Dempster's conditioning and marginalization. Suppose $m_{X,Y}$ is a BPA for (X, Y). He defines the corresponding conditional BPA $m_{Y|x}$, where $x \in \Omega_X$ as follows. Suppose $m_{X=x}$ is a deterministic BPA for X such that $m_{X=x}(\{x\}) = 1$. Then $m_{Y|x}$ is defined as $(m_{X,Y} \oplus m_{X=x})^{\downarrow X}$. He then discusses the problem of going from conditionals to joint and argues that there isn't a unique joint associated with a group of conditionals, e.g., $\{m_{Y|x}\}_{x \in \Omega_X}$. Smets' conditional embedding is discussed whereby a conditional BPA $m_{Y|x}$ for Y is embedded into a BPA $m_{x,Y}$ for (X, Y) (details of Smets' conditional embedding are discussed in Sect. 3). Next, BPA $m_{Y|X}$ for (X, Y) is constructed from conditional embeddings $m_{x,Y}$ for $x \in \Omega_X$ as follows:

$$m_{Y|X} = \oplus\{m_{x,Y} : x \in \Omega_X\}. \tag{1}$$

---

[1] This phenomenon has been observed, e.g., in [11,16], and [12]. An example is given in [10].

Equation (1) implicitly assumes that the conditionally embedded BPAs $m_{x,Y}$ are distinct. Almond claims this assumption is unrealistic except for the case where we start from conditional BPAs $m_{Y|x}$ that are Bayesian.

Xu and Smets [19] discuss conditionals $m_{Y|a}$ for Y when proposition a is observed, where $\emptyset \neq a \in 2^{\Omega_X}$. Let $m_{a,Y}$ denote the BPA for (X,Y) after conditional embedding of $m_{Y|a}$. [1] and [19] discuss Dempster's combination of all such conditionals:

$$\oplus \{m_{a,Y} : \emptyset \neq a \in 2^{\Omega_X}\}. \tag{2}$$

While it may be reasonable to assume that $m_{x,Y}$ for $x \in \Omega_X$ are distinct as in Eq. (1), assuming that all BPAs $m_{a,Y}$ for $\emptyset \neq a \in 2^{\Omega_X}$ are distinct may be unreasonable. The focus of [19] is on computing marginals.

We do not start with a joint BPA when constructing a directed graphical belief function model. Instead, we construct a joint BPA using priors and conditionals. In this context, the current definitions in the literature are not helpful. What exactly is a conditional BPA? What are their properties? This is the primary goal of this article.

An outline of the remainder of the paper is as follows. In Sect. 2, we review the basics of D-S theory. In Sect. 3, we define conditional belief functions, and state some properties. Also, we describe where conditionals come from, including Smets' conditional embedding. We describe Almond's captain's problem [1], a directed graphical belief function model with several examples of conditionals. In Sect. 4, we conclude with a summary.

2   Basics of D-S Theory of Belief Functions

This section sketches the basics of the D-S theory of belief functions [4,13].

Knowledge is represented by basic probability assignments, belief functions, plausibility functions, commonality functions, credal sets, etc. Here we focus only on basic probability assignments and commonality functions.

Consider a set s of variables. For each $X \in s$, let Ω_X denote its finite state space, and let Ω_s denote $\times_{X \in s} \Omega_X$. Let 2^{Ω_s} denote the set of all subsets of Ω_s. A *basic probability assignment* (BPA) m for s is a function $m : 2^{\Omega_s} \to [0,1]$ such that

$$m(\emptyset) = 0, \text{ and } \sum_{\emptyset \neq a \in 2^{\Omega_s}} m(a) = 1. \tag{3}$$

m represents some knowledge about variables in s, and we say the *domain* of m is s. $m(a)$ is the probability assigned to the proposition represented by the subset a of Ω_s. Subsets a such that $m(a) > 0$ are called *focal elements* of m. If m has only one focal element (with probability 1), we say m is *deterministic*. If the focal element of a deterministic BPA is Ω_s, we say m is *vacuous*.

The knowledge encoded in a BPA m can be represented by a corresponding commonality function. The *commonality function* (CF) Q_m corresponding to BPA m for s is such that for all $a \in 2^{\Omega_s}$,

$$Q_m(a) = \sum_{b \supseteq a} m(b). \tag{4}$$

$Q_m(\mathsf{a})$ represents the probability mass that could move to every state in a. Q_m has exactly the same information as m. Given a CF Q for s, we can recover the corresponding BPA m_Q for s as follows [13]: For all $\mathsf{a} \in 2^{\Omega_s}$,

$$m_Q(\mathsf{a}) = \sum_{\mathsf{b} \in 2^{\Omega_s}: \mathsf{b} \supseteq \mathsf{a}} (-1)^{|\mathsf{b} \setminus \mathsf{a}|} Q(\mathsf{b}). \tag{5}$$

Thus, $Q : 2^{\Omega_s} \to [0,1]$ is a CF for s if and only if

$$Q(\emptyset) = 1 \tag{6}$$

$$\sum_{\mathsf{b} \in 2^{\Omega_s}: \mathsf{b} \supseteq \mathsf{a}} (-1)^{|\mathsf{b} \setminus \mathsf{a}|} Q(\mathsf{b}) \geq 0 \quad \text{for all } \emptyset \neq \mathsf{a} \in 2^{\Omega_s}, \text{ and} \tag{7}$$

$$\sum_{\emptyset \neq \mathsf{a} \in 2^{\Omega_s}} (-1)^{|\mathsf{a}|+1} Q(\mathsf{a}) = 1. \tag{8}$$

Equation (6) follows from Eq. (4), Eq. (7) corresponds to non-negativity of BPA values, and Eq. (8) corresponds to the second equation in Eq. (3).

There are two basic inference operators in the D-S theory, marginalization and combination.

Suppose m is a BPA for a set of variables r with state space $\Omega_r = \times_{X \in r} \Omega_X$ and suppose $s \subseteq r$. The marginalization operator transforms a BPA m for r to a BPA $m^{\downarrow s}$ for s by eliminating variables in $r \setminus s$. Projection of states means dropping some coordinates. If $(x, y) \in \Omega_{X,Y}$, then $(x, y)^{\downarrow X} = x$. Projection of subset of states is achieved by projecting every state in the subset. Suppose $\mathsf{a} \in 2^{\Omega_{X,Y}}$. Then, $\mathsf{a}^{\downarrow X} = \{x \in 2^{\Omega_X} : (x, y) \in \mathsf{a}\}$. Suppose m is a BPA for r. Then, the marginal of m for $s \subseteq r$, denoted by $m^{\downarrow s}$, is a BPA for s such that for each $\mathsf{a} \in 2^{\Omega_s}$,

$$m^{\downarrow s}(\mathsf{a}) = \sum_{\mathsf{b} \in 2^{\Omega_r}: \mathsf{b}^{\downarrow s} = \mathsf{a}} m(\mathsf{b}). \tag{9}$$

Dempster's combination rule is described using commonality functions. Consider two distinct BPAs m_1 for r and m_2 for s, and let Q_1 and Q_2 denote the corresponding commonality functions. Then, as showed in [13], for all $\emptyset \neq \mathsf{a} \in 2^{\Omega_{r \cup s}}$

$$(Q_1 \oplus Q_2)(\mathsf{a}) = K^{-1} Q_1(\mathsf{a}^{\downarrow r}) Q_2(\mathsf{a}^{\downarrow s}), \tag{10}$$

where K is a normalization constant defined as follows:

$$K = \sum_{\emptyset \neq \mathsf{a} \in \Omega_{r \cup s}} (-1)^{|\mathsf{a}|+1} Q_1(\mathsf{a}^{\downarrow r}) Q_2(\mathsf{a}^{\downarrow s}). \tag{11}$$

$(1 - K)$ can be regarded as a measure of *conflict* between m_1 and m_2. If $K = 1$, there is no conflict, and if $K = 0$, there is total conflict and Dempster's combination $Q_1 \oplus Q_2$ is undefined.

It is easy to show that Dempster's combination is commutative and associative: $m_1 \oplus m_2 = m_2 \oplus m_1$, and $(m_1 \oplus m_2) \oplus m_3 = m_1 \oplus (m_2 \oplus m_3)$.

There is an important property satisfied by marginalization and Dempster's combination rule called the *local computation* property [17]. Suppose m_1 is a

BPA for r and m_2 is a BPA for s (subsets r and s may not be disjoint) and suppose $X \in r$ and $X \notin s$. Then,

$$(m_1 \oplus m_2)^{\downarrow (r \cup s) \setminus \{X\}} = (m_1)^{\downarrow r \setminus \{X\}} \oplus m_2 \tag{12}$$

This property is the basis of computing marginals of joint belief functions. [6] describes an implementation of a local computation algorithm for computing marginals of graphical belief function models.

Next, we define the removal operator, which is motivated by the following situation in probability theory. Suppose $P_{X,Y}$ is a joint probability mass function (PMF) for (X, Y), and we need to compute the conditional probability table (CPT) $P_{Y|X}$. We know that $P_{X,Y} = P_X \otimes P_{Y|X}$, where $P_X = (P_{X,Y})^{\downarrow X}$ is the marginal PMF for X, and $\otimes$ is the probabilistic combination operator pointwise multiplication followed by normalization. This suggests that $P_{Y|X} = P_{X,Y} \oslash P_X$, where $\oslash$ is the inverse combination operator, pointwise division followed by normalization. If $P_X(x) = 0$, then $P_{X,Y}(x, y)$ must also be zero, and we can consider $0/0$ as undefined (using the symbol $0/0 = ?$) or define it as 1. Thus, if we regard combination $\otimes$ as aggregation of knowledge, then $\oslash$ can be regarded as removal of knowledge, and computing a CPT $P_{Y|X}$ is removing P_X from $P_{X,Y}$.

As we saw in Eq. (10), Dempster's combination is pointwise multiplication of CFs followed by normalization. Thus, removal in the D-S theory can be defined as pointwise division of CFs followed by normalization. Formally, suppose $Q_{X,Y}$ is a joint CF for (X, Y), and let $Q_X = (Q_{X,Y})^{\downarrow X}$ denote the marginal CF for X. Then, we define removal of Q_X from $Q_{X,Y}$ as follows: For all $\emptyset \neq \mathsf{a} \in 2^{\Omega_{X,Y}}$,

$$(Q_{X,Y} \ominus Q_X)(\mathsf{a}) = K^{-1} Q_{X,Y}(\mathsf{a})/Q_X(\mathsf{a}^{\downarrow X}), \tag{13}$$

where K is a normalization constant given by:

$$K = \sum_{\emptyset \neq \mathsf{a} \in 2^{\Omega_{X,Y}}} (-1)^{|\mathsf{a}|+1} Q_{X,Y}(\mathsf{a})/Q_X(\mathsf{a}^{\downarrow X}) \tag{14}$$

As in the probabilistic case, if $Q_X(\mathsf{a}^{\downarrow X}) = 0$, then $Q_{X,Y}(\mathsf{a})$ must also be 0, and we can define $0/0$ as 1.

Unlike probability theory, if we start with an arbitrary joint CF $Q_{X,Y}$, then $Q_{X,Y} \ominus Q_X$ may fail to be a CF because the corresponding BPA has negative masses adding to 1^2. In the next section, we state a proposition that characterizes when removal results in a well-defined CF.

3   Conditional Belief Functions

This section defines a conditional belief function similar to a conditional probability table in probability theory without starting from a joint distribution. Our task is constructing a joint using conditional belief functions as in a graphical model. We begin with the probabilistic case.

---

[2] An example is given in [10].

Suppose P_X denotes a PMF of X, and we wish to construct a joint PMF $P_{X,Y}$ of (X, Y) such that P_X is the marginal of $P_{X,Y}$ for X (as is typically done in a probabilistic graphical model). One way to do this is to define a PMF of Y for each $x \in \Omega_X$ such that[3] $P_X(x) > 0$. Let $P_{Y|x} : \Omega_Y \rightarrow [0, 1]$ denote a PMF of Y when X is known to be x, i.e., for all $y \in \Omega_Y$, $P_{Y|x}(y) \geq 0$ and $\sum_{y \in \Omega_Y} P_{Y|x}(y) = 1$. We can embed all PMFs $P_{Y|x}$ of Y for each $x \in \Omega_X$ into a function $P_{Y|X} : \Omega_{X,Y} \rightarrow [0, 1]$ such that $P_{Y|X}(x, y) = P_{Y|x}(y)$. In the Bayesian network literature, the function $P_{Y|X}$ is called a CPT. The joint PMF $P_{X,Y}$ of (X, Y) can now be defined as $P_{X,Y}(x, y) = P_X(x) \cdot P_{Y|X}(x, y)$. Some observations:

1. Notice that if we marginalize the CPT $P_{Y|X}$ to X, then we get a potential that is identically 1 for all values of $x \in \Omega_X$, which is the vacuous potential in probability theory.
2. If we consider probabilistic combination operator $\otimes$ as pointwise multiplication followed by normalization, then we can write $P_{X,Y} = P_X \otimes P_{Y|X}$. The normalization constant is 1 for this combination.
3. It follows from the first observation that the marginal of $P_{X,Y}$ for X is P_X. So, the CPT $P_{Y|X}$ is used to extend P_X to $P_{X,Y}$ such that the marginal $(P_{X,Y})^{\downarrow X} = P_X$.

A formal definition of a conditional belief function for Y given X in the D-S theory follows.

Definition 1. *Suppose $m_{Y|X}$ is a BPA for (X, Y), where X and Y are distinct variables. We say $m_{Y|X}$ is a conditional BPA for Y given X if and only if*

1. *$(m_{Y|X})^{\downarrow X}$ is a vacuous BPA for X, and*
2. *for any BPA m_X for X, m_X and $m_{Y|X}$ are distinct. Thus, $m_X \oplus m_{Y|X}$ is a BPA for (X, Y).*

The first condition says that $m_{Y|X}$ tells us nothing about X. We will refer to the BPA $m_X \oplus m_{Y|X}$ as the *joint* BPA for (X, Y) and denote it by $m_{X,Y}$. It follows from the local computation property (Eq. (12)) that $(m_{X,Y})^{\downarrow X} = (m_X \oplus m_{Y|X})^{\downarrow X} = m_X \oplus (m_{Y|X})^{\downarrow X} = m_X$. Thus, the second condition says the conditional $m_{Y|X}$ allows us to *extend* any BPA m_X for X to a joint BPA $m_{X,Y}$ for (X, Y) without changing its marginal for X. Notice that m_X and $m_{Y|X}$ are non-conflicting, i.e., the normalization constant K in $m_X \oplus m_{Y|X}$ is 1 (Eq. (11)).

Given a conditional BPA $m_{Y|X}$ for Y given X, we will refer to Y as the *head* of the conditional, and X as the *tail*. A conditional describes the dependency between the head and tail variables. Although we have defined a conditional BPA with the head and tail being single variables, the definition generalizes when the head and tail are disjoint subsets of variables.

---

[3] If $P_X(x) = 0$, then the conditional has no effect on the joint, and $0/0$ can be left undefined, or defined as 1.

Definition 2. *Suppose r and s are disjoint subsets of variables, and $m_{s|r}$ is a BPA for $r \cup s$. We say $m_{s|r}$ is a conditional BPA for s given r if and only if*

1. *$(m_{s|r})^{\downarrow r}$ is a vacuous BPA for r, and*
2. *for any BPA m_r for r, m_r and $m_{s|r}$ are distinct. Thus, $m_r \oplus m_{s|r}$ is a BPA for $r \cup s$.*

In a directed graphical belief function model, we have a conditional associated with each variable X in the model. The head of the associated condition is X, and the tail consists of the parents of X. For variables with no parents, we have priors associated with such variables. For convenience, we can consider priors as conditionals with empty tails. For such BPAs, the first condition in the definition is trivially true as the sum of the probability masses in a BPA is 1.

Properties of Conditionals. The following lemma was stated in [16] where conditionals were defined using an inverse of the combination operator called removal. Here we prove the same results using the definition of conditionals above that include only combination and marginalization.

Lemma 1. *Suppose r, s, and t are disjoint subsets of variables. Let m_r denote a BPA for r, $m_{s|r}$ denote a conditional BPA with head s and tail r, etc. Then, the following statements are true.*

1. *$m_r \oplus m_{s|r} \oplus m_{t|r \cup s} = m_{r \cup s \cup t}$.*
2. *$m_{s|r} \oplus m_{t|r \cup s} = m_{s \cup t|r}$.*
3. *Suppose $s' \subseteq s$. Then, $(m_{s|r})^{\downarrow r \cup s'} = m_{s'|r}$.*
4. *$(m_{s|r} \oplus m_{t|r \cup s})^{\downarrow r \cup t} = m_{t|r}$.*

Proof. 1. m_r, $m_{s|r}$, and $m_{t|r \cup s}$ are all distinct by definition of conditionals. Thus,
$$m_r \oplus m_{s|r} \oplus m_{t|r \cup s} = (m_r \oplus m_{s|r}) \oplus m_{t|r \cup s} = m_{r \cup s} \oplus m_{t|r \cup s} = m_{r \cup s \cup t}.$$
2. Let ι_r denote the vacuous BPA for r. Using the local computation property,
$$(m_{s|r} \oplus m_{t|r \cup s})^{\downarrow r} = ((m_{s|r} \oplus m_{t|r \cup s})^{\downarrow r \cup s})^{\downarrow r} = (m_{s|r} \oplus (m_{t|r \cup s})^{\downarrow r \cup s})^{\downarrow r}$$
$$= (m_{s|r} \oplus \iota_{r \cup s})^{\downarrow r} = (m_{s|r})^{\downarrow r} = \iota_r.$$

Suppose m_r is a BPA for r. Then, it follows from Statement 1 that $m_r \oplus (m_{s|r} \oplus m_{t|r \cup s}) = m_{r \cup s \cup t}$.
3. First, notice that $((m_{s|r})^{\downarrow r \cup s'})^{\downarrow r} = (m_{s|r})^{\downarrow r} = \iota_r$. Suppose m_r is a BPA for r. As m_r and $m_{s|r}$ are distinct, m_r and $m_{s'|r}$ are distinct. Thus, $m_r \oplus (m_{s|r})^{\downarrow r \cup s'} = m_{r \cup s'}$.
4. Using the local computation property,
$$((m_{s|r} \oplus m_{t|r \cup s})^{\downarrow r \cup t})^{\downarrow r} = ((m_{s|r} \oplus m_{t|r \cup s})^{\downarrow r \cup s})^{\downarrow r} = (m_{s|r} \oplus (m_{t|r \cup s})^{\downarrow r \cup s})^{\downarrow r}$$
$$= ((m_{s|r} \oplus \iota_{r \cup s})^{\downarrow r} = (m_{s|r})^{\downarrow r} = \iota_r.$$

Suppose m_r is a BPA for r. As m_r, $m_{s|r}$, and $m_{t|r \cup s}$ are all distinct,
$$m_r \oplus (m_{s|r} \oplus m_{t|r \cup s})^{\downarrow r \cup t} = (m_r \oplus m_{s|r} \oplus m_{t|r \cup s})^{\downarrow r \cup t} = (m_{r \cup s \cup t})^{\downarrow r \cup t} = m_{r \cup t}.$$

$\square$

Where do Conditionals Come From? A conditional BPA $m_{r|s}$ describes the relationship between the variables in r and s. One source of conditionals is Smets' conditional embedding [18]. To describe conditional embedding, consider the case of two variables X and Y. To describe the dependency between X and Y, suppose that when $X = x$, our belief in Y is described by a BPA $m_{Y|x}$ for Y. Thus, $m_{Y|x} : 2^{\Omega_Y} \to [0,1]$ such that $\sum_{a \in 2^{\Omega_Y}} m_{Y|x}(a) = 1$. The BPA $m_{Y|x}$ for Y needs to be embedded into a BPA $m_{x,Y}$ for (X,Y) such that

1. $m_{x,Y}$ is a conditional BPA for (X,Y), i.e., $(m_{x,Y})^{\downarrow X}$ is vacuous BPA for X, and
2. when we add the belief that $X = x$ and marginalize the result to Y, we obtain $m_{Y|x}$.

One way to do this is to take each focal element $b \in 2^{\Omega_Y}$ of $m_{Y|x}$, and convert it to the corresponding focal element

$$(\{x\} \times b) \cup ((\Omega_X \setminus \{x\}) \times \Omega_Y) \in 2^{\Omega_{X,Y}} \tag{15}$$

of BPA $m_{x,Y}$ for (X,Y) with the same mass. It is easy to confirm that this method of embedding satisfies both conditions mentioned above. If we have several distinct conditionals, e.g., $m_{Y|x_1}$, $m_{Y|x_2}$, etc., where x_1, and x_2 are distinct values of X, then we do conditional embedding of each of these BPAs and then combine the embeddings by Dempster's combination rule to obtain $m_{Y|X}$. An example of conditional embedding follows.

Example 1 (Conditional embedding). Consider binary variables X and Y, with $\Omega_X = \{x, \bar{x}\}$ and $\Omega_Y = \{y, \bar{y}\}$. Suppose we have a BPA $m_{Y|x}$ for Y given $X = x$ as follows:

$$m_{Y|x}(y) = 0.8, \ m_{Y|x}(\Omega_Y) = 0.2,$$

then its conditional embedding into the conditional BPA $m_{x,Y}$ for (X,Y) is as follows:

$$m_{x,Y}(\{(x,y), (\bar{x}, y), (\bar{x}, \bar{y})\}) = 0.8, \ m_{x,Y}(\Omega_{X,Y}) = 0.2.$$

Similarly, if we have a BPA $m_{Y|\bar{x}}$ for Y given $X = \bar{x}$ as follows:

$$m_{Y|\bar{x}}(\bar{y}) = 0.3, \ m_{Y|\bar{x}}(\Omega_Y) = 0.7,$$

then its conditional embedding into the conditional BPA $m_{\bar{x},Y}$ for (X,Y) is as follows:

$$m_{\bar{x},Y}(\{(x,y), (x, \bar{y}), (\bar{x}, \bar{y})\}) = 0.3, \ m_{\bar{x},Y}(\Omega_{X,Y}) = 0.7.$$

Assuming we have these two BPAs, and their corresponding embeddings, it is clear that the two BPA $m_{x,Y}$ and $m_{\bar{x},Y}$ are distinct, and can be combined with Dempster's rule of combination, resulting in the conditional BPA $m_{Y|X} = m_{x,Y} \oplus m_{\bar{x},Y}$ for (X,Y). $m_{Y|X}$ has the following properties. First, $(m_{Y|X})^{\downarrow X} = \iota_X$, where ι_X denotes the vacuous BPA for X. Second, if we combine $m_{Y|X}$ with deterministic BPA $m_{X=x}(\{x\}) = 1$ for X, and marginalize the combination to Y, then we get $m_{Y|x}$, i.e., $(m_{Y|X} \oplus m_{X=x})^{\downarrow Y} = m_{Y|x}$. Third, $(m_{Y|X} \oplus m_{X=\bar{x}})^{\downarrow Y} = m_{Y|\bar{x}}$. $m_{Y|X}$ is the belief function equivalent of CPT $P_{Y|X}$ in probability theory. $\square$

In probability theory, a joint distribution $P_{X,Y}$ can always be factored into marginal $P_X = (P_{X,Y})^{\downarrow X}$ and a conditional $P_{Y|X}$ such that $P_{X,Y} = P_X \otimes P_{Y|X}$. This is not true in the D-S theory. The following proposition describes when a joint belief function can be factored into a marginal and a conditional.

Proposition 1. *Suppose $m_{X,Y}$ is a BPA for $\{X,Y\}$ with corresponding CF $Q_{m_{X,Y}}$. Let m_X denote the marginal of $m_{X,Y}$ for X, i.e., $m_X = (m_{X,Y})^{\downarrow X}$. Then, $Q_{m_{X,Y}} \ominus Q_{m_X}$ is a CF if and only if there exists a BPA m for $\{X,Y\}$ such that $m_{X,Y} = m_X \oplus m$, and m is a conditional for Y given X.*

A proof of this proposition can be found in [8]. The proposition states that if we remove BPA m_X from $m_{X,Y}$ such that m_X is included in $m_{X,Y}$ in the sense that $m_{X,Y}$ is Dempster's combination of the marginal m_X for X and a conditional m for Y given X, then such removal always results in a well-defined CF.

Smets' conditional embedding is only one way to obtain conditionals. Black and Laskey [2] propose other methods to get conditionals. The following example from [1], called the captain's problem, has many examples of conditionals. The description of Almond's captain's problem is taken from [9].

Example 2 (Captain's problem). A ship's captain is concerned about how many days his ship may be delayed before arrival at a destination. The arrival delay is the sum of departure delay and sailing delay. Departure delay may be a result of maintenance (at most one day), loading delay (at most one day), or a forecast of bad weather (at most one day). Sailing delays may result from bad weather (at most one day) and whether repairs are needed at sea (at most one day). If maintenance is done before sailing, chances of repairs at sea are less likely. The forecast is 80% reliable. The captain knows the loading delay and whether maintenance is done before departure.

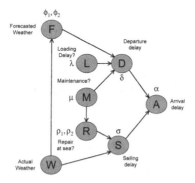

Fig. 1. The directed acyclic graph for the captain's problem. The Greek alphabets adjacent to a variable denote the prior or conditional or evidence associated with the variable.

Table 1. The variables, their state spaces, and associated conditionals in the captain's problem.

Variable	Name	State space, Ω	Associated conditional
W	Actual weather	$\{g_w, b_w\}$	vacuous for W
F	Forecasted weather	$\{g_f, b_f\}$	ϕ_1 for $F|W$
L	Loading delay?	$\{t_l, f_l\}$	λ for L
M	Maintenance done?	$\{t_m, f_m\}$	μ for M
R	Repair at sea needed?	$\{t_r, f_r\}$	ρ_1, ρ_2 for R given $M = t_m, t_f$, resp.
D	Departure delay (in days)	$\{0, 1, 2, 3\}$	δ for $D|\{F, L, M\}$
S	Sailing delay (in days)	$\{0, 1, 2, 3\}$	σ for $S|\{W, R\}$
A	Arrival delay (in days)	$\{0, 1, 2, 3, 4, 5, 6\}$	α for $A|\{D, S\}$

Table 1 describes the variables, their state spaces, and associated conditionals, and Fig. 1 shows the directed acyclic graph associated with this problem. The details of some of the conditional BPAs are as follows.

1. Weather forecast is 80% accurate. ϕ_1 is a conditional BPA for F given W.

$$\phi_1(\{(g_w, g_f), (b_w, b_f)\}) = 0.8, \ \phi_1(\Omega_{W,F}) = 0.2.$$

2. Bad weather and repair at sea each adds a day to sailing delay. This proposition is true 90% of the time. σ is a conditional for S given (W, R).

$$\sigma(\{(g_w, f_r, 0), (b_w, f_r, 1), (g_w, t_r, 1), (b_w, t_r, 2)\}) = 0.9, \ \sigma(\Omega_{W,R,S}) = 0.1.$$

3. Departure delay may be a result of maintenance (at most 1 day), loading delay (at most 1 day), or a forecast of bad weather (at most 1 day). δ is a deterministic conditional BPA for D given $\{F, L, M\}$.

$$\delta(\{(g_f, f_l, f_m, 0), (b_f, f_l, f_m, 1), (g_f, t_l, f_m, 1), (g_f, f_l, t_m, 1),$$
$$(b_f, t_l, f_m, 2), (b_f, f_l, t_m, 2), (g_f, t_l, t_m, 2), (b_f, t_l, t_m, 3)\}) = 1.$$

4. The arrival delay is the sum of departure delay and sailing delay. α is a deterministic conditional BPA for A given $\{D, S\}$.

$$\alpha(\{(0, 0, 0), (0, 1, 1), (0, 2, 2), (0, 3, 3), (1, 0, 1), (1, 1, 2), (1, 2, 3), (1, 3, 4),$$
$$(2, 0, 2), (2, 1, 3), (2, 2, 4), (2, 3, 5), (3, 0, 3), (3, 1, 4), (3, 2, 5), (3, 3, 6)\}) = 1.$$

4   Summary and Conclusions

We have explicitly defined conditionals in the D-S theory using only the marginalization and Dempster's combination operators. The main goal of the definition is to enable the construction of directed graphical belief function models. Conditional belief functions are also defined in [16] using an inverse of

Dempster's combination operator called *removal*. Since Dempster's combination is pointwise multiplication of commonality functions followed by normalization, removal consists of division of commonality functions followed by normalization. Thus, $m_{Y|X} = m_{X,Y} \ominus m_X$. One issue with this definition is that a conditional BPA is defined starting from a joint BPA, which is not useful in constructing a joint BPA. Another issue is that if m_X is not already included in $m_{X,Y}$, the removal operation may result in a BPA with negative masses. We have stated some properties of conditionals given in [16] and these properties remain valid using our definition. Smets' conditional embedding [18] is one way to obtain conditionals. There are other ways to obtain conditionals, and some examples of conditionals are described using Almond's captain's problem [1].

Acknowledgments. This study was supported by the Czech Science Foundation Grant No. 19-06569S to the first two authors and by the Ronald G. Harper Professorship at the University of Kansas to the third author.

References

1. Almond, R.G.: Graphical Belief Modeling. Chapman & Hall, London, UK (1995)
2. Black, P.K., Laskey, K.B.: Hierarchical evidence and belief functions. In: Shachter, R.D., Levitt, T.S., Kanal, L.N., Lemmer, J.F. (eds.) Uncertainty in Artificial Intelligence 4, Machine Intelligence and Pattern Recognition, vol. 9, pp. 207–215. North-Holland, Amsterdam, Netherlands (1990)
3. Cano, J., Delgado, M., Moral, S.: An axiomatic framework for propagating uncertainty in directed acyclic networks. Int. J. Approximate Reasoning **8**(4), 253–280 (1993)
4. Dempster, A.P.: Upper and lower probabilities induced by a multivalued mapping. Ann. Math. Stat. **38**(2), 325–339 (1967)
5. Fagin, R., Halpern, J.Y.: A new approach to updating beliefs. In: Bonissone, P., Henrion, M., Kanal, L., Lemmer, J. (eds.) Uncertainty in Artificial Intelligence 6, pp. 347–374, North-Holland (1991)
6. Giang, P., Shenoy, S.: The belief function machine: an environment for reasoning with belief functions in Matlab. Working paper, University of Kansas School of Business, Lawrence, KS 66045 (2003). https://pshenoy.ku.edu/Papers/BFM072503.zip
7. Halpern, J.Y., Fagin, R.: Two views of belief: belief as generalized probability and belief as evidence. Artif. Intell. **54**(3), 275–317 (1992)
8. Jiroušek, R., Kratochvíl, V., Shenoy, P.P.: Entropy for evaluation of Dempster-Shafer belief function models. Working Paper 342, University of Kansas School of Business, Lawrence, KS 66045 (2022). https://pshenoy.ku.edu/Papers/WP342.pdf
9. Jiroušek, R., Kratochvíl, V., Shenoy, P.P.: Computing the decomposable entropy of graphical belief function models. In: Studený, M., Ay, N., Coletti, G., Kleiter, G.D., Shenoy, P.P. (eds.) Proceedings of the 12th Workshop on Uncertain Processing (WUPES 2022), pp. 111–122. MatfyzPress, Prague, Czechia (2022). https://pshenoy.ku.edu/Papers/WUPES22a.pdf
10. Jiroušek, R., Shenoy, P.P.: Compositional models in valuation-based systems. Int. J. Approximate Reasoning **55**(1), 277–293 (2014)

11. Kong, A.: Multivariate belief functions and graphical models. Ph.D. thesis, Department of Statistics, Harvard University, Cambridge, Massachusetts (1986)
12. Lauritzen, S.L., Jensen, F.V.: Local computation with valuations from a commutative semigroup. Ann. Math. Artif. Intell. **21**(1), 51–69 (1997)
13. Shafer, G.: A Mathematical Theory of Evidence. Princeton University Press, Princeton (1976)
14. Shafer, G.: Belief functions and parametric models. J. Roy. Stat. Soc. B **44**(3), 322–352 (1982)
15. Shafer, G.: An axiomatic study of computation in hypertrees. Working Paper 232, University of Kansas School of Business, Lawrence, KS 66045 (1991). http://glennshafer.com/assets/downloads/hypertrees_91WP232.pdf
16. Shenoy, P.P.: Conditional independence in valuation-based systems. Int. J. Approximate Reasoning **10**(3), 203–234 (1994)
17. Shenoy, P.P., Shafer, G.: Axioms for probability and belief-function propagation. In: Shachter, R.D., Levitt, T., Lemmer, J.F., Kanal, L.N. (eds.) Uncertainty in Artificial Intelligence 4, Machine Intelligence and Pattern Recognition Series, vol. 9, pp. 169–198. North-Holland, Amsterdam, Netherlands (1990)
18. Smets, P.: Un modele mathematico-statistique simulant le processus du diagnostic medical. Ph.D. thesis, Free University of Brussels (1978)
19. Xu, H., Smets, P.: Reasoning in evidential networks with conditional belief functions. Int. J. Approximate Reasoning **14**(2–3), 155–185 (1996)

Valid Inferential Models Offer Performance and Probativeness Assurances

Leonardo Cella$^{(\boxtimes)}$ and Ryan Martin

Department of Statistics, North Carolina State University, Raleigh, NC 27695, USA
{lolivei,rgmarti3}@ncsu.edu

Abstract. Bayesians and frequentists are now largely focused on developing methods that *perform* well in a frequentist sense. But the widely-publicized replication crisis suggests that performance guarantees are not enough for good science. In addition to reliably detecting hypotheses that are incompatible with data, users require methods that can *probe* for hypotheses that are actually supported by the data. In this paper, we demonstrate that valid inferential models achieve both performance and probativeness properties. We also draw important connections between inferential models and Deborah Mayo's severe testing.

Keywords: Bayesian · Frequentist · p-value · Possibility measure · Severity

1 Introduction

Important decisions affecting our everyday experiences are increasingly data-driven. But is data helping us make better decisions? The widely-publicized *replication crisis* in science raises serious concerns, e.g., the American Statistical Association's president commissioned a formal *Statement on Statistical Significance and Replicability*.[1] The lack of any clear guidance in that statement reveals that there are important and fundamental questions concerning the foundations of statistical inference that remain unanswered:

> *Should probability enter to capture degrees of belief about claims? ... Or to ensure we won't reach mistaken interpretations of data too often in the long run of experience?* (Mayo 2018, p. xi)

The two distinct roles of probability above correspond to the classical frequentist and Bayesian schools of statistical inference, which have two fundamentally different priorities, referred to here as *performance* and *probativeness*, respectively. Over the last 50+ years, however, the lines between the two perspectives

---

[1] https://magazine.amstat.org/blog/2021/08/01/task-force-statement-p-value/.

R. Martin—Partially supported by the U.S. National Science Foundation, SES–205122.

S. Le Hégarat-Mascle et al. (Eds.): BELIEF 2022, LNAI 13506, pp. 219–228, 2022.
https://doi.org/10.1007/978-3-031-17801-6_21

and their distinct priorities have been blurred. Indeed, both Bayesians and frequentists now focus almost exclusively on performance. Such considerations are genuinely important for the logic of statistical inference:

> *even if an empirical frequency-based view of probability is not used directly as a basis for inference, it is unacceptable if a procedure... of representing uncertain knowledge would, if used repeatedly, give systematically misleading conclusions* (Reid and Cox 2015, p. 295).

As the replication crisis has taught us, there is more to inference than achieving, say, Type I and II error rate control. Beyond performance, we are also concerned with probativeness, i.e., can methods probe for hypotheses that are genuinely supported by the observed data? Modern statistical methods cannot achieve both performance and probativeness objectives, so a fully satisfactory framework for scientific inferences requires new perspectives.

To set the scene, denote the observable data by Y. The statistical model for Y will be denoted by $\mathsf{P}_{Y|\theta}$, where $\theta \in \Theta$ is an unknown model parameter. Note that the setup here is quite general: Y, θ, or both can be scalars, vectors, or something else. We focus here on the typical case where *no genuine prior information is available/assumed.* So, given only the model $\{\mathsf{P}_{Y|\theta} : \theta \in \Theta\}$ and the observed data $Y = y$, the goal is to quantify uncertainty about the unknown θ for the purpose of making inference. For concreteness, we will interpret "making inference" as making (data-driven) judgments about hypotheses concerning θ. Let $\mathcal{H}$ denote a collection of subsets of Θ, containing the singletons and closed under complementation, and associate $H \in \mathcal{H}$ with a hypothesis about θ.

Section 2.1 briefly describes the Bayesian vs. frequentist *two-theory problem* in our context of hypothesis testing. There we justify our above claim that modern statistical methods fail to meet both the performance and probativeness objectives. This includes the default-prior Bayes solution that aims to strike a balance between the two theories. What holds the default-prior Bayes solution back from meeting the performance and probativeness objectives is its lack of calibration, which is directly related to the constraint that the posterior distribution be a precise probability. Fortunately, the relatively new *inferential model* (IM) framework, reviewed briefly in Sect. 2.2 below, is able to achieve greater flexibility by embracing a certain degree of imprecision in its construction. Our main contribution here, in Sect. 3, is to highlight the IM's ability to simultaneously achieve both *performance* and *probativeness*. Two illustrations are presented in Sect. 4 and some concluding remarks are given in Sect. 5.

The probativeness conclusion is a direct consequence of the IM output's imprecision. That the additional flexibility of imprecision creates opportunities for more nuanced judgments is one of the motivations for accounting for imprecision, so this is no big surprise. But our contribution here is valuable for several reasons. First, the statistical community is aware of this need to see beyond basic performance criteria (e.g., Mayo 2018), but no clear, general, and easy-to-follow guidance has been offered. What we are suggesting here, however, is simple: *just follow the general theory of valid IMs and you get both performance and probativeness assurances.* Second, it showcases the importance of the role of belief functions and imprecise probability more generally, by reinforcing the key point that imprecision is *not* due to an inadequate formulation of the problem, but, rather, an essential part of the complete solution.

2   Background

2.1   Two-Theory Problem

In a nutshell, the two dominant schools of thought in statistics are as follows.

BAYESIAN. Uncertainty is quantified directly through specification of a prior probability distribution for θ, representing the data analyst's *a priori* degrees of belief. Bayes's theorem is then used to update the prior to a data-dependent posterior distribution for θ. The posterior probability of a hypothesis H represents the analyst's degree of belief in the truthfulness of H, given data, and would be essential for inference concerning H. That is, the magnitudes of the posterior probabilities naturally drive the data analyst's judgments about which hypotheses are supported by the data and which are not.

FREQUENTIST. Uncertainty is quantified indirectly through the use of reliable procedures that control error rates. Consider, e.g., a p-value for testing a hypothesis H. What makes such a p-value meaningful is that, by construction, it tends to be not-small when H is true. Therefore, observing a small p-value gives the data analyst reason to doubt the truthfulness of H:

> *The force with which such a conclusion is supported is logically that of the simple disjunction: Either an exceptionally rare chance has occurred, or* [the hypothesis] *is not true* (Fisher 1973, p. 42).

The p-value *does not* represent the "probability of H" in any sense. So, a not-small (resp. small) p-value cannot be interpreted as direct support for H (resp. H^c) or any sub-hypothesis thereof.

The point is that, at least in principle, Bayesians focus on probativeness whereas frequentists focus on performance. But the line between frequentist and modern Bayesian practice is not so clear. Even Bayesians typically assume little or no prior information, as we have assumed here, so default priors are the norm (e.g., Berger 2006; Jeffreys 1946). But with a default prior, the "degree of belief" interpretation or the posterior probabilities is lost,

> [Bayes's theorem] *does not create real probabilities from hypothetical probabilities* (Fraser 2014, p. 249)

and, along with it, the probative nature of inferences based on them,

> *...any serious mathematician would surely ask how you could use* [Bayes's theorem] *with one premise missing by making up an ingredient and thinking that the conclusions of the* [theorem] *were still available* (Fraser 2011, p. 329).

The default-prior Bayes posterior probabilities could still have performance assurances *if* they were suitably calibrated. But the *false confidence theorem* of Balch et al. (2019) shows that this is not the case: there exists false hypotheses to which the posterior distribution tends to assign large probabilities. This implies that inferences based on the magnitudes of default-prior Bayes posterior probabilities can be "systematically misleading" (cf. Reid and Cox). This is perhaps why modern Bayesian analysis focuses less on the posterior probabilities and more on the performance of procedures (tests and credible sets) derived from the posterior. Hence modern Bayesians and frequentists are not so different.

The key take-away message is as follows. Frequentist methods focus on detecting incompatibility between data and hypotheses (performance), so they do not offer any guidance on how to identify hypotheses actually supported by the data (probativeness). Default-prior Bayesian methods are effectively no different, so this critique applies to them too. More specifically, the default-prior Bayes posterior probabilities lack the calibration necessary to reliably check for either incompatibility or support. Therefore, neither of the dominant schools of thought in statistical inference are able to simultaneously achieve both the performance and probativeness objectives.

2.2   Inferential Models Overview

Inferential models (IMs) were first developed in Martin and Liu (2013, 2015) to balance the Bayesians' desire for belief assignments and the frequentists' desire for error rate control. A key distinction between IMs and the familiar Bayesian and frequentist frameworks is that the output is an *imprecise probability* or, more specifically, a *necessity–possibility measure* pair. This imprecision, however, is not the result of an inability to precisely specify a model, etc., it is a necessary condition for inference to be *valid* in the sense defined in Sect. 3 below.

> *Possibility is an entirely different idea from probability, and it is sometimes, we maintain, a more efficient and powerful uncertainty variable, able to perform semantic tasks which the other cannot* (Shackle 1961, p. 103).

The false confidence theorem establishes that validity cannot be achieved via ordinary probability. More recently it has been shown that the possibility-theoretic formulation is key to achieving the relevant performance-related properties.

The original IM construction put forward in Martin and Liu (2013) relied on suitable random sets, whereas Liu and Martin (2021) recently offered a direct construction using possibility measures. The latter construction starts by associating data Y and unknown parameter θ with an unobservable auxiliary variable U with known distribution P_U via the formula

$$A(Y, \theta, U) = 0, \quad U \sim \mathsf{P}_U.$$

Let π denote a plausibility contour on the U-space such that the corresponding possibility measure is *consistent* with P_U in the sense that $\mathsf{P}_U(B) \leq \sup_{u \in B} \pi(u)$ for all subsets B. Now define its extension to Θ as

$$\pi_y(\vartheta) = \sup_{u:A(y,\vartheta,u)=0} \pi(u), \quad \vartheta \in \Theta.$$

Assuming this is a genuine/normal contour, then it defines an IM for θ having the mathematical form of a possibility measure with upper probability

$$\overline{\Pi}_y(H) = \sup_{\vartheta \in H} \pi_y(\vartheta), \quad H \subseteq \Theta,$$

and lower probability $\underline{\Pi}_y(H) = 1 - \overline{\Pi}_y(H^c)$. The IM's output is meaningful thanks to the properties it satisfies, which we discuss in Sect. 3. The performance-related properties have been the focus in previous work, but it is interesting that the performance properties together with the inherent imprecision in the IM's possibilistic output leads to probativeness properties too.

3 Two P's in the Same Pod

3.1 Performance

As discussed above, the property that gets the most attention in the statistics literature is *performance*, i.e., procedures developed for the purpose of making inference-related decisions (e.g., accept or reject a hypothesis) have error rate control guarantees. This is genuinely important: if statistical methods are not even reliable, then they have no hope of helping to advance science.

Our main result here, which is not new, is that procedures derived from a valid IM achieve the desired performance-related properties. As presented in the reference given in Sect. 2.2 above, we say that an IM without lower and upper probability output $y \mapsto (\underline{\Pi}_y, \overline{\Pi}_y)$ is *valid* if

$$\sup_{\theta \in H} \mathsf{P}_{Y|\theta}\{\overline{\Pi}_Y(H) \leq \alpha\} \leq \alpha, \quad \text{for all } \alpha \in [0,1] \text{ and all } H \in \mathcal{H}. \tag{1}$$

This means that, with respect to the model $\mathsf{P}_{Y|\theta}$, it is a relatively rare event that the IM assigns relatively small upper probability to a true hypothesis about θ. Property (1) closely resembles the defining stochastically-no-smaller-than-uniform property of p-values. As such, Fisher's "logical disjunction" argument also applies to the valid IM's output, giving it objective meaning.

Although we are not aware of Fisher ever making such a statement, we believe that Fisher's disdain for the Neyman-style behavioral approach to statistical inference at least partially stemmed from the fact that such properties would be immediate consequences of the calibration needed for his "disjunction" argument to apply. So if Fisher's calibration is satisfied, then Neyman's error rate control is a corollary. Indeed, if the IM with output $(\underline{\Pi}_y, \overline{\Pi}_y)$, with corresponding plausibility contour π_y, is valid in the sense of (1), then

- for any fixed $\alpha \in (0,1)$, the test "reject H if and only if $\overline{\Pi}_y(H) \leq \alpha$" controls the frequentist Type I error probability at level α, and

- for any fixed $\alpha \in (0, 1)$, the set $C_\alpha(y) = \{\vartheta : \pi_y(\vartheta) > \alpha\}$ is a $100(1 - \alpha)\%$ confidence set, i.e., its frequentist coverage probability at least $1 - \alpha$.

These claims are almost immediate consequences of (1); see, e.g., Martin (2021) for a proof. Therefore, valid IMs offer performance guarantees.

Here it is also worth briefly pointing out that the connection between valid IMs and performance guarantees is even more fundamental. It was recently shown in Martin (2021) that *every* procedure with provable frequentist performance guarantees has, working behind the scenes, a valid IM with the form of a possibility measure. So, not only does the IM framework offer performance guarantees, it is really the only framework that does so. This also highlights the deep connections between frequentist inference and possibility theory.

3.2   Probativeness

The literature on IMs has largely focused on performance, i.e., that (1) implies that the output is suitably calibrated which leads to the results quoted in Sect. 3.1 above. While the IM output does, as discussed above, represent lower and upper probabilities, or degrees of necessity/support and possibility, a clear explanation of their post-data interpretation, and why non-additivity is valuable, has yet to be given. This section aims to fill that gap.

Standard performance metrics, such as Type I and Type II error probabilities, are not data-dependent and, therefore, cannot directly speak to whether the actual observed data offer any direct support to a particular hypothesis. The IM output returns both lower and upper probabilities but, so far, the literature has largely only focused on one of these, typically the upper probability. Perhaps the lower probability will be of some value after all.

Suppose that the data y is such that $\overline{\Pi}_y(H)$ is relatively large, i.e., the data are not incompatible with the hypothesis H. If, instead, $\overline{\Pi}_y(H)$ were small, then we can apply all of what we are about to describe to H^c instead of H. If we determine that the data are not incompatible with H, then a natural follow-up question is to ask if the data actually *support* the hypothesis H or any proper subset, say, $H' \subset H$. For this, we propose to consider the lower probability

$$\underline{\Pi}_y(H') = 1 - \overline{\Pi}_y(H'^c) = 1 - \sup\{\pi_y(\vartheta) : \vartheta \notin H'\},$$

where the right-most expression is exclusive to the case where the IM output takes the form of a possibility measure, as we consider here. Coincidentally or not, Shafer (1976, Ch. 11) refers to the lower probability function, $H \mapsto \underline{\Pi}_y(H)$, as a *support function*, which is consistent with how we propose to use it here. If $\overline{\Pi}_y(H)$ is not small, then $\underline{\Pi}_y(H') \leq \overline{\Pi}_y(H)$ can be small or (relatively) large, and its magnitude determines the extent to which the data supports the truthfulness of H', beyond just compatibility or plausibility. On the one hand, if $\underline{\Pi}_y(H')$ is small and $\overline{\Pi}_y(H)$ is relatively large, then H' is plausible—or not incompatible—with the data y but there is little direct support in y for its truthfulness. This corresponds to a case with relatively large "don't know" in the sense of Dempster

(2008). On the other hand, if both $\underline{\mathit{\Pi}}_y(H')$ and $\overline{\mathit{\Pi}}_y(H)$ are relatively large, then y is not only compatible with H, it also directly supports H'.

What makes the "if $\underline{\mathit{\Pi}}_y(H')$ is relatively large, then infer H'" judgment warranted? Readers familiar with imprecise probability might be surprised by this question—this is precisely what lower probabilities are designed for—but remember that $\underline{\mathit{\Pi}}_y$ is not a subjective assessment of the data analyst's degrees of belief. So the data analyst should require, à la Reid and Cox, that their IM will tend not to lead them to erroneous judgments. Like $\underline{\mathit{\Pi}}_y$ is the dual to $\overline{\mathit{\Pi}}_y$, there is a corresponding dual to the validity property (1):

$$\sup_{\theta \notin H} \mathsf{P}_{Y|\theta}\{\underline{\mathit{\Pi}}_Y(H) > 1 - \alpha\} \le \alpha, \quad \text{for all } \alpha \in [0,1] \text{ and all } H \in \mathcal{H}. \qquad (2)$$

It is easy to verify that (2) and (1) are equivalent, but it is worth considering both versions because, while the latter refers primarily to assessments of compatibility between data and hypotheses, the former is relevant to judgments about when data actually support a certain hypothesis.

In Sect. 1, we remarked that there have been recent efforts by statisticians to supplement the standard p-values, etc. with measures designed to *probe* for hypotheses supported by the data. In particular, Mayo (2018) proposes a so-called *severity* measure but only gives one concrete example. If we extrapolate her suggestion beyond that one example, then it boils down to what we described above. That is, the map $H' \mapsto \underline{\mathit{\Pi}}_y(H')$ on subcollections of $\mathcal{H}$ can be used to probe for hypotheses that are actually supported by the data.

There is, however, a minor difference between ours and Mayo's perspective. On the one hand, Mayo is thinking in terms of a specific test of a particular hypothesis, so her severity measure is intended to describe how severe the test is, how deep that tests probes for actual support in the data beyond just compatibility or lack thereof. On the other hand, we are thinking in terms of big-picture uncertainty quantification. In light of the fundamental connection between valid IMs and frequentist inference, perhaps it is no surprise that Mayo's proposal, despite coming from a slightly different perspective, ends up directly aligning with what the valid IM does automatically; see Sect. 4.1. It is now clear that probabitiveness is inherent in the valid IM—no supplements needed!

4   Illustrations

4.1   Normal Mean

Mayo (2018, p. 142) describes a hypothetical water plant where the water it discharges is intended to be roughly 150° Fahrenheit. More specifically, under ideal settings, water temperature measurements ought to be normally distributed with mean 150° and standard deviation 10°. To test the water plant's settings, a sample $Y = (Y_1, \ldots, Y_n)$ of $n = 100$ water temperature measurements are taken; then the sample mean, $\bar{Y}$, is $\mathsf{N}(150, 1)$. Since water temperatures higher than 150° might damage the ecosystem, of primary interest are hypotheses $H_\vartheta =$

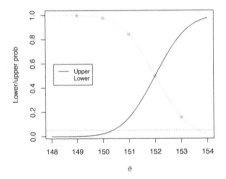

Fig. 1. Results of the valid IM applied to Mayo's normal mean example; the red dots correspond to the values in Table 3.1 of Mayo (2018) (Color figure online)

$(-\infty, \vartheta]$ for ϑ near 150. For hypotheses of this form, the "optimal" IM (Martin and Liu 2013, Sect. 4.3) has upper probability

$$\overline{\Pi}_y(H_\vartheta) = 1 - \Phi(\bar{y} - \vartheta), \quad \vartheta \in \mathbb{R},$$

where Φ denotes the standard normal distribution function.

Suppose we observe $\bar{y} = 152$, which is potentially incompatible with the hypothesis H_{150}. Indeed, a plot of the upper probability is shown in Fig. 1(a) and we see that, at $\vartheta = 150$, the upper probability is smaller than 0.05, so we would be inclined to reject the hypothesis $\theta \leq 150$. To probe for support of subsets of the alternative hypothesis, we also plot the lower probability

$$\underline{\Pi}_y(H_\vartheta^c) = \Phi(\bar{y} - \vartheta), \quad \vartheta \in \mathbb{R},$$

and we see that there is, in fact, non-negligible support in the data for, say, $H_{151}^c = (151, \infty)$. These results agree exactly with the analysis presented in Mayo (2018) based on her supplement of the ordinary p-value with a severity measure. Mayo elaborates on this example in a couple different ways but, for the sake of space, suffice it to say that our analysis perfectly agrees with hers.

4.2   Bivariate Normal Correlation

Suppose that Y consists of n independent and identically distributed pairs $Y_i = (Y_{1,i}, Y_{2,i})$ having a bivariate normal distribution with zero means, unit variances, and correlation $\theta \in [-1, 1]$. Let $\mathsf{P}_{Y|\theta}$ denote the corresponding joint distribution. An asymptotic pivot based on the maximum likelihood estimator, $\hat{\theta}$, can be constructed and the corresponding Wald test would look very similar to that in Sect. 4.1. This bivariate normal correlation problem, however, corresponds to one of those "curved exponential families" where $\hat{\theta}$ is not a sufficient statistic so some efficiency is lost in the Wald test for finite n. So we take a different approach here, which extends us beyond the cases Mayo considers.

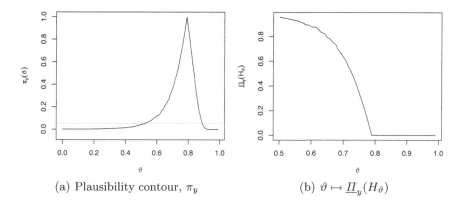

(a) Plausibility contour, π_y (b) $\vartheta \mapsto \underline{\mathit{\Pi}}_y(H_\vartheta)$

Fig. 2. Results of the valid IM applied to Efron's law school admissions data.

Let $\vartheta \mapsto L_y(\vartheta)$ denote the likelihood function for θ based on data y. Following Martin (2015, 2018), a valid IM can be constructed based on the relative likelihood, $r_y(\vartheta) = L_y(\vartheta)/L_y(\hat{\theta})$, with plausibility contour function

$$\pi_y(\vartheta) = \mathsf{P}_{Y|\vartheta}\{r_Y(\vartheta) \leq r_y(\vartheta)\}, \quad \vartheta \in [-1, 1].$$

This resembles the p-value function for a suitable likelihood ratio test. The IM's output, $(\underline{\mathit{\Pi}}_y, \overline{\mathit{\Pi}}_y)$, is determined by optimizing the contour function.

As an illustration of the ideas presented above, consider the law school admissions data analyzed in Efron (1982), which consists of $n = 15$ data pairs with $Y_1 = $ LSAT scores and $Y_2 = $ undergrad GPA. For our analysis, we standardize these so that the mean zero–unit variance is appropriate. Of course, this standardization has no effect on the correlation, which is our object of interest. In this case, the sample correlation is 0.776; the maximum likelihood estimator, which has no closed-form expression, is $\hat{\theta} = 0.789$. A plot of the plausibility contour π_y for this data is shown in Fig. 2(a). The horizontal line at $\alpha = 0.05$ determines the 95% plausibility interval, which is an exact 95% confidence interval. It is clear that the data shows virtually no support for $\theta = 0$, but there is some marginal support for the hypothesis $H = (0.5, 1]$. To probe this further, consider the class of sub-hypotheses $H_\vartheta = (\vartheta, 1]$, $\vartheta > 0.5$. A plot of the function $\vartheta \mapsto \underline{\mathit{\Pi}}_y(H_\vartheta)$ is shown in Fig. 2(b). As expected from Panel (a), the latter function is decreasing in ϑ and we clearly see no support for H_ϑ as soon as $\vartheta \geq \hat{\theta}$. But there is non-negligible support for H_ϑ with ϑ less than, say, 0.65–0.70.

5   Conclusion

Here we showed that there is more to the IM framework than what has been presented in the existing literature. Specifically, the validity property, together with its inherent imprecision implies both performance and probativeness assurances. This is of special interest to the belief function/possibility theory community as it showcases the fundamental importance of its brand of imprecision.

We also identified a connection between IMs and Mayo's severe testing framework. This is beneficial to severe testers, as the IM construction exposed in Sect. 2.2 provides a general recipe for assessing severity in a wide range of modern applications. We also find it attractive that the IM framework has this notion of probativeness built in, as opposed to being an add-on to classical testing. Illustrations in cases beyond the simple, low dimensional problems above will be reported elsewhere, as well as the extension of the notion of probativeness/severity to statistical learning problems.

References

Balch, M.S., Martin, R., Ferson, S.: Satellite conjunction analysis and the false confidence theorem. Proc. Roy. Soc. A **475**(2227), 2018.0565 (2019)

Berger, J.: The case for objective Bayesian analysis. Bayesian Anal. **1**(3), 385–402 (2006)

Dempster, A.P.: The Dempster-Shafer calculus for statisticians. Internat. J. Approx. Reason. **48**(2), 365–377 (2008)

Efron, B.: The Jackknife, the Bootstrap and other Resampling Plans. CBMS-NSF Regional Conference Series in Applied Mathematics, vol. 38. Society for Industrial and Applied Mathematics (SIAM), Philadelphia (1982)

Fisher, R.A.: Statistical Methods and Scientific Inference, 3rd edn. Hafner Press, New York (1973)

Fraser, D.A.S.: Rejoinder: "Is Bayes posterior just quick and dirty confidence?". Statist. Sci. **26**(3), 329–331 (2011)

Fraser, D.A.S.: Why does statistics have two theories? In: Lin, X., Genest, C., Banks, D.L., Molenberghs, G., Scott, D.W., Wang, J.-L. (eds.) Past, Present, and Future of Statistical Science, chap. 22. Chapman & Hall/CRC Press (2014)

Jeffreys, H.: An invariant form for the prior probability in estimation problems. Proc. Roy. Soc. London Ser. A **186**, 453–461 (1946)

Liu, C., Martin, R.: Inferential models and possibility measures. In: Handbook of Bayesian, Fiducial, and Frequentist Inference arXiv:2008.06874 (2021, to appear)

Martin, R.: Plausibility functions and exact frequentist inference. J. Amer. Statist. Assoc. **110**(512), 1552–1561 (2015)

Martin, R.: On an inferential model construction using generalized associations. J. Statist. Plann. Inference **195**, 105–115 (2018)

Martin, R.: An imprecise-probabilistic characterization of frequentist statistical inference (2021). https://researchers.one/articles/21.01.00002

Martin, R., Liu, C.: Inferential models: a framework for prior-free posterior probabilistic inference. J. Amer. Statist. Assoc. **108**(501), 301–313 (2013)

Martin, R., Liu, C.: Inferential Models: Reasoning with Uncertainty. Monographs on Statistics and Applied Probability, vol. 147. CRC Press, Boca Raton (2015)

Mayo, D.G.: Statistical Inference as Severe Testing. Cambridge University Press, Cambridge (2018)

Reid, N., Cox, D.R.: On some principles of statistical inference. Int. Stat. Rev. **83**(2), 293–308 (2015)

Shackle, G.L.S.: Decision Order and Time in Human Affairs. Cambridge University Press, Cambridge (1961)

Shafer, G.: A Mathematical Theory of Evidence. Princeton University Press, Princeton (1976)

Links with Other Uncertainty Theories

A Qualitative Counterpart of Belief Functions with Application to Uncertainty Propagation in Safety Cases

Yassir Idmessaoud[1(✉)], Didier Dubois[2], and Jérémie Guiochet[1]

[1] LAAS-CNRS, University of Toulouse, Toulouse, France
{yassir.id-messaoud,jeremie.guiochet}@laas.fr
[2] IRIT, University of Toulouse, Toulouse, France
dubois@irit.fr

Abstract. Critical systems such as those developed in the aerospace, railway or automotive industries need official documents to certify their safety via convincing arguments. However, informal tools used in certification documents seldom cover the uncertainty that pervades safety cases. Several works use quantitative approaches based on belief functions to model and propagate confidence/uncertainty in the argument structures (particularly those using goal structuring notation). However the numerical uncertainty information is often a naive encoding of qualitative expert inputs. In this paper, we outline a qualitative substitute to Dempster-Shafer theory and suggest new qualitative confidence propagation models. We also propose a more faithful encoding of expert inputs.

Keywords: Goal Structuring Notation · Argument structures · Confidence elicitation · Dempster-Shafer theory · Qualitative capacities

1 Introduction

As the use of artificial intelligence (AI) in systems increases, the need for safety assessment methods in the latter is also increasing. However, the lack of confidence can jeopardize the social acceptance of these systems and therefore their existence. Several approaches are used to assess confidence/uncertainty in such systems (especially, the safety critical ones).

Many papers addressing the assessment of safety of systems rely on the graphical representation of an argument structure like GSN (Goal Structuring Notation), plus quantitative representations of uncertainty. Typically, probability theory is often used in Bayesian network models of GSNs. In order to address the issue of incomplete information, Dempster-Shafer theory (DST) is also proposed. In the latter case, argument trees can be modelled in classical logic using if-then rules [6,7].

However the quantification of uncertainty is often problematic, when it relies on expert assessments. In many cases, experts supply qualitative assessments using linguistic values like *probable, very probable, unlikely*, etc., which are then translated into numbers on the $[0,1]$ scale. This translation is somewhat arbitrary. So, a legitimate question is whether a purely qualitative approach to uncertainty, that would be a counterpart to the belief function approach, could be promising. The idea is to avoid the quantitative encoding of qualitative estimates. It makes all the more sense as numerical degrees of belief obtained via uncertainty propagation are often translated back to the qualitative scale, so as to make the results more palatable. So it is legitimate to investigate a qualitative approach.

This paper is a first step in this direction. It is structured as follows. Section 2 presents theoretical background on qualitative capacities that can be viewed as a qualitative counterpart of belief functions, based on [5]. Section 3 deals with the elicitation of qualitative capacities, based on an existing method where linguistic term scales were mapped to belief functions. Section 4 use qualitative belief measures on classical inference patterns. Section 5 recalls the a graphical representation called Goal Structuring Notation (GSN), dedicated to argument structures for safety cases. This section applies the qualitative uncertainty propagation method from premises to conclusions of several types arguments. In Sect. 6, a preliminary comparison of qualitative and quantitative uncertainty propagation is proposed via an example.

2   From Belief Functions to Qualitative Capacities

As a generalization of probability theory, Dempster-Shafer theory [8](DST) offers tools to model and propagate both aleatory (due to random events) and epistemic (due to incomplete information) uncertainty.

A mass function, or basic belief assignment (BBA), is a probability distribution over the power set of the universe of possibilities (W), known as the *frame of discernment*. Formally, a mass function $m : 2^W \rightarrow [0,1]$ is such that $\sum_{E \subseteq W} m(E) = 1$, and $m(\emptyset) = 0$. Any subset E of W such as $m(E) > 0$ is called a focal set of m. $m(E)$ quantifies the probability that we only know that the truth lies in E; in particular $m(W)$ quantifies the amount of ignorance.

A mass assignment induces a so-called belief function $Bel : 2^W \rightarrow [0,1]$, defined by: $Bel(A) = \sum_{E \subseteq A} m(E)$. It represents the sum of all the masses supporting a statement A. The degree of belief in the negation $\neg A$ of the statement A is called *disbelief*: $Disb(A) = Bel(\neg A)$; the value $Uncer(A) = 1 - Bel(A) - Disb(A)$ quantifies the lack of information about A.

The *conjunctive rule of combination* combines multiple pieces of evidence (represented by mass functions m_i, with $i = 1, 2$) coming from independent sources of information: $m_\cap = m_1 \otimes m_2$ such that: $m_\cap(A) = \sum_{E_1 \cap E_2 = A} m_1(E_1) \cdot m_2(E_2)$. In DST, an additional step eliminates conflict that may exist by means of a normalization factor (dividing $m_\cap$ by $1 - m_\cap(\emptyset)$). This is Dempster rule of combination [8], which is associative.

In contrast, we outline the qualitative approach in [3–5]. Let L be a finite totally ordered set representing certainty levels. A qualitative capacity (q-capacity, for short) is a function $\gamma : 2^W \rightarrow L$ such that:

$\gamma(\emptyset) = 0$; $\gamma(W) = 1$; $A \subseteq B \Rightarrow \gamma(A) \leq \gamma(B)$. *Any* q-capacity can be put in the form:

$$\gamma(A) = \max_{\emptyset \neq B \subseteq A} \rho(B), \forall A \subseteq W, \tag{1}$$

where ρ is formally a basic possibility assignment (BΠA) [3], namely, a possibility distribution $\rho : 2^W \rightarrow L$ on the power set of W, such that $\max_{B \subseteq W} \rho(B) = 1$ and $\rho(\emptyset) = 0$. The value $\rho(B)$ is the strength of piece of evidence B. Several BΠA's can generate the same γ, the least of which is the qualitative Moebius transform (QMT) of γ such that:

$$\gamma_{\#}(A) = \begin{cases} \gamma(A) \text{ if } \gamma(A) > \gamma(A \setminus \{w\}), \forall w \in A; \\ 0 \text{ otherwise.} \end{cases} \tag{2}$$

The value $\gamma(A)$ (resp. $\gamma(\neg A)$) qualifies the support in favor of (resp. against) A, i.e. belief (resp. disbelief) in A using an element in the qualitative scale L. The pair $(\gamma(A), \gamma(\neg A))$ thus describes our epistemic stance with respect to A in terms of belief and disbelief, ranging from no information (i.e., $(0, 0)$), to full conflicting information (i.e., $(1, 1)$), from full belief (i.e., $(1, 0)$) to full disbelief (i.e., $(0, 1)$). This is more general than possibility theory where the case $(1, 1)$ is not allowed.

Figure 1 presents the credibility and information orderings on pairs (belief, disbelief) including extreme cases [5]. A proposition A is at least as credible as B if $\gamma(A) \geq \gamma(B)$ and $\gamma(\neg A) \leq \gamma(\neg B)$ (solid arrows from B to A), thus ranging from certainty of falsity $(0, 1)$ up to certainty of truth $(1, 0)$. A proposition A is at least as informed as B if $\gamma(A) \geq \gamma(B)$ and $\gamma(\neg A) \geq \gamma(\neg B)$ (dotted arrows from B to A), thus ranging from ignorance $((0, 0)$, no information) up to conflict $((1, 1)$, full contradictory information). In this situation, the amount of evidence supporting the conclusion is equal to the one rejecting it. The set $L \times L$ is then equipped with a bilattice structure. In order to qualitatively combine pieces of evidence represented by possibilistic mass functions, i.e., BΠA's ρ_i, coming from several sources of information, the qualitative counterpart of the conjunctive rule of combination for belief functions is: $\rho_{\cap} = \rho_1 \oplus \rho_2$ such that:

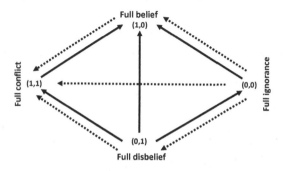

Fig. 1. Evolution of certainty and information in pairs (belief, disbelief)

$$\rho_\cap(A) = \max_{E_1 \cap E_2 = A} \{\min[\rho_1(E_1), \rho_2(E_2)]\} \tag{3}$$

Due to the use of the (idempotent) minimum operation, the combined pieces of evidence are not supposed to be independent. The result is not always a BΠA, strictly speaking. First we may have that $\rho_\cap(A) < 1$ for all A. So we must add the condition $\rho_\cap(W) = 1$. This will not occur if we restrict to non-dogmatic BΠA's such that $\rho_i(W) = 1$, which we assume in this paper. Besides, we may have that $\rho_\cap(\emptyset) > 0$, indicating conflict between the pieces of evidence.

3   Expert Elicitation Approach

In order to elicit qualitative capacities, we borrow from a methodology by Cyra and Gorski [2]. Two types of information are collected from experts about a statement A: A so-called *decision* and a level of *confidence* associated to it. Then, these pieces of information are numerically encoded, and transformed to belief and disbelief degrees in the sense of Shafer (see also [7]). More precisely:

– The decision index denoted by $Dec(A)$, describes which side the assessor leans towards, i.e., acceptance or rejection of A. It is associated with a bipolar scale $D = \{0_D = d_{-k}, d_{k-1}, \dots, d_0 = e, d_1, \dots d_k = 1_D\}$ with $2k + 1$ values, the bottom of which (0_D) expressing rejection, the top (1_D) acceptance, and the midpoint (e) a neutral position. Here we assume $k = 2$.
– The confidence index denoted by $Conf(A)$ reflects the amount of information an assessor possesses to support the decision. It uses a positive uni-polar scale K with $k + 1$ values (the top 1_K expresses full confidence, the bottom 0_K is neutral - no information). For $k = 2$: these levels mean: lack of confidence ($C_0 = 0_K$), moderate confidence (C_1) and full confidence ($C_2 = 1_K$).

The bipolar scale D is equipped with an order-reversing map ν_D such that $\nu_D(d_{-i}) = d_i$. Especially we have that $\nu_D(Dec(A)) = Dec(\neg A)$. The unipolar scale K is isomorphic to the positive part of D. This assumption makes K and D commensurate. K is equipped with an order-reversing map ν_K such that: $\nu_K(C_i) = C_{k-i}$.

In order to switch from a $(Dec(A), Conf(A))$ pair to $(\gamma(A), \gamma(\neg A))$, we use a transformation that maps $D \times K$ to the belief-disbelief scale $L \times L$ containing pairs $(\gamma(A), \gamma(\neg A))$. The scale L has the same number of elements as K (i.e., 3 here). The mapping $f : D \times K \to L \times L$: must satisfy some conditions [5]:

– If the expert declares lack of confidence, the result is $f(Dec(A), 0) = (0, 0)$, whatever the trend expressed on the decision scale.
– If the expert is fully confident, then $f(1, 1) = (\gamma(A), \gamma(\neg A)) = (1, 0)$, $f(0, 1) = (0, 1)$, $f(e, 1) = (1, 1)$. Indeed, for the latter, there is a total conflict: the expert is maximally informed ($Conf(A) = 1$), and cannot decide between A and its negation ($Dec(A) = e$).
– $\max(\gamma(A), \gamma(\neg A)) = Conf(A)$: the belief in A or its negation cannot be stronger than the confidence.

- if $Dec(A)$ is the midpoint of D, then $\gamma(A) = \gamma(\neg A)(= Conf(A))$ (no reason to take side).
- if $Dec(A)$ is less than the midpoint of D, then $\gamma(A) < \gamma(\neg A) = Conf(A)$, and the smaller $Dec(A)$, the smaller $\gamma(A)$.
- if $Dec(A)$ is greater than the midpoint of D, then $\gamma(A) = Conf(A) > \gamma(\neg A)$, and the greater $Dec(A)$, the smaller $\gamma(\neg A)$.

These conditions lead to propose the following translation formulas [5]:

if $Dec(A) < e$, $\gamma(A) = \min[\nu_K(Dec(\neg A)), Conf(A)]$ and $\gamma(\neg A) = Conf(A)$;

if $Dec(A) > e$, $\gamma(A) = Conf(A)$ and $\gamma(\neg A) = \min[\nu_K(Dec(A)), Conf(A)]$;

if $Dec(A) = Dec(\neg A) = e$, $\gamma(A) = \gamma(\neg A) = Conf(A)$.

In Table 1, we grouped all possible $(Dec, Conf)$ pairs on premises with their appropriate counterparts $(\gamma(A), \gamma(\neg A)) \in L \times L$, using the formulas above. We can notice an anti-symmetry between belief and disbelief degrees regarding the central column ($D_0 = e$: no decision). We also notice that when no information is available (C_0: Lack of confidence), no matter what choice is made the degrees of belief and disbelief take a minimal value. On the other hand, in the case of a fully informed expert (C_2: Full confidence) the decision value varies from rejection to acceptance.

Table 1. Values from $(Dec, Conf)$ to $(Bel, Disb)$ pairs on premises

$Conf$	Dec				
	D_{-2} (Rej)	D_{-1} (Opp)	$D_0(ND)$	D_1 (Tol)	D_2 (Acc)
C_0 (Lack of confidence)	$(0, 0)$	$(0, 0)$	$(0, 0)$	$(0, 0)$	$(0, 0)$
C_1 (Moderate confidence)	$(0, \lambda)$	(λ, λ)	(λ, λ)	(λ, λ)	$(\lambda, 0)$
C_2 (Full confidence)	$(0, 1)$	$(\lambda, 1)$	$(1, 1)$	$(1, \lambda)$	$(1, 0)$

4   Logical Inference for Qualitative Capacities

Logical reasoning and numerical belief functions are not often put together. An approach to reasoning with Dempster rule of combination was proposed in [1]. In this approach each formula in a knowledge base is viewed as a simple support function and combined with other formulas in the knowledge base. Besides, the application of belief functions to argument structures has been studied in [2,7,10] to build models for uncertainty propagation. For instance in [7], we assign mass functions to logical expressions such as facts p_i, $\neg p_i$, and rules $p_i \Rightarrow C$, $\neg p_i \Rightarrow \neg C$, $(\wedge_{i=1}^n p_i) \Rightarrow C$ and $(\wedge_{i=1}^n \neg p_i) \Rightarrow \neg C$, in order to deduce the belief on the conclusions C and $\neg C$. Here we develop the same approach, albeit using qualitative capacities.

The simplest pattern is modus ponens, i.e. inferring C from p and $p \Rightarrow C$. We assume two BΠA's ρ_p on $\{p, \neg p\}$ with values in L, say $\rho_p(p) = \alpha_p, \rho_p(\neg p) =$

$\overline{\alpha}_p, \rho_p(\top) = 1$, and a simple support function $\rho_\Rightarrow$ with $\rho_\Rightarrow(p \Rightarrow C) = \beta_\Rightarrow$, $\rho_\Rightarrow(\top) = 1$, where $\top$ stands for the tautology. The capacity γ_C is obtained via projection for the conclusion C by the q-conjunctive rule has a $B\Pi A$ ρ_C such that: $\gamma_C(C) = \rho_C(p \wedge C) = \min(\rho_p(p), \rho_\Rightarrow(p \Rightarrow C)) = \min(\alpha_p, \beta_\Rightarrow), \gamma_C(\neg C) = 0$.

If $p = \bigwedge_{i=1}^n p_i$, then the above formula holds with $\rho_p(p) = \min_{i=1}^n \rho_p^i(p_i)$. If there is also a $B\Pi A$ $\rho_\Leftarrow$ assigning a weight $\beta_\Leftarrow$ to the reversed implication $\neg p \Rightarrow \neg C$ there is an additional weight on $\neg C$ via the combination $\rho_p \oplus \rho_\Rightarrow \oplus \rho_\Leftarrow$ using Eq. (3) and projection on C's universe.

$$\gamma_C(\neg C) = \rho_C(\neg p \wedge \neg C) = \min(\rho_p(\neg p), \rho_\Leftarrow(p \Leftarrow C)) = \min(\overline{\alpha}_p, \beta_\Leftarrow).$$

Consider the case with more than one premise. Suppose we have to merge $B\Pi A$'s ρ_p^i on p_i, $\rho_\Rightarrow^i$, $\rho_\Leftarrow^i, i = 1, \ldots n$. As in its quantitative counterpart, the $B\Pi A$ pertaining to the conclusion C obtained from this fusion may assign a mass to the contradiction. Conflict always appears when four items are merged of the form: p_i and $p_i \Rightarrow C$ with $\neg p_j$ and $\neg p_j \Rightarrow \neg C, j \neq i$, whose conjunction is a contradiction $\emptyset$ with mass:

$$\rho_C^{ij}(\emptyset) = \min[\rho_C^i(p_i \wedge C), \rho_C^j(\neg p_j \wedge \neg C)]$$
$$= \min[\rho_p^i(p_i), \rho_\Rightarrow^i(p_i \Rightarrow C), \rho_p^j(\neg p_j), \rho_\Rightarrow^j(\neg p_j \Rightarrow \neg C)]]$$

For two premises, the final mass on contradiction is $\rho_C(\emptyset) = \max(\rho_C^{12}(\emptyset), \rho_C^{21}(\emptyset))$. Besides, using (1) we get: $\gamma_C(C) = \max[\rho_C(p_1 \wedge C), \rho_C(p_2 \wedge C)] \geq \rho_C(\emptyset)$ and $\gamma_C(\neg C) = \max[\rho_C(\neg p_1 \wedge \neg C), \rho_C(\neg p_2 \wedge \neg C)] \geq \rho_C(\emptyset)$.

5   Application to Safety Cases

Goal structuring notation (GSN) is a graphical notation/language which represents argument structures (i.e., safety and assurance cases) in the form of directed acyclic graphs (directed trees or arborescences). It breaks down a top claim, called "goal", into elementary sub-goals following a specific strategy and in accordance with a particular context. Each sub-goal is associated with pieces of evidence, called solutions, which support the conclusion. Despite the fact that it presents all the evidence supporting the safety of the system, GSN fails to show how premises support the conclusion and the confidence that can be given to them. Both questions bring uncertainty to arguments, which may affect their merits. To address this issue, confidence propagation schemes were proposed to complement GSN patterns.

Some approaches use DST to model and propagate confidence in GSN patterns in the literature [2,7,9]. These papers consider a number of argument types and associate confidence propagation formulas to each of them. In practice, they also devise transformation formulas that turn uncertainty assessments of experts (on a qualitative scale) about premises to numerical belief and disbelief degrees. This transformation is a source of uncertainty. Indeed, qualitative inputs are often naively translated into equidistant values in the unit interval. Therefore, the qualitative approach to uncertainty developed in [3–5], and the elicitation and the inference methods of Sects. 3 and 4 may lead to more robust confidence assessment approaches.

Here, we use the argument types defined in [7]. An argument type describes the interaction between premises to support a conclusion. This type of interaction is either a conjunction (C-Arg), a disjunction (D-Arg), or a combination of both (H-Arg, a hybrid type). We can translate each argument type into logical expressions often called rules. Since we use only implication to describe links between the universe of premises ($W_p = \{p, \neg p\}$) and that of the conclusion ($W_C = \{C, \neg C\}$), two kinds of rules are used: *direct rules*, which model the acceptance of the conclusion (C), and *reverse rules* which model its rejection ($\neg C$). Then, to each rule, we assigned a simple support function (a mass on the rule, and another on the tautology). We also assigned masses to the premises and their negation. Finally the propagation formulas, for each type, are obtained using the qualitative combination rule (3). Below, we recall our argument types and associate to each of them to qualitative uncertainty propagation formula.

Simple Argument (S-Arg): This argument describes the case of a conclusion (C) supported by a single premise (p), hence the name "simple". If the premise is true, then so is the conclusion: $p \Rightarrow C$. Note that only the information about the acceptance of the conclusion can be inferred in this situation. Since we work on a three-state paradigm (belief, disbelief and uncertainty), the reverse rule $\neg p \Rightarrow \neg C$ is introduced to add conditions for the possible denial of the conclusion. Then, we associate to the direct and reverse rules simple $B\Pi A$'s (resp., $\rho_\Rightarrow$ and $\rho_\Leftarrow$), and a $B\Pi A$ on the premise space, as done above. We can prove:

$$\text{S-Arg}: \begin{cases} \gamma_C(C) & = \min[\gamma_p(p), \gamma_\Rightarrow(p \Rightarrow C)] \\ \gamma_C(\neg C) & = \min[\gamma_p(\neg p), \gamma_\Leftarrow(\neg p \Rightarrow \neg C)] \end{cases} \tag{4}$$

We can notice that the belief $\gamma_C(C)$ depends only on the direct rule and the acceptance of the premise, while the disbelief $\gamma_C(\neg C)$ only depends on the reverse rule and the disbelief of the premise.

Conjunctive Argument (C-Arg): This argument type describes the situation when two premises or more are jointly needed to support a conclusion. We formally defined its direct and reverse rules (resp.) by: $(\wedge_{i=1}^n p_i) \Rightarrow C$ and $\wedge_{i=1}^n(\neg p_i \Rightarrow \neg C)$. Following the same reasoning of the previous argument type, we put a simple $B\Pi A$ on each rule ($\rho_\Rightarrow$ and $\rho_\Leftarrow^i$), and another $B\Pi A$ on each premise (ρ_p^i). Then we combine them with the rule of combination ($\rho = \rho_r \oplus \rho_p$, with $\rho_p = \rho_p^1 \oplus ... \oplus \rho_p^n$ and $\rho_r = \rho_\Rightarrow \oplus (\oplus_{i=1}^n \rho_\Leftarrow^i)$) and get:

$$\text{C-Arg}: \begin{cases} \gamma_C(C) & = \min\{\min_{i=1}^n \gamma_p^i(p_i), \gamma_\Rightarrow([\wedge_{i=1}^n p_i] \Rightarrow C)\} \\ \gamma_C(\neg C) & = \max_{i=1}^n\{\min[\gamma_p^i(\neg p_i), \gamma_\Leftarrow^i(\neg p_i \Rightarrow \neg C)]\} \end{cases} \tag{5}$$

In the formulas of the quantitative approach [7] operations $a + b - ab$ and ab replace max, min, respectively, thus highlighting the similarity between the results obtained from each model. Indeed, we can notice that the C-Arg, like its quantitative counterpart, favors the propagation of the premise with the least strength (minimal belief, with a maximal disbelief degree).

Disjunctive Argument (D-Arg): In this situation, each premise can support alone the whole conclusion. Formally, the direct and reverse rules are defined as follows: $\wedge_{i=1}^{n}(p_i \Rightarrow C)$ and $(\wedge_{i=1}^{n}\neg p_i) \Rightarrow \neg C$. The calculation of $\gamma_C(C)$ and $\gamma_C(\neg C)$ is identical to the one above, swapping the two expressions:

$$\text{D-Arg} : \begin{cases} \gamma_C(C) = \max_{i=1}^{n}\{\min[\gamma_p^i(p_i), \gamma_{\Rightarrow}^i(p_i \Rightarrow C)]\} \\ \gamma_C(\neg C) = \min\{\min_{i=1}^{n}\gamma_p^i(\neg p_i), \gamma_{\Leftarrow}([\wedge_{i=1}^n\neg p_i] \Rightarrow \neg C)\} \end{cases} \quad (6)$$

We can notice that this model, as its quantitative counterpart [7], favors the propagation of the premise with the greatest strength (maximal belief and minimal disbelief degree).

Hybrid Argument (H-Arg): This argument type describes the situation where each premise supports the conclusion to some degree, but their conjunction does it to a larger one. Therefore, all conjunctive and disjunctive rules will be used in this argument type. Thus, we obtain:

$$\text{H-Arg} : \begin{cases} \gamma_C(C) = \max\{\min[\min_{i=1}^{n}\gamma_p^i(p_i), \gamma_{\Rightarrow}([\wedge_{i=1}^n p_i] \Rightarrow C)], \\ \qquad \max_{i=1}^{n}(\min[\gamma_p^i(p_i), \gamma_{\Rightarrow}^i(p_i \Rightarrow C)]\} \\ \gamma_C(\neg C) = \max\{\min[\min_{i=1}^{n}\gamma_p^i(\neg p_i), \gamma_{\Leftarrow}([\wedge_{i=1}^n\neg p_i] \Rightarrow \neg C)], \\ \qquad \max_{i=1}^{n}\min[\gamma_p^i(\neg p_i), \gamma_{\Leftarrow}^i(\neg p_i \Rightarrow \neg C)]\} \end{cases} \quad (7)$$

We can notice that Eq. (7), presents a combination between C-Arg formulas (5), and D-Arg (6). Assuming a maximal belief (=1) (resp. disbelief) on premises, it is enough that the simple direct rules take a null value (resp. the reversed conjunctive one) to get the conjunctive argument type. And conversely, to get the disjunctive argument type, put null values on direct conjunctive and simple reversed rules. The S-Arg, represent a special case when only one premise is available ($n = 1$). In the following, only the H-Arg will be used since it covers the four types.

6   Application Example

On an artificial example (Fig. 3) that displays three argument types (C-Arg, D-Arg and H-Arg), we apply our approach in order to see how each type affects the propagation of uncertainty from premises to the overall goal (conclusion). We also apply the quantitative approach presented in [7] on the same example. To compare results from both approaches, we will use the same decision and confidence scales (see Fig. 2).

Regarding elicitation, we use the evaluation matrix in Fig. 2 to collect expert opinions, and transform them using formulas in Sect. 3 to get belief and disbelief on premises. Regarding the elicitation of belief weight on rules, we benefit from

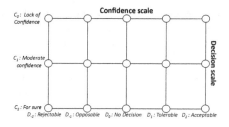

Fig. 2. Evaluation matrix

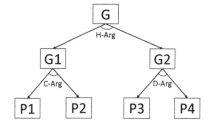

Fig. 3. GSN artificial example

an observation made on the quantitative models [7]. Indeed, we notice that under some assumptions for the premises, the value of the conclusion is the value of the rule. For instance, assuming full support (resp. positive or negative) on all premises gives the value of the conjunctive rule (resp. direct and reversed): $\gamma_C(C) = \gamma_\Rightarrow([\wedge_{i=1}^n p_i] \Rightarrow C)$ or $\gamma_C(\neg C) = \gamma_\Leftarrow([\wedge_{i=1}^n \neg p_i] \Rightarrow \neg C)$. On the other hand, assuming a total support (resp. positive or negative) on one premise (p_i) and total ignorance on the other gives the value of the appropriate disjunctive rule: $\gamma_C(C) = \gamma_\Rightarrow^i(p_i \Rightarrow C)$ or $\gamma_C(\neg C) = \gamma_\Leftarrow^i(\neg p_i \Rightarrow \neg C)$. So, we will use the same Table 1 to transform the assessment to rules. However, to avoid the negation of rules, the assessor can only choose between the positive decision (from *"no decision"* to *"acceptable"*) for direct rules; only negative decisions (from *"rejectable"* to *"no decision"*) for the reversed ones. Indeed, rules can only infer uncertainty on one side of the decision scale.

The example in Fig. 3, presents a top-goal (G) supported by two sub-goals (G1) and (G2) through a hybrid argument type (H-Arg). Each one of them is also supported, respectively, by two premises. Goal (G1) is supported by the premises (P1) and (P2) related by a conjunctive argument type (C-Arg). On the other hand, goal (G2) is supported by the premises (P3) and (P4) related by a disjunctive argument type (D-Arg). For simplicity, we set all masses on rules to their maximal values (=1). Then, we use four settings with different premise values and compute the confidence in the top goal.

Table 2. Pairs (decision, confidence) according to both qualitative (Qual.) and quantitative [7] (Quant.) methods for the example (see Fig. 2 for the meaning of symbols)

	Meth.	1st	2nd	3th	4th
P_1	-	$(Opp;C_2)$	$(Tol;C_2)$	$(Tol;C_2)$	$(Opp;C_2)$
P_2	-	$(Tol;C_2)$	$(Tol;C_2)$	$(Tol;C_2)$	$(Opp;C_2)$
P_3	-	$(Tol;C_2)$	$(Tol;C_2)$	$(Tol;C_2)$	$(Opp;C_2)$
P_4	-	$(Tol;C_2)$	$(Opp;C_2)$	$(Tol;C_2)$	$(Opp;C_2)$
G	Quant.	$(ND;C_0)$	$(Tol;C_1)$	$(Tol;C_1)$	$(Opp;C_1)$
	Qual.	$(ND;C_2)$	$(Tol;C_2)$	$(Tol;C_2)$	$(Opp;C_2)$

In general, we can see from Table 2 that both approaches give close results which fit well with our expectations. The only difference is in the confidence values. We can say that, in this case the qualitative approach gives results with higher levels of confidence than the quantitative one.

We notice from Table 2 that the first case gives a *"no decision"*. This result is explained by the fact that we end up with two opposite judgments in the H-Arg (conflict situation) due to C-Arg that propagates the premise with least strength (opposable) to G_1 (G_2: tolerable). On the contrary, in the 2nd column, we get a *"tolerable"* decision, because the D-Arg favors the propagation of the premise with the greatest weight (tolerable) to G_2 (G_1: tolerable). In the 3th and 4th columns we can notice, as expected, that the top goal keeps the same decision as premises respectively: *"tolerable"* and *"opposable"*.

The difference in the degree of confidence between qualitative and quantitative approaches is due to the nature of the operations used. For example, the C-Arg favors the propagation of the weakest premise (weaker belief and stronger disbelief). In the quantitative setting, we use the product and the probabilistic sum. And in the qualitative case, we use min and max, which does not model attenuation or reinforcement effects in case of independent pieces of information. This is one limitation of the qualitative approach.

7   Conclusion

In this paper, we propose a qualitative confidence assessment approach. We provide formulas to propagate confidence in GSN from the premises to the top-goal using qualitative mass functions. Each of these functions is collected from experts in the form of a decision and the associated confidence degree, and then transformed into a q-capacity. By sticking to qualitative values, the possible arbitrariness of the transformation of expert opinions into quantitative values (used in some previous approaches) is eliminated. Furthermore, it seems that the qualitative approach gives results similar to the quantitative one in [7]. However, more experiments are needed to confirm this conclusion.

References

1. Chatalic, P., Dubois, D., Prade, H.: An approach to approximate reasoning based on Dempster rule of combination. Int. J. Expert Syst. Res. Appl. **1**, 67–85 (1987)
2. Cyra, L., Górski, J.: Support for argument structures review and assessment. Reliab. Eng. Syst. Saf. **96**(1), 26–37 (2011)
3. Dubois, D., Faux, F., Prade, H., Rico, A.: A possibilistic counterpart to Shafer evidence theory. In: IEEE International Conference on Fuzzy Systems (FUZZIEEE) (2019)
4. Dubois, D., Faux, F., Prade, H., Rico, A.: Qualitative capacities and their informational comparison. In: Proceedings 12th Conference of the European Society for Fuzzy Logic and Technology (EUSFLAT 2021), pp. 19–24 (2021)
5. Dubois, D., Faux, F., Prade, H., Rico, A.: Qualitative capacities: basic notions and potential applications. Int. J. Approximate Reasoning (2022, in press)

6. Idmessaoud, Y., Dubois, D., Guiochet, J.: Belief functions for safety arguments confidence estimation: a comparative study. In: Davis, J., Tabia, K. (eds.) SUM 2020. LNCS (LNAI), vol. 12322, pp. 141–155. Springer, Cham (2020). https://doi.org/10.1007/978-3-030-58449-8_10

7. Idmessaoud, Y., Dubois, D., Guiochet, J.: Quantifying confidence of safety cases with belief functions. In: Denœux, T., Lefèvre, E., Liu, Z., Pichon, F. (eds.) BELIEF 2021. LNCS (LNAI), vol. 12915, pp. 269–278. Springer, Cham (2021). https://doi.org/10.1007/978-3-030-88601-1_27

8. Shafer, G.: A Mathematical Theory of Evidence. Princeton University Press (1976)

9. Wang, R., Guiochet, J., Motet, G., Schön, W.: D-S theory for argument confidence assessment. In: Vejnarová, J., Kratochvíl, V. (eds.) BELIEF 2016. LNCS (LNAI), vol. 9861, pp. 190–200. Springer, Cham (2016). https://doi.org/10.1007/978-3-319-45559-4_20

10. Wang, R., Guiochet, J., Motet, G., Schön, W.: Safety case confidence propagation based on Dempster-Shafer theory. Int. J. Approximate Reasoning **107**, 46–64 (2019)

The Extension of Dempster's Combination Rule Based on Generalized Credal Sets

Andrey G. Bronevich$^{(\boxtimes)}$ (ID) and Igor N. Rozenberg (ID)

JSC "Research and Design Institute for Information Technology,
Signalling and Telecommunications on Railway Transport", Orlikov per.5,
building 1, 107996 Moscow, Russia
brone@mail.ru

Abstract. Dempster's aggregation rule plays an important role in the theory of belief functions. Recently, the concept of a generalized credal set has been introduced that allows us to model uncertainty caused by imprecision, conflict and contradiction in information. This concept generalizes in some sense constructions used in evidence theory and imprecise probabilities. In this paper, we show that Dempster's rule can be viewed as the average conditional of a given plausibility function w.r.t. a random set. This allows us to define the extension of this rule into a wider family of uncertainty models presented by generalized credal sets.

Keywords: Belief functions · Conditional generalized credal sets · Dempster's combination rule

1 Introduction

Although there are some common features between evidence theory [8,10] and imprecise probabilities [1,12], some researchers [2,12] argue that the basic constructions of evidence theory, like Dempster's combination rule, do not have a solid justification based on the probabilistic interpretation. On the other hand, the theory of belief functions has strong capabilities providing merging information from several sources with the conflict management. This is possible, because we can model the contradiction in the information if we allow the belief function to assign a strictly positive value to the empty set [11]. The analogous idea is exploited in generalized credal sets [4–7], where sets of probability measures are given with the corresponding amount of contradiction or likelihood.

In the paper, we consider the result of Dempster's rule as the average conditional plausibility function describing the first source of information, given a random set describing the second source of information. Using this interpretation, we extend Dempster's rule for the case, when the first source of information is described by a generalized credal set.

The paper has the following structure. Sections 2 and 3 give us the basic notions concerning belief functions and imprecise probabilities. In Sect. 4 and 5,

S. Le Hégarat-Mascle et al. (Eds.): BELIEF 2022, LNAI 13506, pp. 242–250, 2022.
https://doi.org/10.1007/978-3-031-17801-6_23

we give the necessary information about generalized credal sets concerning their representation and updating information based on them. In Sect. 6, we analyze the expression for Dempster's rule, that helps us to view it as the average conditional w.r.t. a random set. We finish our paper with the conclusion about the obtained results.

2   Basic Notions Concerning Monotone Measures and Belief Functions

Let X be a finite set and let 2^X be the powerset of X. A set function $\mu : 2^X \to [0,1]$ is called a *monotone measure* [9] if 1) $\mu(\emptyset) = 0$, $\mu(X) = 1$; 2) $\mu(A) \leqslant \mu(B)$ if $A \subseteq B$. The set of all monotone measures is denoted by $M_{mon}(X)$.

We will use the following relations and operations on $M_{mon}(X)$:

1) $\mu_1 \leqslant \mu_2$ for $\mu_1, \mu_2 \in M_{mon}(X)$ if $\mu_1(A) \leqslant \mu_2(A)$ for all $A \in 2^X$;
2) $\mu = a\mu_1 + (1-a)\mu_2$ for $a \in [0,1]$ and $\mu_1, \mu_2 \in M_{mon}(X)$ if $\mu(A) = a\mu_1(A) + (1-a)\mu_2(A)$;
3) μ^d is called the *dual* of $\mu \in M_{mon}(X)$ if $\mu^d(A) = 1 - \mu(A^c)$ for all $A \in 2^X$, where A^c denotes the complement of A.

A $Bel \in M_{mon}(X)$ is called a *belief function* if there is a $m : 2^X \to [0,1]$ called the *basic belief assignment* (bba) such that

1) $\sum_{B \in 2^X} m(B) = 1$;
2) $Bel(A) = \sum_{B \subseteq A | B \in 2^X} m(B)$ for every $A \in 2^X$.

The dual Pl of a belief function Bel is called the *plausibility function*. It can be computed through the bba m as

$$Pl(A) = \sum_{A \cap B \neq \emptyset | B \in 2^X} m(B), \quad A \in 2^X.$$

Note that the belief function Bel, and the corresponding m and Pl are equivalent representations of the same uncertain information. If $Bel(\emptyset) = 0$ (or equivalently $m(\emptyset) = 0$, $Pl(X) = 1$), Bel, m, Pl are called *normalized*. With the help of $m(\emptyset) > 0$, we can model the contradiction in information, and this subject will be discussed later.

Consider an arbitrary belief function Bel with the bba m. An $A \in 2^X$ is called a *focal element* if $m(A) > 0$. The set of all focal elements for Bel is called the *body of evidence*. If the body of evidence contains only one focal element, then Bel is called *categorical*. The categorical belief function with the focal element $B \in 2^X$ is denoted by $\eta_{\langle B \rangle}$ and it is easy to see that $\eta_{\langle B \rangle}(A) = 1$ if $B \subseteq A$ and $\eta_{\langle B \rangle}(A) = 0$ otherwise. Every belief function Bel with the bba m can be represented as a convex sum of categorical belief functions as

$$Bel = \sum_{B \in 2^X} m(B)\eta_{\langle B \rangle}.$$

A normalized belief function is called *Bayesian* if its body of evidence consists of singletons. Formally, every Bayesian belief function is a probability measure. We will use the following notations:

- $M_{bel}(X)$ is the set of all normalized belief functions on 2^X;
- $M_{pr}(X)$ is the set of all (normalized) probability measures on 2^X.

Assume that $\mathcal{M} \subseteq M_{mon}(X)$, then $\mathcal{M}^d = \{\mu^d | \mu \in \mathcal{M}\}$. Using this notation, we can denote, for example, the set of all plausibility functions on 2^X by $M_{bel}^d(X)$.

3   Modelling Uncertainty by Belief Functions and Imprecise Probabilities

If we use the probabilistic interpretation of (normalized) belief functions, then every $Bel \in M_{bel}(X)$ defines the *credal set* (the non-empty set of probability measures) defined by

$$\mathbf{P}(Bel) = \{P \in M_{pr}(X) | P \geqslant Bel\}.$$

Therefore, using belief functions, we can model two types of uncertainty [3]: conflict that is inherent to probability measures and non-specificity that characterizes the possible choice of a probability measure in $\mathbf{P}(Bel)$.

More generally, a credal set $\mathbf{P}$ is a convex and closed subset of $M_{pr}(X)$. In other words, if $X = \{x_1, ..., x_n\}$, then $\{(P(\{x_1\}), ..., P(\{x_n\})) | P \in \mathbf{P}\}$ is a convex and closed subset of $\mathbb{R}^n$. Unfortunately, there are credal sets, which are not representable by belief functions [1,12]. In this case, we can represent credal sets by lower or upper previsions.

Let $\mathcal{F}$ be the set of all real valued functions on X and $\mathcal{K} \subseteq \mathcal{F}$. Then every $\underline{P}:$ $\mathcal{K} \to \mathbb{R}$ can be viewed as a *lower prevision* if $\underline{P}(f) \leqslant \max_{x \in X} f(x)$ for every $f \in \mathcal{K}$. Formally, values of $\underline{P}$ can be considered as lower bounds of expectations. Assume that $P \in M_{pr}(X)$ and $E_P(f) = \sum_{x \in X} f(x)P(\{x\})$. Then a lower prevision $\underline{P} : \mathcal{K} \to \mathbb{R}$ is called *non-contradictory* (consistent) if it defines the non-empty set of probability measures (credal set)

$$\mathbf{P}(\underline{P}) = \{P \in M_{pr}(X) | \forall f \in \mathcal{K} : \underline{P}(f) \leqslant E_P(f)\}.$$

Analogously, we can associate the lower prevision

$$\underline{E}_{\mathbf{P}}(f) = \inf\{E_P(f) | P \in \mathbf{P}\}, \quad f \in \mathcal{F}, \tag{1}$$

with any possible credal set $\mathbf{P} \subseteq M_{pr}(X)$. There is the bijection between credal sets and lower previsions defined by (1). A non-contradictory lower prevision $\underline{P} : \mathcal{K} \to \mathbb{R}$ is called *coherent* if there is a credal set $\mathbf{P} \subseteq M_{pr}(X)$ such that $\underline{P}(f) = \underline{E}_{\mathbf{P}}(f)$ for all $f \in \mathcal{K}$.

It is possible to transform every non-contradictory lower prevision $\underline{P} : \mathcal{K} \to \mathbb{R}$ to the coherent lower prevision $\underline{P}' : \mathcal{K} \to \mathbb{R}$ computed as $\underline{P}'(f) = \underline{E}_{\mathbf{P}(\underline{P})}(f)$, $f \in \mathcal{K}$. This transform is called the *natural* extension.

We can also model uncertainty using upper previsions. In this case, the functional $\overline{P} : \mathcal{K} \to \mathbb{R}$ can be viewed as an *upper prevision* if $\overline{P}(f) \geqslant \min_{x \in X} f(x)$; values of $\overline{P}$ can be viewed as upper bounds of expectations and $\overline{P}$ is non-contradictory if it defines the credal set

$$\mathbf{P}(\overline{P}) = \{P \in M_{pr}(X) | \forall f \in \mathcal{K} : \overline{P}(f) \geqslant E_P(f)\}.$$

It is easy to see that if $\underline{P} : \mathcal{K} \to \mathbb{R}$ is an non-contradictory lower prevision, then the upper prevision $\overline{P} : \mathcal{K}' \to \mathbb{R}$, where $\mathcal{K}' = -f | f \in \mathcal{K}$, defined by $\overline{P}(f) = -\underline{P}(-f)$, $f \in \mathcal{K}'$, defines the same credal set, i.e. $\mathbf{P}(\overline{P}) = \mathbf{P}(\underline{P})$. This allows us to say that uncertainty models based on lower and upper probabilities, and convex, closed credal sets have the same expressive capabilities. We use the analogous terminology and notations for upper previsions: if $\mathbf{P} \subseteq M_{pr}(X)$, then the corresponding coherent upper prevision on $\mathcal{F}$ is $\overline{E}_{\mathbf{P}}(f) = \sup\{E_P(f) | P \in \mathbf{P}\}$, $f \in \mathcal{F}$. An upper non-contradictory prevision $\overline{P} : \mathcal{K} \to \mathbb{R}$ is coherent if $\overline{P}(f) = \overline{E}_{\mathbf{P}(\overline{P})}(f)$ for all $f \in \mathcal{K}$. We can define the natural extension for upper previsions analogously as for lower previsions.

4   Contradictory Upper Previsions and Generalized Credal Sets

Definition 1. *Let $\overline{P} : \mathcal{K} \to \mathbb{R}$ be an upper prevision. The amount of contradiction in $\overline{P}$ is the real number computed as*

$$Con(\overline{P}) = \inf\{a \in [0,1] | \overline{P} = (1-a)\overline{P}^{(1)} + a\overline{P}^{(2)}\},$$

where the disaggregation $\overline{P} = (1-a)\overline{P}^{(1)} + a\overline{P}^{(2)}$ is produced over all possible $\underline{P}^{(i)} : \mathcal{K} \to \mathbb{R}$, $i = 1, 2$, where $\overline{P}^{(1)}$ is an non-contradictory upper prevision, and $\overline{P}^{(2)}$ is a contradictory upper prevision.

An upper prevision $\overline{P} : \mathcal{K} \to \mathbb{R}$ is called fully contradictory iff $Con(\overline{P}) = 1$. In the next, we will consider mostly two cases, when $\mathcal{K} = \mathcal{F}$ or $\mathcal{K} = \{1_A\}_{A \in 2^X}$, where 1_A is the characteristic function of the set A. If $\mathcal{K} = \mathcal{F}$, then the fully contradictory upper prevision is $\overline{P}(f) = \min_{x \in X} f(x)$. In the case, when $\mathcal{K} = \{1_A\}_{A \in 2^X}$, we describe uncertainty information by monotone measures called *upper probabilities*, and their values are viewed as upper bounds of probabilities. Therefore, every $\mu \in M_{mon}(X)$ can be an upper probability, and it can be described by the upper prevision $\overline{P}(1_A) = \mu(A)$ for all $A \in 2^X$. Analogously, if $\mathcal{K} = \{1_A\}_{A \in 2^X}$, then the fully contradiction is described by monotone measure $\overline{P}(1_A) = \min_{x \in X} 1_A(x) = \eta_{\langle X \rangle}(A)$, $A \in 2^X$.

Assume that $X = \{x_1, ..., x_n\}$ and consider monotone measures of the type:

$$P = a_0 \eta_{\langle X \rangle} + \sum_{i=1}^{n} a_i \eta_{\langle \{x_i\} \rangle}, \tag{2}$$

where $a_i \in [0,1]$, $i = 0, ..., n$, $\sum_{i=0}^{n} a_i = 1$. If they are viewed as upper probabilities, then $Con(P) = a_0$, and in the case, when $a_0 < 1$, we have the following disaggregation: $P = (1-a_0)P^{(1)} + a_0\eta_{\langle X\rangle}$, where $P^{(1)} = \frac{1}{1-a_0}\sum_{i=1}^{n} a_i\eta_{\langle\{x_i\}\rangle}$ is a probability measure and $\eta_{\langle X\rangle}$ describes the contradiction in P. The same information can be described by an upper prevision $\overline{P}(f) = (1-a_0)E_{P^{(1)}}(f) + a_0\min_{x\in X} f(x)$, or

$$\overline{P}(f) = \sum_{i=1}^{n} a_i f(x_i) + a_0 \min_{x\in X} f(x). \tag{3}$$

Notice that the right part of (3) is the Choquet integral w.r.t. P, and we further denote the $\overline{P}(f)$ in (3) by $E_P(f)$. We see that the meaningful information in (2) is contained in the probability measure $P^{(1)}$ and the term $a_0\eta_{\langle X\rangle}$ only gives us the degree of contradiction. Therefore, using monotone measures from (2), we can model conflict and contradiction in information. The set of all such measures on 2^X is denoted by $M_{cpr}(X)$.

Definition 2. *An non-empty subset* $\mathbf{P} \subseteq M_{cpr}(X)$ *is called a lower generalized credal set (LG-credal set) if*

1) $\mathbf{P}$ *is convex and closed;*
2) $P_1 \leqslant P_2$ *for* $P_1 \in M_{cpr}(X)$ *and* $P_2 \in \mathbf{P}$ *implies* $P_1 \in \mathbf{P}$.

Every $P = a_0\eta_{\langle X\rangle} + \sum_{i=1}^{n} a_i\eta_{\langle\{x_i\}\rangle}$ in $M_{cpr}(X)$ can be represented as a point $(a_1, ..., a_n)$ in $\mathbb{R}^n$. Therefore, we can write $P = (a_1, ..., a_n)$ for the measure defined by (2). Then we see that a LG-credal set is represented as a convex and closed subset of $\{(a_1, ..., a_n) \in \mathbb{R}^n | a_i \geqslant 0, i = 1, ..., n; \sum_{i=1}^{n} a_i \leqslant 1\}$.

Additionally, it is easy to show that $P_1 \leqslant P_2$ for measures $P_j = (a_1^{(j)}, ..., a_n^{(j)})$, $j = 1,2$, defined by (2) iff $a_i^{(1)} \leqslant a_i^{(2)}$, $i = 1, ..., n$.

When we use generalized credal sets, we assume that all useful information is contained in their profiles. Let $\mathbf{P} \subseteq M_{cpr}(X)$ is a LG-credal set, then its *profile* denoted by $profile(\mathbf{P})$ is the set of all maximal elements in $\mathbf{P}$ w.r.t. $\leqslant$. Obviously, the profile of a LG-credal set $\mathbf{P} \subseteq M_{cpr}(X)$ defines it uniquely since $\mathbf{P} = \{P \in M_{cpr}(X)|\exists Q \in profile(\mathbf{P}) : P \leqslant Q\}$. Every usual credal set $\mathbf{P}' \subseteq M_{pr}(X)$ can be represented as the LG-credal set $\mathbf{P} \subseteq M_{cpr}(X)$ with $profile(\mathbf{P}) = \mathbf{P}'$.

The amount of contradiction in a LG-credal set $\mathbf{P} \subseteq M_{cpr}(X)$ is defined as

$$Con(\mathbf{P}) = \inf\{Con(P)|P \in \mathbf{P}\}.$$

Every upper prevision $\overline{P} : \mathcal{K} \to \mathbb{R}$ can be described by a LG-credal set

$$\mathbf{P}(\overline{P}) = \{P \in M_{cpr}|\forall f \in \mathcal{K} : E_P(f) \leqslant \overline{P}(f)\}. \tag{4}$$

Proposition 1. *Let* $\overline{P} : \mathcal{K} \to \mathbb{R}$ *be a an upper prevision, and let* $\mathbf{P}(\overline{P})$ *be the LG-credal set defined by (4), then* $Con(\mathbf{P}(\overline{P})) = Con(\overline{P})$.

An upper prevision $\overline{P} : \mathcal{K} \to \mathbb{R}$ is called *coherent w.r.t. LG-credal sets* (LG-coherent) if there is a LG-credal set $\mathbf{P} \subseteq M_{cpr}(X)$ such that $\overline{P}(f) = \overline{E}_{\mathbf{P}}(f)$ for all $f \in \mathcal{K}$, where $\overline{E}_{\mathbf{P}}(f) = \sup\{E_P(f)|P \in \mathbf{P}\}$. Every $\overline{P} : \mathcal{K} \to \mathbb{R}$ can be transformed to the LG-coherent upper prevision $\overline{P'} : \mathcal{K} \to \mathbb{R}$ as $\overline{P'}(f) = \overline{E}_{\mathbf{P}(\overline{P})}(f)$ for all $f \in \mathcal{K}$. The last transformation is called *the natural extension based on LG-credal sets*.

In [5], a reader can find the necessary and sufficient conditions, when an upper prevision $\overline{P} : \mathcal{F} \to \mathbb{R}$ is LG-coherent. We recall one simple property of $\overline{P}$: $\overline{P}(f + c) = \overline{P}(f) + c$ for every $f \in \mathcal{F}$ and $c \in \mathbb{R}$. This allows us to define such functionals only on $f \in \mathcal{F}$ normalized from below, i.e. functions $f \in \mathcal{F}$ such that $\min\limits_{x \in X} f(x) = 0$. Therefore, any function $f \in \mathcal{F}$ can be normalized: $\underline{f}(x) = f(x) - \min\limits_{x \in X} f(x), x \in X$, and $\overline{P}(f) = \overline{P}(\underline{f}) + \min\limits_{x \in X} f(x)$ for every $f \in \mathcal{F}$. The set of all normalized functions in $\mathcal{F}$ is denoted by $\underline{\mathcal{F}}$.

5   Updating Information Based on LG-Credal Sets

Updating information based on LG-credal sets consists of two stages:

1) the conjunction of two information sources;
2) the contradiction correction.

The conjunction of two LG-credal sets is simply their intersection. Let us describe the first stage. Assume that a prior information is described by a LG-credal set $\mathbf{P} \subseteq M_{cpr}(X)$ and we need to find the description of the posterior information given $B \in 2^X$. The occurrence of an event B can be described by the upper probability $\eta^d_{\langle B \rangle}$ or by the LG-credal set $\mathbf{P}(\eta^d_{\langle B \rangle})$ that consists of $a_0 \eta_{\langle X \rangle} + \sum_{i=1}^{n} a_i \eta_{\langle \{x_i\} \rangle}$ in $M_{cpr}(X)$, in which $a_i = 0$ if $x_i \notin B$.

Lemma 1. *Consider a LG-credal set* $\mathbf{P} \subseteq M_{cpr}(X)$ *with a* $profile(\mathbf{P}) = \{P\}$*, where* $P = a_0 \eta_{\langle X \rangle} + \sum_{i=1}^{n} a_i \eta_{\langle \{x_i\} \rangle}$*. Then* $\mathbf{P}_B = \mathbf{P} \cap \mathbf{P}(\eta^d_{\langle B \rangle})$*, where* $profile(\mathbf{P}_B) = \{P_B\}$ *and* $P_B = b_0 \eta_{\langle X \rangle} + \sum_{i=1}^{n} b_i \eta_{\langle \{x_i\} \rangle}$ *is such that* $b_0 = a_0 + \sum_{x_i \in X \setminus B} a_i$*,* $b_i = a_i$ *if* $x_i \in B$*, and* $b_i = 0$ *if* $x_i \in X \setminus B$*.*

Proposition 2. *Assume that we use the notations from Lemma 1,* $\mathbf{P} \subseteq M_{cpr}(X)$ *is a LG-credal set and* $\mathbf{P}_B = \mathbf{P} \cap \mathbf{P}(\eta^d_{\langle B \rangle})$*. Then* $\mathbf{P}_B = \{P_B|P \in \mathbf{P}\}$*.*

Proposition 2 implies that $profile(\mathbf{P}_B) \subseteq \{P_B|P \in profile(\mathbf{P})\}$. The way of finding the description of $\mathbf{P}_B$ for $\mathbf{P}$ with a finite number of extreme points can be found in [6]. The next proposition shows, how the conditioning is produced for coherent upper previsions w.r.t. LG-credal sets.

Proposition 3. *Let* $\overline{P} : \underline{\mathcal{F}} \to \mathbb{R}$ *be a LG-coherent upper prevision and* $B \in 2^X$*. Let us denote* $\mathbf{P} = \mathbf{P}(\overline{P})$ *and* $\overline{P}_B(f) = \overline{P}(1_B f)$*,* $f \in \underline{\mathcal{F}}$*. Then* $\mathbf{P}_B = \mathbf{P}(\overline{P}_B)$*. In addition,* $\overline{P}_B$ *is also the LG-coherent upper prevision.*

Proposition 3 can be extended to LG-coherent upper probabilities as follows. Assume that $\mu \in M_{mon}(X)$ is an LG-coherent upper probability. Then its conditioning μ_B given $B \in 2^X$ is $\mu_B(A) = \mu(A \cap B) + (1 - \mu(B))\eta_{\langle X \rangle}(A)$, $A \in 2^X$. In addition, μ_B is an LG-coherent upper probability. Formally, the formula for μ_B repeats the result formulated in Proposition 3, since $\mu_B(A) = \mu(A \cap B)$, $A \in 2^X \backslash \{X\}$.

The contradiction correction should be produced before the moment, when we need to make a decision. During this stage, we transform a LG-credal set to the usual model of imprecise probabilities (credal set) without contradiction. In [6], two possible transformations are introduced that guarantee usual conditioning used for probability measures:

1) $\Phi_1(\mathbf{P}) = \{\varphi(P)|P \in \mathbf{P}, Con(P) = Con(\mathbf{P})\}$;
2) $\Phi_2(\mathbf{P}) = \{\varphi(P)|P \in profile(\mathbf{P})\}$;

where $\varphi(P)$ is defined for all $P \in M_{cpr}(X)$ with $Con(P) < 1$ such that

$$\varphi(P) = \frac{1}{1-a_0} \sum_{i=1}^{n} a_i \eta_{\langle \{x_i\} \rangle} \text{ if } P = a_0 \eta_{\langle X \rangle} + \sum_{i=1}^{n} a_i \eta_{\langle \{x_i\} \rangle}.$$

Assume that information is described by a plausibility function Pl on 2^X. Consider how to define its conditioning given $B \in 2^X$. In this case, Pl_B is defined by

$$Pl_B(A) = Pl(A \cap B) + (1 - Pl(B))\eta_{\langle X \rangle}(A), \ A \in 2^X.$$

In [6], we show that after the contradiction correction, the updated information is described by the plausibility function $Pl_{|B}(A) = Pl(A \cap B)/Pl(B)$, $A \in 2^X$, after both possible transformations Φ_i, $i = 1, 2$.

6   Generalized Credal Sets and Dempster's Rule

Consider firstly the *conjunctive rule* [11] that can be applied to so called independent sources of information, given by bbas $m_i : 2^X \rightarrow [0, 1]$ and corresponding plausibility functions Pl_i, $i = 1, 2$. Then their conjunction has the bba

$$m(D) = \sum_{C, B \in 2^X | C \cap B = D} m_1(C)m_2(B), \ D \in 2^X,$$

and the corresponding belief and plausibility functions are defined as

$$Bel = \sum_{C, B \in 2^X} m_1(C)m_2(B)\eta_{\langle C \cap B \rangle}, \ Pl = \sum_{C, B \in 2^X} m_1(C)m_2(B)\eta^d_{\langle C \cap B \rangle}.$$

Since

$$Pl(A) = \sum_{C, B \in 2^X} m_1(C)m_2(B)\eta^d_{\langle C \cap B \rangle}(A) = \sum_{C, B \in 2^X} m_1(C)m_2(B)\eta^d_{\langle C \rangle}(A \cap B) =$$
$$\sum_{B \in 2^X} m_2(B) \sum_{C \in 2^X} m_1(C)\eta^d_{\langle C \rangle}(A \cap B) = \sum_{B \in 2^X} m_2(B)Pl_1(A \cap B).$$

In the last expression, the set function $(Pl_1)_B(A) = Pl_1(A \cap B)$, $A \in 2^X$, can be considered as the conditioning of Pl given B. Therefore, we can write

$$Pl = \sum_{B \in 2^X} m_2(B)(Pl_1)_B. \tag{5}$$

It should noticed in the transferable belief model [11], the fully contradiction is modelled by the set function $\eta_{\langle\emptyset\rangle}$ that is identical to one if we represent information by belief functions, and the set function $\eta_{\langle\emptyset\rangle}^d$ that is identical to zero if we represent information by plausibility functions. This implies that amount of contradiction in Pl is $Con(Pl) = 1 - \sum_{B \in 2^X} m(B)Pl_1(B)$. We can express the amount of contradiction in Pl using bbas as $Con(Pl) = \sum_{B,C \in 2^X | B \cap C = \emptyset} m_1(B)m_2(C)$. If we produce the contradiction correction in Pl using Φ_1 and Φ_2, then in both cases, we get the plausibility function generated by Dempster's aggregation rule.

The expression (5) can be viewed as the average conditional of Pl_1 w.r.t. a random set Ξ defined by probabilities $\Pr(\Xi = B) = m_2(B)$, i.e. we can use the notation $PL = (Pl_1)_\Xi$. This interpretation allows us to extend the average conditionals w.r.t. a random set Ξ for an arbitrary LG-credal set $\mathbf{P} \subseteq M_{cpr}(X)$ as $\mathbf{P}_\Xi = \sum_{B \in 2^X} \Pr(\Xi = B)\mathbf{P}_B$ or for a LG-coherent upper prevision $\overline{P} : \mathcal{F} \to \mathbb{R}$ as $\overline{P}_\Xi = \sum_{B \in 2^X} \Pr(\Xi = B)\overline{P}_B$.

7 Conclusion

As we see from Sect. 6, Dempster's rule of aggregation has the probabilistic interpretation based on generalized credal sets, and this allows to investigate its applicability in some cases. It is also possible to extend Dempster's rule for the case, when both sources of information are described by generalized credal sets using approximations of generalized credal sets, but this problem is the topic for the future research.

References

1. Augustin, T., Coolen, F.P.A., de Cooman, G., Troffaes, M.C.M. (eds.): Introduction to Imprecise Probabilities. Wiley, New York (2014)
2. Brodzik, A.K., Enders, R.H.: A case of combination of evidence in the Dempster-Shafer theory inconsistent with evaluation of probabilities. CoRR arXiv:1107.0082 (2011)
3. Bronevich, A.G., Klir, G.J.: Measures of uncertainty for imprecise probabilities: an axiomatic approach. Int. J. Approximate Reasoning **51**(4), 365–390 (2010)
4. Bronevich, A.G., Rozenberg, I.N.: Incoherence correction and decision making based on generalized credal sets. In: Antonucci, A., Cholvy, L., Papini, O. (eds.) ECSQARU 2017. LNCS (LNAI), vol. 10369, pp. 271–281. Springer, Cham (2017). https://doi.org/10.1007/978-3-319-61581-3_25
5. Bronevich, A.G., Rozenberg, I.N.: Modelling uncertainty with generalized credal sets: application to conjunction and decision. Int. J. Gen. Syst. **27**(1), 67–96 (2018)

6. Bronevich, A.G., Rozenberg, I.N.: Conditioning of imprecise probabilities based on generalized credal sets. In: Kern-Isberner, G., Ognjanović, Z. (eds.) ECSQARU 2019. LNCS (LNAI), vol. 11726, pp. 374–384. Springer, Cham (2019). https://doi.org/10.1007/978-3-030-29765-7_31

7. Bronevich, A.G., Rozenberg, I.N.: The contradiction between belief functions: its description, measurement, and correction based on generalized credal sets. Int. J. Approximate Reasoning **112**, 119–139 (2019)

8. Dempster, A.P.: Upper and lower probabilities induced by a multivalued mapping. Ann. Math. Stat. **38**, 325–339 (1967)

9. Denneberg, D.: Non-additive Measure and Integral. Kluwer, Dordrecht (1997)

10. Shafer, G.: A Mathematical Theory of Evidence. Princeton University Press, Princeton (1976)

11. Smets, P.: Analyzing the combination of conflicting belief functions. Inf. Fusion **8**(4), 387–412 (2007)

12. Walley, P.: Statistical Reasoning with Imprecise Probabilities. Chapman and Hall, London (1991)

A Correspondence Between Credal Partitions and Fuzzy Orthopartitions

Stefania Boffa$^{(\boxtimes)}$ ⓘ and Davide Ciucci ⓘ

Dipartimento di Informatica, Sistemistica e Comunicazione,
Università degli Studi di Milano-Bicocca, Viale Sarca 336, 20126 Milano, Italy
{stefania.boffa,davide.ciucci}@unimib.it

Abstract. This work highlights the connections between fuzzy ortho-partitions and credal partitions, which are both mathematical structures. It is shown that fuzzy orthopartitions are a more general way to represent partitions with uncertainty than credal partitions.

Keywords: Fuzzy orthopartitions · Credal partitions · Bayesian bbas

1 Introduction

Credal partitions are relevant structures in evidential clustering used to represent partitions in cases of partial knowledge concerning the membership of elements [1] to classes. Assuming that $C = \{C_1, \ldots, C_n\}$ is a standard partition of a universe $U = \{u_1, \ldots, u_l\}$, a credal partition is mathematically defined as a collection $m = \{m_1, \ldots, m_l\}$ of basic belief assignments. By a basic belief assignment (bba), we mean a function $m_i : 2^C \to [0,1]$ verifying the condition $\sum_{A \subseteq C} m_i(A) = 1$[1]. Let $A \subseteq C$, $m_i(A)$ called *mass of belief*, quantifies the evidence supporting the claim "u_i belongs to a block of A" [2,3]. Credal partitions subsume the concept of fuzzy probabilistic partitions, which are composed of all Bayesian bbas, i.e., bbas assigning a non-zero degree only to the singletons of 2^U [4].

Fuzzy orthopartitions have been introduced in [5,6] to model (fuzzy) Ruspini partitions [7] including uncertainty, and are also a generalization of orthopartitions based on classical sets [8]. Mathematically, fuzzy orthopartitions are defined as collections of Intuitionistic Fuzzy Sets (IFS) satisfying a specific list of axioms[2]. Each IFS (μ_i, ν_i) of a fuzzy orthopartition represents a class to which elements belong with a degree of [0,1] that is not precisely known: given $u \in U$, then $\mu_i(u)$ and $\nu_i(u)$ are respectively the *degrees of membership* and *non-membership* of u to the class i, and $h_i(u) = 1 - (\mu_i(u) + \nu_i(u))$ is the *degree of indeterminacy* (or *uncertainty*) of u to the class i. We can view a fuzzy orthopartition where all degrees of indeterminacy equal 0 as a Ruspini partition,

---

[1] We also assume here that m_i is normalized, namely $m_i(\emptyset) = 0$.

[2] By an intuitionistic fuzzy set on a universe U, we mean a pair of functions $\mu_i : U \to [0,1]$ and $\nu_i : U \to [0,1]$ verifying the condition $\mu_i(u) + \nu_i(u) \leq 1$ for each $u \in U$.

© The Author(s), under exclusive license to Springer Nature Switzerland AG 2022
S. Le Hégarat-Mascle et al. (Eds.): BELIEF 2022, LNAI 13506, pp. 251–260, 2022.
https://doi.org/10.1007/978-3-031-17801-6_24

which is formally a collection of fuzzy sets $\pi_1, \ldots, \pi_n : U \to [0,1]$ such that $\pi_1(u) + \ldots + \pi_n(u) = 1$ for each $u \in U$.

Here, to a fuzzy orthopartition we attach the following semantics. Let (μ_i, ν_i) be an IFS on U, $\mu_i(u)$ and $\nu_i(u)$ are respectively the degrees of belief that "u belongs to the class i" and "u does not belong to the class i". Moreover, according to this interpretation, $pl_i(u) = 1 - \nu_i(u)$ is the degree of plausibility that "u belongs to the class i". Therefore, fuzzy orthopartitions like credal partitions, are understood as extensions of fuzzy probabilistic partitions.

This work principally explores the connections between fuzzy orthopartitions and credal partitions. The article is organized as follows. The first section recalls the notion of fuzzy orthopartitions and focuses on their interpretation based on evidence theory. Moreover, we recall how fuzzy orthopartitions as well as credal partitions, are a generalization of fuzzy probabilistic partitions. Section 3 is composed of two subsections. In Subsect. 3.1, a fuzzy orthopartition O is mapped into a class of credal partitions denoted with $\mathcal{F}(O)$. We show that $\mathcal{F}(O)$ can be empty, made of one or infinite credal partitions. In reverse, Subsect. 3.2 aims to assign a fuzzy orthopartition to each credal partition. This correspondence leads to an equivalence relation on the set of all credal partitions (i.e., we say that two credal partitions are equivalent if and only if they correspond to the same fuzzy orthopartition). For these reasons, fuzzy orthopartitions can be considered more general than credal partitions. In the last section, we present the conclusions of this work.

2   Fuzzy Orthopartitions

In the sequel, we consider a standard partition $C = \{C_1, \ldots, C_n\}$ of the universe $U = \{u_1, \ldots, u_l\}$[3]. Furthermore, we respectively use the symbols O and m to indicate the fuzzy orthopartition $\{(\mu_1, \nu_1), \ldots, (\mu_n, \nu_n)\}$ of U and the credal partition $\{m_1, \ldots, m_l\}$ of U, which represent C.

Definition 1. *Let* $O = \{(\mu_1, \nu_1), \ldots, (\mu_n, \nu_n)\}$ *be a family of IFSs of* U. *Then,* O *is a fuzzy orthopartition of* U *if and only if the following properties hold for each* $u \in U$:

a) $\sum_{i=1}^{n} \mu_i(u) \leq 1$,
b) $\mu_i(u) + h_j(u) \leq 1$, $\forall i \neq j$,
c) $\sum_{i=1}^{n} \mu_i(u) + h_i(u) \geq 1$,
d) $\forall i \in \{1, \ldots, n\}$ *with* $h_i(u) > 0$, $\exists j \neq i$ *such that* $h_j(u) > 0$.

Each IFS (μ_i, ν_i) of a fuzzy orthopartition O describes the belief and plausibility relating to the belonging of the elements to the class C_i. Then, $\mu_i(u)$ and $\nu_i(u)$ are respectively interpreted as the *degrees of belief* attached to the claims the true class of the object u is C_i and the true class of the object u is not C_i. Moreover, $pl_i(u) = 1 - \nu_i(u)$ represents the *degree of plausibility* that u belongs to C_i and $h_i(u) = pl_i(u) - \mu_i(u)$.

---

[3] Of course, we need to suppose that $2 \leq n \leq l$.

Let us underline that all properties of Definition 1 are fundamental to consider fuzzy orthopartitions as an extension of standard partitions, and more generally of the concept of orthopartitions based on classical sets as explained in the following remark (see [5,6] for more details).

Remark 1. Let O be a fuzzy orthopartition of U where $\mu_1, \ldots, \mu_n, \nu_1, \ldots, \nu_n$ are Boolean functions (i.e., $\mu_i(u), \nu_i(u) \in \{0, 1\}$ for each $u \in U$). Then, we have proved in [6] that O is also a crisp orthopartition as defined in [8]. Therefore, in this specific case, the fuzzy orthopartition O is understood as a partition where the membership class of some elements is known with certainty (i.e., we can find $i \in \{1, \ldots, n\}$ such that $\mu_i(u) = 1$), whereas the membership class of the remaining elements is completely unknown (i.e., $\mu_i(u) = 0$ for each $i \in \{1, \ldots, n\}$).

In the sequel, we show that a fuzzy orthopartition where all degrees of belief and plausibility coincide, specifies a fuzzy probabilistic partition[4].

Definition 2. *Let O be a fuzzy orthopartition of U, we consider $m_1, \ldots, m_l$: $2^C \to [0, 1]$ such that for $A \in 2^C$,*

$$m_i(A) = \begin{cases} \mu_j(u_i) & \text{if } A = \{C_j\} \text{ with } j \in \{1, \ldots, n\}, \\ 0 & \text{otherwise.} \end{cases} \tag{1}$$

Proposition 1. *Let O be a fuzzy orthopartition of U such that $\mu_i(u) = pl_i(u)$ for each $u \in U$ and $i \in \{1, \ldots, n\}$. Then, $m_1, \ldots, m_l$ given by Definition 2 form a fuzzy probabilistic partition.*

Proof. Let $u_i \in U$, we intend to prove that m_i is a bba, namely $\sum_{A \subseteq C} m_i(A) = 1$. By hypothesis, $h_j(u_i) = 0$ for each $j \in \{1, \ldots, n\}$. Then, by Definition 1 (items (a) and (c)),

$$\mu_1(u_i) + \ldots + \mu_n(u_i) = 1. \tag{2}$$

Moreover, $\sum_{A \subseteq C} m_i(A) = m_i(C_1) + \ldots m_i(C_n)$ by considering that m_i is defined by (1) ($m_i(A) = 0$ for each $A \in 2^C$ that is not a singleton). Using (1) again, we can notice that $m_i(C_1) + \ldots + m_i(C_n)$ coincides with $\mu_1(u_i) + \ldots + \mu_n(u_i)$, which is equal to 1 from (2). Hence, we are sure that m_i is a bba.

Finally, m_i is trivially Bayesian due to Definition 2.

Vice versa, a fuzzy probabilistic partition can be seen as a special fuzzy orthopartition where all degrees of belief and plausibility coincide.

Proposition 2. *Let m be a fuzzy probabilistic partition of U, and let $O = \{(\mu_1, \nu_1), \ldots, (\mu_n, \nu_n)\}$ such that*

$$\mu_i(u_j) = m_j(C_i) \quad \text{and} \quad \nu_i(u_j) = 1 - m_j(C_i) \; \forall i \in \{1, \ldots, n\} \text{ and } j \in \{1, \ldots, l\}. \tag{3}$$

---

[4] Let us recall that a fuzzy probabilistic partition $\{m_1, \ldots, m_l\}$ of U is a collection of Bayesian bbas, namely for each $i \in \{1, \ldots, l\}$, $\sum_{A \subseteq C} m_i(A) = 1$ and $m_i(A) = 0$ for each $A \subseteq C$ that is not a singleton.

Then, O is a fuzzy orthopartition of U such that $h_i(u_j) = 0$ $\forall i \in \{1, \ldots, n\}$ and $j \in \{1, \ldots, l\}$.

Proof. Let $i \in \{1, \ldots, n\}$. By (3), $\mu_i(u) + \nu_i(u) = 1$ for each $u \in U$. This clearly implies that (μ_i, ν_i) is an IFS on U and $h_i(u) = 0$ for each $u \in U$. Moreover, $\{(\mu_1, \nu_1), \ldots, (\mu_n, \nu_n)\}$ satisfies all properties characterizing a fuzzy orthopartition and listed in Definition 1. This can be demonstrated starting from the definition of μ_i and ν_i given by (3) together with the assumptions that m_i is a Bayesian bba , $h_i(u) = 0$ for each $i \in \{1, \ldots, n\}$, and $u \in U$.

Then, the previous propositions lead to a first connection between fuzzy orthopartitions and credal partitions: both coincide with fuzzy probabilistic partitions in special cases. That is, a fuzzy orthopartition where all degrees of uncertainty are 0, can be viewed as a credal partition made of all Bayesian bbas.

Example 1. Let $U = \{u_1, u_2, u_3\}$, consider a credal partition $m = \{m_1, m_2, m_3\}$ and a fuzzy orthopartition $O = \{(\mu_1, \nu_1), (\mu_2, \nu_2)\}$ of U defined in Table 1.

Table 1. Definition of the elements of O and m.

u	$\mu_1(u)$	$\nu_1(u)$	$\mu_2(u)$	$\nu_2(u)$
u_1	0.2	0.8	0.8	0.2
u_2	0.5	0.5	0.5	0.5
u_3	0.6	0.4	0.4	0.6

A	$m_1(A)$	$m_2(A)$	$m_3(A)$
$\emptyset$	0	0	0
C_1	0.2	0.5	0.6
C_2	0.8	0.5	0.4
C	0	0	0

We can notice that O and m are equivalent since they can be obtained one from each other employing Definition 2 and Proposition 2. Indeed, $\mu_1(u_2) = m_2(C_1) = 0.5$, $\mu_2(u_3) = m_3(C_2) = 0.4$, and so on.

3   Connections Between Fuzzy Orthopartition and Credal Partitions

This section explores the connections between fuzzy orthopartitions and credal partitions. They can both be seen as a fuzzy partition under condition of uncertainty. As such, a fuzzy ortho/credal partition can represent several fuzzy partitions, once the uncertainty is solved. More formally, we define the class of all fuzzy partitions that could coincide with a given credal partition.

Definition 3. *Let m be a credal partition of U. We say that a fuzzy probabilistic partition m' is compatible with m if and only if*

$$m_j(\{C_i\}) \leq m'_j(\{C_i\}) \leq \sum_{\{A \mid C_i \in A\}} m_j(A)$$

for each $i \in \{1, \ldots, n\}$ and $j \in \{1, \ldots, l\}$.

Similarly, we can define the class of all fuzzy partitions that could coincide with a given fuzzy orthopartition.

Definition 4. *Let O be a fuzzy orthopartition of U. We say that a fuzzy probabilistic partition m' is compatible with O if and only if $\mu_i(u_j) \leq m'_j(C_i) \leq pl_i(u_j)$, for each $i \in \{1, \ldots, n\}$ and $j \in \{1, \ldots, l\}$.*

Such correspondences naturally arise in a dynamic situation, where knowledge about the membership class of the elements is partial and increases over the time so that fuzzy orthopartitions and credal partitions become fuzzy probabilistic partitions. In this context, a credal partition $\{m_1, \ldots, m_l\}$ is transformed in $\{m'_1, \ldots, m'_l\}$ such that m'_i is a Bayesian bba and $m'_i(\{C_j\})$ belongs to the interval $[m_i(\{C_j\}), \sum_{\{A \mid C_j \in A\}} m_i(A)]$. Therefore, if $A = \{C'_1, \ldots, C'_k\}$ where $C'_1, \ldots, C'_k$ belong to C and $k \geq 2$ (i.e., A is not a singleton), then $m_i(A)$ is distributed among the masses of belief concerning $C'_1, \ldots, C'_k$, i.e., the degrees $m'_i(C'_1), \ldots, m'_i(C'_k)$ supporting the propositions "u_i belongs to C'_1", ..., "u_i belongs to C'_k". Moreover, the limit cases $m'_i(\{C_j\}) = m_i(\{C_j\})$ and $m'_i(\{C_j\}) = \sum_{\{A \mid C_j \in A\}} m_i(A)$ respectively occur when

- "if $C_j \subset A$ and u_i belongs to A, then u_i belongs to $A \setminus \{C_j\}$" and
- "if $C_j \subset A$ and u_i belongs to A, then u_i belongs to C_j".

Similarly, a fuzzy orthopartition $O = \{(\mu_1, \nu_1), \ldots, (\mu_n, \nu_n)\}$ is transformed in a new fuzzy orthopartition $O' = \{(\pi_1, 1 - \pi_1), \ldots, (\pi_n, 1 - \pi_n)\}$ where all degrees of plausibility and belief coincide, and $\mu_i(u_j) \leq \pi_i(u_j) \leq pl_i(u_j)$. By Proposition 1, O' is equivalent to a fuzzy probabilistic partition $\{m'_1, \ldots, m'_l\}$ given by Definition 2 and so, it is true that $\mu_i(u_j) \leq m'_j(C_i) \leq pl_i(u_j)$.

In the sequel, we use the symbols $\mathcal{C}_O$ and $\mathcal{C}_m$ to denote the collections of all fuzzy probabilistic partitions that are respectively compatible with a fuzzy orthopartition O and a credal partition m.

3.1 From a Fuzzy Orthopartition to a Class of Credal Partitions

In this subsection, we associate a given fuzzy orthopartition with a class of credal partitions.

Suppose that O represents a partition C, we are interested in all credal partitions that could correspond to C, according to the uncertainty contained in O. In other words, if O is extracted from a given dataset D, then we intend to determine the characteristics that a credal partition m has to satisfy to be obtained from D and O.

Firstly, the degrees supporting the proposition "u_i belongs to C_j" w.r.t. m and O, have to coincide, i.e., it should hold that $m_j(\{C_i\}) = \mu_i(u_j)$. Secondly, the degrees reflecting the uncertainty about the proposition "u_i belongs to C_j" w.r.t. m and O, have to be equal, i.e., it holds that $\sum_{\{A \mid \{C_j\} \subset A\}} m_i(A) = pl_j(u_i) - \mu_j(u_i)$.

Formally, the class of credal partitions corresponding to a given fuzzy orthopartition is defined as follows. For convenience, we denote the collection of all credal partitions of l bbas with $\mathcal{M}$.

Definition 5. *Let O be a fuzzy orthopartition of U. Then, we put*

$$\mathcal{F}(O) = \{m \in \mathcal{M} \mid m_j(\{C_i\}) = \mu_i(u_j) \text{ and } \sum_{\{A \mid C_i \in A\}} m_j(A) = pl_j(u_i)$$

$$\forall i \in \{1, \ldots, n\} \text{ and } j \in \{1, \ldots, l\}\}. \quad (4)$$

So, according with Eq. (4), given a fuzzy orthopair, for $j \in \{1, \ldots, l\}$, we set $m_j(\{C_1\}), \ldots, m_j(\{C_n\})$ to $\mu_1(u_j), \ldots, \mu_n(u_j)$. Then, it remains to find the value of every $m_j(A)$ with $|A| \geq 2$ so that

$$\sum_{\{A \mid C_1 \in A\}} m_j(A) = pl_1(u_j), \quad \ldots, \quad \sum_{\{A \mid C_n \in A\}} m_j(A) = pl_n(u_j)$$

from (4), and $\sum_{A \subseteq C} m_j(A) = 1$ since m_j must be a bba.

Therefore, if we consider a variable x_A^i for each $A \subseteq C$ such that $\{C_i\} \subset A$ and $i \in \{1, \ldots, n\}$, then the values of $\{m_j(A) \mid |A| \geq 2\}$ form a solution of the following system:

$$S_j = \begin{cases} \mu_1(u_j) + \sum_{\{A \mid \{C_1\} \subset A\}} x_A^1 = pl_1(u_j), \\ \vdots \\ \mu_n(u_j) + \sum_{\{A \mid \{C_n\} \subset A\}} x_A^n = pl_n(u_j), \\ \mu_1(u_j) + \ldots + \mu_n(u_j) + \sum_{x \in \mathcal{A}} x = 1, \end{cases} \quad (5)$$

where $\mathcal{A}$ is a set of variables of $\{x_A^i \mid i \in \{1, \ldots, n\}$ and $|A| \geq 2\}$ such that there exists a one-to-one correspondence between $\mathcal{A}$ and the respective values in $\{m_j(A) \mid |A| \geq 2\}$. Let us notice that S_j is a linear system with $n + 1$ equations and $2^n - n - 1$ variables, then it can admit zero, one or infinite solutions[5]. More precisely, since each m_j is determined by solving S_j, we can say that

- $\mathcal{F}(O) = \emptyset$ if and only if there exists $j \in \{1, \ldots, l\}$ such that S_j is impossible.
- $\mathcal{F}(O)$ has infinite solutions if and only if $S_1, \ldots, S_l$ are non-impossible and at least one of them is indeterminate.
- $\mathcal{F}(O)$ has a unique solution if and only if $S_1, \ldots, S_l$ are determinate.

The next examples show the existence of fuzzy orthopartitions associated with one, none and infinite credal partitions.

Example 2. Let $C = \{C_1, C_2\}$ be a partition of $U = \{u_1, u_2, u_3\}$ represented by the fuzzy orthopartition $O = \{(\mu_1, \nu_1), (\mu_2, \nu_2)\}$, where μ_1, ν_1, μ_2, and ν_2 are determined by Table 2. In order to find $\mathcal{F}(O)$, we need to consider the systems S_1, S_2 and S_3 having 3 equations and one variable. According to (5), we get

$$S_1 = \begin{cases} 0.3 + x_C^1 = 0.6, \\ 0.4 + x_C^1 = 0.7, \\ 0.3 + 0.4 + x_C^1 = 1. \end{cases} \quad S_2 = \begin{cases} 0.4 + x_C^2 = 0.5, \\ 0.5 + x_C^2 = 0.6, \\ 0.4 + 0.5 + x_C^2 = 1. \end{cases} \quad S_3 = \begin{cases} 0 + x_C^3 = 0.6, \\ 0.4 + x_C^3 = 1, \\ 0 + 0.4 + x_C^3 = 1. \end{cases}$$

$$\quad (6)$$

---

[5] We notice that the number of equations is less than or equal to the number of variables in S_j, but this does not imply the existence of a solution. Indeed, if $x = (x_1, \ldots, x_k)$ is a solution of S_j, it must hold that $x_1, \ldots, x_k \geq 0$ (i.e., x_i cannot be any real number).

Table 2. Definition of the elements of O and $\{m_1, m_2, m_3\}$.

u	$\mu_1(u)$	$\nu_1(u)$	$\mu_2(u)$	$\nu_2(u)$
u_1	0.3	0.4	0.4	0.3
u_2	0.4	0.5	0.5	0.4
u_3	0	0.4	0.4	0

A	$m_1(A)$	$m_2(A)$	$m_3(A)$
$\emptyset$	0	0	0
C_1	0.3	0.4	0
C_2	0.4	0.5	0.4
C	0.3	0.1	0.6

Thus, S_1, S_2 and S_3 have a unique solution: $x_C^1 = 0.3, x_C^2 = 0.1$, and $x_C^3 = 0.6$. Consequently, $\mathcal{F}(O) = \{\{m_1, m_2, m_3\}\}$, where m_1, m_2 and m_3 are defined by Table 2.

Example 3. Let $C = \{C_1, C_2\}$ be a partition of $U = \{u_1, u_2, u_3\}$ represented by the fuzzy orthopartition $O = \{(\mu_1, \nu_1), (\mu_2, \nu_2)\}$, where μ_1, ν_1, μ_2, and ν_2 are determined by Table 3.

Table 3. Definition of the elements of the fuzzy orthopartition of Example 3.

u	$\mu_1(u)$	$\nu_1(u)$	$\mu_2(u)$	$\nu_2(u)$
u_1	0.3	0.5	0.2	0.3
u_2	0.5	0.2	0.4	0.4
u_3	0	0.4	0.4	0

Then, the systems S_1, S_2 and S_3 associated with O have 3 equations and one variable:

$$S_1 = \begin{cases} 0.3 + x_C^1 = 0.5, \\ 0.2 + x_C^1 = 0.7, \\ 0.3 + 0.2 + x_C^1 = 1. \end{cases} \quad S_2 = \begin{cases} 0.5 + x_C^2 = 0.8, \\ 0.4 + x_C^2 = 0.6, \\ 0 + 0.4 + x_C^2 = 1. \end{cases} \quad S_3 = \begin{cases} 0 + x_C^3 = 0.6, \\ 0.4 + x_C^3 = 1, \\ 0 + 0.4 + x_C^3 = 1. \end{cases}$$
(7)

We can immediately see that S_1, S_2 and S_3 have no solutions, Hence, $\mathcal{F}(O) = \emptyset$.

Example 4. Let $C = \{C_1, C_2, C_3, C_4\}$ be a partition of $U = \{u_1, u_2, u_3, u_4\}$ represented by the fuzzy orthopartition $O = \{(\mu_1, \nu_1), (\mu_2, \nu_2), (\mu_3, \nu_3), (\mu_4, \nu_4)\}$, where $\mu_1, \nu_1, \mu_2, \nu_2, \mu_3, \nu_3, \mu_4$, and ν_4 are determined by Table 4. We can

Table 4. Definition of the elements of the fuzzy orthopartition of Example 4.

u	$\mu_1(u)$	$\nu_1(u)$	$\mu_2(u)$	$\nu_2(u)$	$\mu_3(u)$	$\nu_3(u)$	$\mu_4(u)$	$\nu_4(u)$
u_1	0.1	0.5	0.1	0.7	0.25	0.45	0.15	0.55
u_2	0.15	0.55	0.1	0.5	0.1	0.7	0.25	0.45
u_3	0.3	0.5	0.2	0.7	0.2	0.65	0.1	0.75
u_4	0.2	0.7	0.1	0.75	0.3	0.5	0.2	0.65

verify that the systems S_1, S_2, S_3 and S_4 associated with O by using (5), have infinite solutions. For example, the system S_1 together with two of its solutions $m_1 : 2^C \rightarrow [0,1]$ and $m_2 : 2^C \rightarrow [0,1]$ are defined by (8) and (9), where $C^* = \{\{C_1\}, \{C_2\}, \{C_1, C_2, C_3\}, \{C_1, C_2, C_4\}\}$ and $C^{**} = \{\{C_1\}, \{C_2\}, \{C_1, C_3\}, \{C_1, C_4\}\}$.

$$S_1 = \begin{cases} 0.1 + \sum_{\{A \mid \{C_1\} \subset A\}} x_A^1 = 0.5, \\ 0.1 + \sum_{\{A \mid \{C_2\} \subset A\}} x_A^1 = 0.3, \\ 0.25 + \sum_{\{A \mid \{C_3\} \subset A\}} x_A^1 = 0.55, \\ 0.15 + \sum_{\{A \mid \{C_4\} \subset A\}} x_A^1 = 0.45, \\ 0.1 + 0.1 + 0.25 + 0.15 + \sum_{x \in A} x = 1. \end{cases} \qquad (8)$$

$$m_1(A) = \begin{cases} 0.1 & \text{if } A \in C^*, \\ 0.2 & \text{if } A = \{C_1, C_3, C_4\}, \\ 0.25 & \text{if } A = \{C_3\}, \\ 0.15 & \text{if } A = \{C_4\}, \\ 0 & \text{otherwise}, \end{cases} \qquad m_1'(A) = \begin{cases} 0.1 & \text{if } A \in C^{**}, \\ 0.2 & \text{if } A = \{C_1, C_2, C_3, C_4\}, \\ 0.25 & \text{if } A = \{C_3\}, \\ 0.15 & \text{if } A = \{C_4\}, \\ 0 & \text{otherwise}, \end{cases}$$
$$(9)$$

Hence, we can conclude that $\mathcal{F}(O)$ is made of infinite credal partitions.

As explained at the beginning of this section, a fuzzy orthopartition O represents the set of compatible fuzzy probabilistic partition C_O. The next theorem shows that the same set of fuzzy partitions is obtained as those compatible with any credal partition $m \in \mathcal{F}(O)$.

Theorem 1. *Let O be a fuzzy orthopartition of U, and let $m \in \mathcal{F}(O)$. Then, $C_m = C_O$.*

Proof. If $m' \in C_m$ then $m_i(\{C_j\}) \leq m_i'(\{C_j\}) \leq \sum_{\{A|C_j \in A\}} m_i(A)$ from Definition 3. Moreover, $m \in \mathcal{F}(O)$ implies that $m_i(\{C_j\}) = \mu_j(u_i)$ and $\sum_{\{A|C_j \in A\}} m_i(A) \leq pl_j(u_i)$ from Definition 5. Then, $C_m \subseteq C_O$ clearly follows from Definition 4. The case $C_O \subseteq C_m$ is symmetric and omitted.

3.2 From a Credal Partition to a Fuzzy Orthopartition

In this subsection, we explain how to associate a fuzzy orthopartition to a given credal partition. The meaning of such correspondence is dual to that exhibited in Subsect. 3.1. Given a credal partition m, we intend to consider a fuzzy orthopartition O such that both m and O can be seen as the generalization of an initial partition in the same conditions of uncertainty/knowledge.

Definition 6. *Let $m \in \mathcal{M}$. Then, we consider $O_m = \{(\mu_1, \nu_1), \dots, (\mu_n, \nu_n)\}$ such that for each $i \in \{1, \dots, l\}$ and $j \in \{1, \dots, n\}$, of course, we get $pl_j(u_i) = \sum_{\{A \mid C_j \in A\}} m_i(A)$.*

$$\mu_j(u_i) = m_i(C_j) \quad and \quad \nu_j(u_i) = 1 - \sum_{\{A \mid C_j \in A\}} m_i(A). \qquad (10)$$

We can prove that the set of pairs assigned to a given credal partition (by means of (10)) is a fuzzy orthopartition.

Theorem 2. *Let $m \in \mathcal{M}$. Then, O_m is a fuzzy orthopartition.*

Proof. First of all, we need to verify that (μ_i, ν_i) is an IFS. By Definition 6,

$$\mu_j(u_i) + \nu_j(u_i) = m_i(C_j) + 1 - \sum_{\{A \mid C_j \in A\}} m_i(A).$$

Since $m_i(C_j) \leq \sum_{\{A \mid C_j \in A\}} m_i(A)$, we have $\mu_j(u_i) + \nu_j(u_i) \leq 1$. The latter implies that (μ_i, ν_i) is an IFS. Then, we intend to prove that all properties of Definition 1 hold for O_m.

a) Let $u_i \in U$. By hypothesis, m_i is a bba. Then, $m_i(C_1) + \ldots + m_i(C_n) \leq 1$. Thus, $\mu_1(u_i) + \ldots + \mu_n(u_i) \leq 1$ by considering that $\mu_j(u_i) = m_i(C_j)$ for each $j \in \{1, \ldots, n\}$. Then, Property (a) of Definition 1 holds for O_m.

b) Let $u_i \in U$ and let $j, k \in \{1, \ldots, n\}$ with $j \neq k$, we want to prove that $\mu_j(u_i) + h_k(u_i) \leq 1$.
   By Definition 6, we get $\mu_j(u_i) + h_k(u_i) = \mu_j(u_i) + 1 - \mu_k(u_i) - \nu_k(u_i) = m_i(C_j) + 1 - m_i(C_k) - (1 - \sum_{\{A \mid C_k \in A\}} m_i(A)) = m_i(C_j) + \sum_{\{A \mid \{C_k\} \subset A\}} m_i(A)$. Since m_i is a bba, it must be $m_i(C_j) + \sum_{\{A \mid \{C_k\} \subset A\}} m_i(A) \leq 1$. Finally, we can conclude that $\mu_j(u_i) + h_k(u_i) \leq 1$ and so, Property (b) of Definition 1 holds for O_m.

c) Let $u_i \in U$. Then, $(\mu_1(u_i) + h_1(u_i)) + \ldots + (\mu_n(u_i) + h_n(u_i))$ is equal to $\sum_{\{A \mid C_1 \in A\}} m_i(A) + \ldots + \sum_{\{A \mid C_n \in A\}} m_i(A)$ due to Definition 6. Moreover, the latter is greater than or equal to $\sum_{A \subseteq C} m_i(A)$, which is 1 because m_i is a bba. Then, Property (c) of Definition 1 holds for O_m.

d) Firstly, by Definition 6, we can observe that $h_j(u_i) = \sum_{\{A \mid \{C_j\} \subset A\}} m_i(A)$ for each $i \in \{1, \ldots, l\}$ and $j \in \{1, \ldots, n\}$.
   Given $u_i \in U$, we suppose that there exists $j \in \{1, \ldots, n\}$ such that $h_j(u_i) \geq 0$. Equivalently, $\sum_{\{A \mid \{C_j\} \subset A\}} m_i(A) > 0$. Hence, there exists a subset A' of C such that $\{C_j\} \subset A'$ and $m_i(A') > 0$. This implies that $h_k(u_i) = \sum_{\{A \mid \{C_k\} \subset A\}} m_i(A) > 0$ when we consider $C_k \in A'$ such that $C_k \neq C_j$. Then, Property (d) of Definition 1 holds for O_m.

By Definition 6, we can observe that different credal partitions can correspond to a same fuzzy orthopartition. Therefore, we can define an equivalence relation on $\mathcal{M}$ as follows: let $m, m' \in \mathcal{M}$, $m \sim m'$ if and only if $O_m = O_{m'}$. This section ends by connecting the fuzzy orthopartitions and credal partitions provided by Definitions 5 and 6.

Theorem 3. *Let O be a fuzzy orthopartition of U and let $m \in \mathcal{F}(O)$. Then, $O_m = O$.*

Proof. The thesis clearly follows from both Definitions 5 and 6.

In other words, we can start from a fuzzy orthopartition O, consider $\mathcal{F}(O)$ and obtain O again by applying Definition 6 to any credal partition in $\mathcal{F}(O)$. Lastly, it is important to notice that the previous theorem allows us to rewrite Equation (4) as $\mathcal{F}(O) = \{m \in \mathcal{M} \mid O_m = O\}$ and see O_m as a fuzzy orthopartition of U verifying $\mathcal{C}_{O_m} = \mathcal{C}_m$ (this result is dual to that provided by Theorem 1).

4   Conclusions

We explored the connections between fuzzy orthopartitions and credal partitions. In summary, we have shown that fuzzy orthopartitions and credal partitions coincide when both are fuzzy probabilistic partitions. In the general case, fuzzy orthopartitions are a more general construct than credal partitions. Indeed, fuzzy orthopartitions are partitioned into three non-empty classes: $\{O \mid \mathcal{F}(O) = \emptyset\}$, $\{O \mid \mathcal{F}(O) = 1\}$, and $\{O \mid \mathcal{F}(O)$ is infinite$\}$. Finally, we have considered equivalence classes on $\mathcal{M}$ composed of all credal partitions corresponding to the same fuzzy orthopartition. We remark that we have focused on normalized bbas, however, our results can be extended in the more general case and this will enable us to deal with the presence of outliers in clustering applications. Indeed, in the case of fuzzy orthopartitions, outliers can be managed considering an IFS (μ_0, ν_0) such that $\mu_0(u_i) = 0$ and $\nu_0(u_i) = k \in (0, 1]$ if u_i is an outlier, and $\nu_0(u_i) = 1$ otherwise. So, the correspondence shown in this paper will be generalized by modifying Definitions 5 and 6. As future developments, besides the application of fuzzy orthopartitions to clustering, we will compare the existing measures of uncertainty in both settings.

References

1. Denœux, T., Masson, M.-H.: EVCLUS: evidential clustering of proximity data. IEEE Trans. Syst. Man Cybern. Part B (Cybern.) **34**(1), 95–109 (2004)
2. Shafer, G.: A Mathematical Theory of Evidence. Princeton University Press, Princeton (1976)
3. Smets, P., Kennes, R.: The transferable belief model. Artif. Intell. **66**(2), 191–234 (1994)
4. Bezdek, J.C., Keller, J., Krisnapuram, R., Pal, N.: Fuzzy Models and Algorithms for Pattern Recognition and Image Processing, vol. 4. Springer, New York (1999). https://doi.org/10.1007/b106267
5. Boffa, S., Ciucci, D.: Logical entropy and aggregation of fuzzy orthopartitions. Fuzzy Sets Syst. (2022, accepted)
6. Boffa, S., Ciucci, D.: Fuzzy orthopartitions and their logical entropy. In: CEUR Workshop Proceedings, p. 3074. WILF (2021)
7. Ruspini, E.H.: A new approach to clustering. Inf. Control **15**(1), 22–32 (1969)
8. Campagner, A., Ciucci, D.: Orthopartitions and soft clustering: soft mutual information measures for clustering validation. Knowl.-Based Syst. **180**, 51–61 (2019)

Toward Updating Belief Functions over Belnap-Dunn Logic

Sabine Frittella[1], Ondrej Majer[2], and Sajad Nazari[1(✉)]

[1] INSA Centre Val de Loire, Univ. Orléans, LIFO EA 4022, Orléans, France
{sabine.frittella,sajad.nazari}@insa-cvl.fr
[2] Institute of Philosophy, Czech Academy of Sciences, Prague, Czech Republic
majer@flu.cas.cz

Abstract. [3] and [7] generalize the notion of probability measures and belief functions to Belnap-Dunn (BD) logic, respectively. This work aims at providing an alternative way to treat contradictory information by relying on a logic that was introduced to reason about incomplete and contradictory information rather than on classical logic. In this article, we study how to update belief functions over BD logic with new pieces of information. We present a first approach via a frame semantics of BD logic. This frame semantics relying on sets, we can use Bayesian update and Dempster-Shafer combination rule over powerset algebras to define their corresponding updates within the framework of BD logic.

Keywords: Belief functions · Belnap-Dunn logic · Bayesian update · Dempster-Shafer combination rule

1 Introduction

Belief functions were introduced to generalise the notion of probabilities to situations with incomplete information. They can be used to encode the information given by one or many pieces of evidence. They were first introduced on Boolean algebras [8], that is within the framework of classical reasoning. Dempster-Shafer theory uses the equivalent representation of belief functions via their mass functions to propose a method to aggregate the information conveyed by two belief functions. Since, the theory of belief functions has been developed on distributive lattices [1,5,9] and finite lattices [4]. Notice that the definition of belief functions (see Definition 3) relies only on the lattice structure of the algebra. However, its dual notion, plausibility, is usually defined by combining a belief function and the classical negation. In this framework, belief is interpreted as a lower bound on the probability of an event and plausibility as an upper bound. Similarly, Dempster-Shafer combination rule also relies strongly on the underlying algebra being Boolean. In this article, we discuss the interpretation of belief

The research of Sabine Frittella and Sajad Nazari was funded by the grant ANR JCJC 2019, project PRELAP (ANR-19-CE48-0006).

and plausibility functions within the framework of Belnap-Dunn logic rather than through classical logic, because it allows us to reason with incomplete and contradictory information.

Belnap-Dunn logic was introduced to reason about information rather than about truth. In classical logic a statement p is either *true* or *false*, meaning that p is *true* (resp. false) iff the statement p is true (resp. false) in the world. In Belnap-Dunn (BD) logic, a statement p is either "supported by the available information", or "contradicted by the available information", or "neither supported nor contradicted by the available information", or "both supported and contradicted by the available information". These four truth values are respectively denoted **t** (*true*), **f** (*false*), **n** (*neither*), **b** (*both*). [3] introduces belief functions over BD models and presents a logic to reason with them. This work was motivated by the counterintuitive results that Dempster-Shafer (DS) theory can produce in presence of highly contradictory pieces of information. Indeed, the strategy of Dempster-Shafer combination rule is to ignore contradictory information rather than dealing with the contradictions and saying something meaning full about them. Since BD logic is a simple, well-established extension of classical logic that was introduced to formalise reasoning based on incomplete and contradictory information, it appeared natural to adapt Dempster-Shafer theory over BD logic.

The next step is to look at taking new pieces of information into account by updating the belief functions. [6] presents different ways to update belief and plausibility functions in the classical framework. This article expends on [3] and discusses how to adapt and interpret those results within the framework of belief and plausibility functions over Belnap-Dunn logic.

Structure of the Paper. In Sect. 2, we present BD logic and non-standard probabilities over BD logic, then we recall definitions and lemmas about belief and plausibility functions and Dempster-Shafer combination rule, finally we present existing proposals to update belief functions over Boolean algebras. In Sect. 3, we introduce models for belief and plausibility over BD logic. We discuss how to update belief and plausibility when getting a new piece of information. In Sect. 4, we discuss further research.

2   Preliminaries

In this section, we first introduce the necessary definitions about BD logic and non-standard probabilities. Then we recall useful definitions and lemmas about belief and plausibility functions and Dempster-Shafer combination rule. Finally, we present two ways to update belief functions on Boolean algebras.

In the reminder of the paper, we will always work with finite lattices. Recall that a *lattice* is a tuple $\mathcal{L} = \langle L, \vee, \wedge \rangle$, such that $\vee$ and $\wedge$ are binary, commutative, associative, and idempotent operations that satisfy the following rules: $x \vee (x \wedge y) = x$ and $x \wedge (x \vee y) = x$, for all $x, y \in \mathcal{L}$. A lattice is *bounded* if it contains nullary operators $\bot$ and $\top$ that represent respectively its lower and upper bounds, i.e., for every element $x \in \mathcal{L}$, we have $x \vee \top = \top$ and $x \wedge \bot = \bot$. A lattice is *distributive* if

$(x \vee y) \wedge z = (x \wedge z) \vee (y \wedge z)$ holds for all $x, y, z \in \mathcal{L}$. A *(bounded) De Morgan algebra* is a (bounded) distributive lattice equipped with an additional unary operation $\neg$ such that $\neg\neg x = x$ and $\neg(x \wedge y) = \neg x \vee \neg y$ for all $x, y \in \mathcal{L}$. A *Boolean algebra* is a bounded De Morgan algebra that satisfies the law of excluded middle ($p \vee \neg p = \top$) and the principle of explosion ($p \wedge \neg p = \bot$). A function $\mu : \mathcal{P}(S) \to [0, 1]$ is called a *(finitely additive) probability measure* if it satisfies the following properties: (i) $\mu(S) = 1$, and (ii) $\mu(A \cup B) = \mu(A) + \mu(B)$ for A, B disjoint.

2.1 Belnap-Dunn Logic

BD logic mentioned in the introduction was introduced by Nuel Belnap in [2]. His main aim was to design a logical system capable of dealing with inconsistent or/and incomplete information. The language $\mathscr{L}_{\mathsf{BD}}$ of BD logic is defined via the following grammar over a finite set of propositions Prop:

$$\varphi := p \in \mathtt{Prop} \mid \neg\varphi \mid \varphi \wedge \varphi \mid \varphi \vee \varphi \mid \bot \mid \top.$$

The constants $\bot$ and $\top$ are not a standard part of the signature of $\mathscr{L}_{\mathsf{BD}}$, we include them for the sake of simplicity. Most of the results can be straightforwardly adapted for the more general framework without the constants. Semantics for BD logic can be provided in two equivalent ways. One possibility is to evaluate the formulas directly into the set $\{\mathbf{t}, \mathbf{f}, \mathbf{n}, \mathbf{b}\}$. We will use the second option based on the idea of independence of positive and negative information. In particular, a lack of positive support for a claim does not automatically mean a support for its negation. This is formally represented by two valuations representing positive (negative) support, respectively.

Definition 1 (Belnap-Dunn models). *A* Belnap-Dunn model (BD model) *is a tuple* $\mathfrak{M} = \langle S, v^+, v^- \rangle$ *where* $S \neq \varnothing$ *is a finite set of states and* v^+, v^- : Prop $\to \mathcal{P}(S)$ *are positive, negative valuation functions respectively.*

The valuations are extended to the corresponding satisfaction relations. Each of them evaluates compound formulas in a way which is in fact classical: e.g. a conjunction is supported positively in a state iff each of its conjuncts is, and it is supported negatively if at least one of the conjuncts is.

Definition 2 (Frame semantics for BD). *Let* $\varphi, \varphi' \in \mathscr{L}_{\mathsf{BD}}$, $\mathfrak{M} = \langle W, v^+, v^- \rangle$ *a* BD *model and* $w \in S$. *Then* $\vDash^+$ *and* $\vDash^-$ *are defined as follows.*

$$w \vDash^+ p \text{ iff } w \in v^+(p) \qquad\qquad w \vDash^- p \text{ iff } w \in v^-(p)$$
$$w \vDash^+ \neg\varphi \text{ iff } w \vDash^- \varphi \qquad\qquad w \vDash^- \neg\varphi \text{ iff } w \vDash^+ \varphi$$
$$w \vDash^+ \varphi \wedge \varphi' \text{ iff } w \vDash^+ \varphi \text{ and } w \vDash^+ \varphi' \quad w \vDash^- \varphi \wedge \varphi' \text{ iff } w \vDash^- \varphi \text{ or } w \vDash^- \varphi'$$
$$w \vDash^+ \varphi \vee \varphi' \text{ iff } w \vDash^+ \varphi \text{ or } w \vDash^+ \varphi' \quad w \vDash^- \varphi \vee \varphi' \text{ iff } w \vDash^- \varphi \text{ and } w \vDash^- \varphi'$$

We will make use of the notions of the *positive extension of a formula* (the set of states supporting it): $|\varphi|^+ = \{s \in S \mid s \vDash^+ \varphi\}$, and analogously the *negative extension of a formula* $|\varphi|^- = \{s \in S \mid s \vDash^- \varphi\}$. Notice that in general a negative extension is not a set theoretical complement of the corresponding positive extension, i.e. $|\varphi|^+ \cup |\varphi|^- \subsetneq S$ (some states might support neither φ nor $\neg\varphi$) and $|\varphi|^+ \cap |\varphi|^- \neq \varnothing$ (states in the intersection support both φ and $\neg\varphi$). Moreover, positive and negative extensions are mutually definable: $|\neg\varphi|^+ = |\varphi|^-$.

When we define a measure on the formulas of some language, it is natural to require that equivalent formulas have the same value (the probability of an event should not depend on its name). Another option which is technically more convenient is to work with equivalence classes of formulas directly. Two formulas are equivalent, denoted $\varphi \sim \psi$, if they are mutually derivable with respect to the axiomatisation of BD logic (see e.g. [3, p. 5]). Formally, this structure is called the Lindenbaum algebra of the logic. For BD logic, it is the free De Morgan algebra generated by the set of atomic variables Prop. It is defined as $\mathcal{L}_{\mathsf{BD}} = (L, \wedge', \neg')$ such that L is the set of equivalence classes $[\varphi] = \{\psi \mid \psi \sim \varphi\}$, $[\varphi] \wedge' [\psi] = [\varphi \wedge \psi]$ and $\neg'[\varphi] = [\neg\varphi]$. In what follows we will not distinguish between a formula and its equivalence class.

Information might be incomplete or contradictory, but usually it is also uncertain. There were several attempts to propose a probabilistic version of BD logic, which extends the basic idea of independence of positive and negative information to the probabilistic case. We build on the framework of non-standard probabilities presented in [7]. We will use the notion of a *probabilistic BD model*, $\mathfrak{M} = \langle S, v^+, v^-, \mu \rangle$, which is a BD model equipped with a probability measure μ on S. The probability of a statement expressed by a formula φ is defined as the (classical) measure of its positive extension: $p(\varphi) = \mu(|\varphi|^+)$. Although the non-standard probability is defined using classical measure on a Boolean algebra, it satisfies axioms weaker than the Kolomogorovian ones. In particular, the additivity axiom does not hold, it is replaced by a weaker principle called inclusion/exclusion : $p(\varphi \vee \psi) = p(\varphi) + p(\psi) - p(\varphi \wedge \psi)$. As a consequence, some classical principles are not valid any more: it might happen that $p(\varphi) + p(\neg\varphi) < 1$ (probabilistic information is incomplete) and $p(\varphi \wedge \neg\varphi) > 0$ (probabilistic information is contradictory).

2.2   Belief and Plausibility Functions

We recall the definitions of belief functions, plausibility functions and mass functions. Usually those definitions are given on Boolean algebras, here, we directly generalise them to lattices.

Definition 3 (Belief function). *Let $\mathcal{L}$ be a a bounded lattice. A function* bel : $\mathcal{L} \to [0,1]$ *is called a* belief function *if the following conditions hold: (i)* bel$(\bot) =$ 0 *and* bel$(\top) = 1$ *; (ii)* bel *is monotone with respect to $\mathcal{L}$, i.e. for every $x, y \in \mathcal{L}$, if $x \leq_{\mathcal{L}} y$, then* bel$(x) \leq$ bel(y) *; (iii)* bel *is weakly totally monotone, i.e. for every $k \geq 1$ and every $a_1, \ldots, a_k \in \mathcal{L}$, it holds that*

$$\mathtt{bel}\left(\bigvee_{1 \leq i \leq k} a_i\right) \geq \sum_{\substack{J \subseteq \{1, \ldots, k\} \\ J \neq \varnothing}} (-1)^{|J|+1} \cdot \mathtt{bel}\left(\bigwedge_{j \in J} a_j\right). \tag{1}$$

Definition 4 (Mass function). *Let $\mathcal{L} \neq \varnothing$ be an arbitrary lattice. A* mass function *on $\mathcal{L}$ is a function $\mathbf{m} : \mathcal{L} \to [0,1]$ such that $\sum_{x \in \mathcal{L}} \mathbf{m}(x) = 1$.*

The following well-known lemma, see for example [9, Theorem 2.8], shows the relation between mass functions and belief functions.

Lemma 1 (Mass function associated to a belief function). *Let $\mathcal{L}$ be a finite lattice and $\mathbf{bel} : \mathcal{L} \to [0,1]$ a belief function. Then, there is a mass function $\mathbf{m}_{\mathbf{bel}} : \mathcal{L} \to [0,1]$, called the* mass function associated to $\mathbf{bel}$, *such that, for every $x \in \mathcal{L}$, $\mathbf{bel}(x) = \sum_{y \leq x} \mathbf{m}_{\mathbf{bel}}(y)$. Conversely, for any mass function $\mathbf{m}$ on the lattice $\mathcal{L}$, the function $\mathbf{bel}_m : \mathcal{L} \to [0,1]$ defined as $\mathbf{bel}_m(x) = \sum_{y \leq x} \mathbf{m}(y)$ is a belief function.*

Definition 5 (Plausibility functions). *Let $\mathcal{L}$ be a bounded lattice. $\mathbf{pl} : \mathcal{L} \to [0,1]$ is called a* plausibility function *if the following conditions hold: (i) $\mathbf{pl}(\bot) = 0$ and $\mathbf{pl}(\top) = 1$; (ii) $\mathbf{pl}$ is monotone with respect to $\mathcal{L}$, (iii) for every $k \geq 1$ and every $a_1, \ldots, a_k \in \mathcal{L}$, it holds that*

$$\mathbf{pl}\left(\bigwedge_{1 \leq i \leq k} a_i\right) \leq \sum_{\substack{J \subseteq \{1, \ldots, k\} \\ J \neq \varnothing}} (-1)^{|J|+1} \cdot \mathbf{pl}\left(\bigvee_{j \in J} a_j\right). \tag{2}$$

The following two lemmas from [3] show the mutual definability of belief, plausibility and their mass functions on De Morgan algebras.

Lemma 2 (Plausibility function associated to a belief function). *Let $\mathcal{L}$ be a bounded De Morgan algebra and $\mathbf{bel} : \mathcal{L} \to [0,1]$ a belief function. Then, the function $\mathbf{pl}_{\mathbf{bel}} : \mathcal{L} \to [0,1]$ such that $\mathbf{pl}_{\mathbf{bel}}(x) = 1 - \mathbf{bel}(\neg x)$ is a plausibility function, called the* plausibility function associated to $\mathbf{bel}$.

Lemma 3 (Mass function associated to a plausibility function). *Let $\mathcal{L}$ be a bounded De Morgan algebra, and $\mathbf{pl} : \mathcal{L} \to [0,1]$ a plausibility function. Then, the function $\mathbf{bel}_{\mathbf{pl}} : \mathcal{L} \to [0,1]$ such that $\mathbf{bel}_{\mathbf{pl}}(x) = 1 - \mathbf{pl}(\neg x)$ is a belief function, called the* belief function associated to $\mathbf{pl}$. *We denote $\mathbf{m}_{\mathbf{pl}}$ the mass function associated to $\mathbf{bel}_{\mathbf{pl}}$ and we call $\mathbf{m}_{\mathbf{pl}}$ the* mass function associated to $\mathbf{pl}$. *Then, $\mathbf{pl}(x) = 1 - \sum_{y \leq \neg x} \mathbf{m}_{\mathbf{pl}}(y)$.*

Notice that, like in the classical case, a mass function $\mathbf{m}$ gives rise to a belief $\mathbf{bel}_m$ and a plausibility $\mathbf{pl}_m$ function such that $\mathbf{pl}_m(x) = 1 - \mathbf{bel}_m(\neg x)$. However, here, since $\neg$ is not a Boolean negation, we cannot prove anymore that $\mathbf{bel}_m(x) \leq \mathbf{pl}_m(x)$. In addition, notice that contrary to the classical case, one cannot rewrite the expression $1 - \sum_{y \leq \neg x} \mathbf{m}_{\mathbf{pl}}(y)$ as $\sum_{y:y \wedge x > \bot} \mathbf{m}_{\mathbf{pl}}(y)$. Indeed, for instance, $\neg x \leq \neg x$, but $\neg x \wedge x \neq \bot$.

Belief functions and their mass functions are used to reason about evidence. Dempster-Shafer combination rule allows merging the information provided by different sources, each source being described by a mass function.

Definition 6 (Dempster-Shafer combination rule). *Let* m_1 *and* m_2 *be two mass functions on* $\mathcal{P}(S)$. *Dempster-Shafer combination rule computes their aggregation* $m_1 \oplus m_2 : \mathcal{P}(S) \to [0,1]$ *as follows.*

$$X \mapsto \begin{cases} 0 & \text{if } X = \varnothing \\[2mm] \dfrac{\sum\{m_1(X_1) \cdot m_2(X_2) \mid X_1 \cap X_2 = X\}}{\sum\{m_1(X_1) \cdot m_2(X_2) \mid X_1 \cap X_2 \neq \varnothing\}} & \text{otherwise.} \end{cases}$$

2.3 Classical Updating of Uncertainty Measures

A probability function μ on a Boolean algebra $\mathcal{P}(S)$ represents the information available about the subsets of S representing events. Observing $B \in \mathcal{P}(S)$ changes the probabilities assigned to the elements of $\mathcal{P}(S)$ and leads to a new probability measure μ_B. There are different strategies to define this new measure. Here we focus on Bayesian updating defined as $\mu_B(C) = \frac{\mu(B \cap C)}{\mu(B)}$, for every $C \in \mathcal{P}(S)$. In this section, we present results from [6]. We recall how Bayesian updating is used to define conditional upper and lower probabilities, and conditional belief and plausibility, and how Dempster-Shafer combination rule can be used to define conditional belief and plausibility. In this section, Pl denotes the plausibility function associated to Bel.

Conditioning Upper and Lower Probabilities. Let $\mathcal{A}$ be a non-empty set of probability measures defined over $\mathcal{P}(S)$. Then its *lower and upper probabilities* (resp. $\mathcal{A}_*$ and $\mathcal{A}^*$) are functions defined on $\mathcal{P}(S)$ as follows, for every $X \in \mathcal{P}(S)$: $\mathcal{A}_*(X) = \inf\{\mu(X) : \mu \in \mathcal{A}\}$ and $\mathcal{A}^*(X) = \sup\{\mu(X) : \mu \in \mathcal{A}\}$. [6] proposes the following way to update a set of probabilities $\mathcal{A}$ using Bayesian update. A priori, observing $B \in \mathcal{P}(S)$ leads to the Bayesian updating μ_B of all probabilities $\mu \in \mathcal{A}$ such that $\mu(B) \neq 0$, that is, the probability measures consistent with that observation. Therefore, the update of $\mathcal{A}$ is defined only if there is at least one $\mu' \in \mathcal{A}$ such that $\mu'(B) > 0$. We denote $\mathcal{A}_B = \{\mu_B : \mu \in \mathcal{A} \text{ and } \mu(B) > 0\}$. We define the *conditional updating of* $\mathcal{A}_*$ *and* $\mathcal{A}^*$ *by* B as follows: $(\mathcal{A}^*)_B := (\mathcal{A}_B)^*$ and $(\mathcal{A}_*)_B := (\mathcal{A}_B)_*$.

Conditioning Belief and Plausibility as Lower and Upper Probabilities. In the classical case, [6] introduces two ways to update belief functions: (1) via the representation of belief functions as lower probabilities (see Theorem 1), and (2) via their associated mass functions (see Proposition 2).

Theorem 1 *[6, Theorem 2.6.1].* *Let* Bel *be a belief function defined on* $\mathcal{P}(S)$ *and* $\mathcal{M}_{\text{Bel}} = \{\mu : \mu(X) \geq \text{Bel}(X), \text{ for all } X \in \mathcal{P}(S)\}$. *Then* Bel $= (\mathcal{M}_{\text{Bel}})_*$ *and* Pl $= (\mathcal{M}_{\text{Bel}})^*$.

The set $\mathcal{M}_{\text{Bel}}$ can be updated when it contains at least one measure such that $\mu(B) > 0$, that is when $\text{Pl}_{\text{Bel}}(B) > 0$.

Definition 7. Let $\mathtt{Bel} : \mathcal{P}(S) \to [0,1]$ be a belief function and $\mathtt{Pl}$ is associated plausibility function such that $\mathtt{Pl}(X) = 1 - \mathtt{Bel}(\overline{X})$ for every $X \in \mathcal{P}(S)$. Then conditioning of $\mathtt{Bel}$ and $\mathtt{Pl}$ on B is defined as follows: $\mathtt{Bel}_B(X) = ((\mathcal{M}_{\mathtt{Bel}})_*)_B(X) = ((\mathcal{M}_{\mathtt{Bel}})_B)_*(X)$ and $\mathtt{Pl}_B(X) = ((\mathcal{M}_{\mathtt{Bel}})^*)_B(X) = ((\mathcal{M}_{\mathtt{Bel}})_B)^*(X)$.

In Sect. 3.3, we will work with models containing belief and plausibility functions that are not interdefinable. The following proposition characterises those functions in terms of lower and upper probabilities.

Proposition 1 (Belief and plausibility as an upper/lower probability). Let f be a function defined on $\mathcal{P}(S)$ and $\mathcal{M}_f = \{\mu : \mu(X) \geq f(X), \text{ for all } X \in \mathcal{P}(S)\}$ and $\mathcal{N}_f = \{\mu : \mu(X) \leq f(X), \text{ for all } X \in \mathcal{P}(S)\}$. Now let $\mathtt{Bel}$ and $\mathtt{Pl}$ be belief and plausibility functions defined independently on $\mathcal{P}(S)$. Then,

$$\mathtt{Pl} = (\mathcal{M}_{\mathtt{Bel}_{\mathtt{Pl}}})^* = (\mathcal{N}_{\mathtt{Pl}})^* \quad \text{and} \quad \mathtt{Bel} = (\mathcal{M}_{\mathtt{Bel}})_* = (\mathcal{N}_{\mathtt{Pl}_{\mathtt{Bel}}})_*$$

where $\mathtt{Bel}_{\mathtt{Pl}}$ and $\mathtt{Pl}_{\mathtt{Bel}}$ are respectively the belief and plausibility associated to $\mathtt{Pl}$ and $\mathtt{Bel}$ (see Lemmas 2 and 3).

Proof. Recall that $\mathtt{Bel}_{\mathtt{Pl}}(X) = 1 - \mathtt{Pl}(\overline{X})$. Based on Theorem 1, we have $\mathtt{Bel}_{\mathtt{Pl}} = (\mathcal{M}_{\mathtt{Bel}_{\mathtt{Pl}}})_*$ and $\mathtt{Pl} = (\mathcal{M}_{\mathtt{Bel}_{\mathtt{Pl}}})^*$. Notice that, for every $X \in \mathcal{P}(S)$,

$$\mu(X) \leq \mathtt{Pl}(X) \iff \mu(X) \leq 1 - \mathtt{Bel}_{\mathtt{Pl}}(\overline{X}) \iff \mathtt{Bel}_{\mathtt{Pl}}(\overline{X}) \leq 1 - \mu(X)$$
$$\iff \mathtt{Bel}_{\mathtt{Pl}}(\overline{X}) \leq \mu(\overline{X}) \iff \mathtt{Bel}_{\mathtt{Pl}}(X) \leq \mu(X)$$
$$\iff \mathtt{Bel}_{\mathtt{Pl}}(X) \leq \mu(X)$$

Therefore, $\mathcal{M}_{\mathtt{Bel}_{\mathtt{Pl}}} = \mathcal{N}_{\mathtt{Pl}}$ and $(\mathcal{M}_{\mathtt{Bel}_{\mathtt{Pl}}})^* = (\mathcal{N}_{\mathtt{Pl}})^*$. The proof for $\mathtt{Bel}$ is similar.

Defining conditioning via lower and upper probabilities is not very practical from a computational perspective. The following theorem gives us an explicit formula.

Theorem 2. Let $\mathtt{Bel} : \mathcal{P}(S) \to [0,1]$ be a belief function. Let $\mathtt{Pl}$ be its associated plausibility function. Suppose that $\mathtt{Pl}(B) > 0$. Then,

$$\mathtt{Bel}_B(X) = \begin{cases} 1 & \text{if } \mathtt{Pl}(\overline{X} \cap B) = 0, \\ \dfrac{\mathtt{Bel}(X \cap B)}{\mathtt{Bel}(X \cap B) + \mathtt{Pl}(\overline{X} \cap B)} & \text{if } \mathtt{Pl}(\overline{X} \cap B) > 0. \end{cases}$$

$$\mathtt{Pl}_B(X) = \dfrac{\mathtt{Pl}(X \cap B)}{\mathtt{Pl}(X \cap B) + \mathtt{Pl}(\overline{X} \cap B)}$$

Proof. [6, Theorem 3.8.2] proves the formula for $\mathtt{Bel}$ and the fact that

$$\mathtt{Pl}_B(X) = \begin{cases} 0 & \text{if } \mathtt{Pl}(X \cap B) = 0, \\ \dfrac{\mathtt{Pl}(X \cap B)}{\mathtt{Pl}(X \cap B) + \mathtt{Pl}(\overline{X} \cap B)} & \text{if } \mathtt{Pl}(X \cap B) > 0. \end{cases}$$

Notice that $\text{Pl}(X \cap B) + \text{Pl}(\overline{X} \cap B) > 0$ for every $X \in \mathcal{P}(S)$. Indeed, since Pl is a plausibility function, we have $\text{Pl}(A \cup C) \leq \text{Pl}(A) + \text{Pl}(C) - \text{Pl}(A \cap C)$ for every $A, C \in \mathcal{P}(S)$. If $A \cap C = \emptyset$, then $\text{Pl}(A \cup C) \leq \text{Pl}(A) + \text{Pl}(C)$. Hence, if $A = \overline{X} \cap B$ and $C = X \cap B$, we have $\text{Pl}(B) \leq \text{Pl}(\overline{X} \cap B) + \text{Pl}(X \cap B)$. Since $\text{Pl}(B) > 0$, we have $\text{Pl}(\overline{X} \cap B) + \text{Pl}(X \cap B) > 0$, for every $X \in \mathcal{P}(S)$. Therefore, if $\text{Pl}(X \cap B) = 0$, we have $\frac{\text{Pl}(X \cap B)}{\text{Pl}(X \cap B) + \text{Pl}(\overline{X} \cap B)} = 0$ as required.

Conditioning Belief and Plausibility via Mass Functions. In this case, observing B is encoded via the mass function m_B as $\text{m}_B(B) = 1$ and 0 otherwise. The update Bel^B of a belief function Bel by B is computed via Dempster-Shafer combination rule and its associated mass function is $\text{m}_{\text{Bel}} \oplus \text{m}_B$. To distinguish between this method and the above method we use Bel^B and Pl^B for the conditional belief and plausibility obtained by the latter approach and we call it DS-conditioning. We have the following explicit formulas for Bel^B and Pl^B.

Proposition 2 *[6, Theorem 3.8.5].* Bel^B *and* Pl^B *are defined if* $\text{Pl}(B) > 0$. *For every* $X \in \mathcal{P}(S)$, $\text{Bel}^B(X) = \frac{\text{Bel}(X \cup \overline{B}) - \text{Bel}(\overline{B})}{1 - \text{Bel}(\overline{B})}$ *and* $\text{Pl}^B(X) = \frac{\text{Pl}(X \cap B)}{\text{Pl}(B)}$.

3   Updating Belief and Plausibility over Belnap-Dunn Logic

3.1   Models for Belief and Plausibility over Belnap-Dunn Logic

We define belief functions on BD logic similarly to how we defined probabilities, that is, we use the notion of positive/negative extension.

Definition 8 (DS models and their associated belief functions). *Let* $\mathcal{L}_{\text{BD}}$ *be the Lindenbaum algebra for* BD *logic over the set of propositional letters* **Prop**. *A* DS model *is a tuple* $\mathcal{M} = \langle S, \mathcal{P}(S), \text{Bel}, v^+, v^- \rangle$ *such that* $\langle S, v^+, v^- \rangle$ *is a* BD *model and* Bel *is a belief function on* $\mathcal{P}(S)$. *We denote* $\text{bel}^+_{\mathcal{M}} : \mathcal{L}_{\text{BD}} \to [0,1]$ *and* $\text{bel}^-_{\mathcal{M}} : \mathcal{L}^{op}_{\text{BD}} \to [0,1]$ *the maps such that, for every* $\varphi \in \mathcal{L}_{\text{BD}}$,

$$\text{bel}^+_{\mathcal{M}}(\varphi) = \text{Bel}(|\varphi|^+) \qquad and \qquad \text{bel}^-_{\mathcal{M}}(\varphi) = \text{Bel}(|\varphi|^-) = \text{Bel}(|\neg\varphi|^+). \quad (3)$$

We drop the subscript whenever there is no ambiguity on the model $\mathcal{M}$ *we are considering.*

Notice that as we are defining belief of a formula via its extension, we obtain mutual definability of positive and negative belief: $\text{bel}^-(\varphi) = \text{bel}^+(\neg\varphi)$. This property mirrors how the negation works in BD logic and in non-standard probabilities. It would be possible to define plausibility analogously to the classical case, that is $\text{Pl}(X) = 1 - \text{Bel}(\overline{X})$. The plausibility of φ would then be equal to the sum of the masses of the sets of states that at least partially support φ, i.e. $\sum\{m(A) \mid A \cap |\varphi|^+ \neq \emptyset\}$. This definition does not have an intuitive interpretation, as in the sum (1) we can take into account sets of states that all positively satisfy both φ and $\neg\varphi$ and (2) we do not take into account sets of states that satisfy neither φ nor $\neg\varphi$. However, not having information about φ, in general, is not an argument to say that it is implausible. Therefore, we introduce models where belief and plausibility are not inter-definable.

Definition 9 ($\mathrm{DS_{pl}}$ models and their associated plausibility functions). *Let $\mathcal{L}_{\mathsf{BD}}$ be the Lindenbaum algebra for BD logic over the set of propositional letters* Prop. *A $\mathrm{DS_{pl}}$ model is a tuple $\mathcal{M} = (S, \mathcal{P}(S), \mathsf{Bel}, \mathsf{Pl}, v^+, v^-)$ such that $(S, \mathcal{P}(S), \mathsf{Bel}, v^+, v^-)$ is a DS model, Pl is a plausibility function on $\mathcal{P}(S)$. We denote $\mathsf{pl}^+_{\mathcal{M}} : \mathcal{L}_{\mathsf{BD}} \to [0,1]$ and $\mathsf{pl}^-_{\mathcal{M}} : \mathcal{L}^{op}_{\mathsf{BD}} \to [0,1]$ the maps such that, for every $\varphi \in \mathcal{L}_{\mathsf{BD}}$, $\mathsf{pl}^+_{\mathcal{M}}(\varphi) = \mathsf{Pl}(|\varphi|^+)$ and $\mathsf{pl}^-_{\mathcal{M}}(\varphi) = \mathsf{Pl}(|\varphi|^-) = \mathsf{Pl}(|\neg\varphi|^+)$. We drop the subscript whenever there is no ambiguity on the model $\mathcal{M}$ we are considering.*

In the standard approach both belief and plausibility use in fact the same information represented by the mass function, but deal with it in a different way. While we can see belief as the amount of information which directly supports the statement in question, plausibility represents the amount of information which does not contradict the statement. As Halpern says: "$Plaus_m(U)$ can be thought of as the sum of the probabilities of the evidence that is compatible with the actual world being in U". ([6], p. 38). This idea is captured in the definition of plausibility via mass function: $\mathsf{pl}(A) = \sum_{A \cap B \neq \varnothing} \mathsf{m}(B)$. We can also see belief and plausibility as approximations, as a lower and an upper bound for the 'true' probability: $\mathsf{bel}(A) \leq p(A) \leq \mathsf{pl}(A)$. While in the classical case all these readings coincide, in the case of BD logic they do not, which gives us several possibilities of defining belief/plausibility pairs (see [3, Sect. 3.3] for a more detailed discussion). Notice that, since A and $\neg A$ are independent elements, if we consider a belief function bel and its associated plausibility function $\mathsf{pl_{bel}}$, then we can have $\mathsf{pl_{bel}}(A) < \mathsf{bel}(A)$. Therefore, asking for $\mathsf{bel}(A) \leq \mathsf{pl}(A)$ usually implies that bel and pl are associated to different mass functions. This can be interpreted as follows: an agent has a fixed set of pieces of evidences, however, they do not read the evidence the same way when they ask themselves "does the evidence strongly convince me that φ is the case?" or "is the evidence coherent with the fact that φ might be the case?". We discuss this more in details in [3, Sect. 3.3.3].

3.2   Updating Belief

A natural question that arises is what is the behaviour of the positive and negative belief functions induced by the above models, when one learns a new piece of information. Learning something about φ means finding a positive or negative piece of information or even a contradictory piece of information about φ. Here, we directly adapt the conditioning on belief function proposed in [6]. Indeed, the belief function Bel in a DS models is defined on a powerset algebra. The non-classical behaviour with respect to the negation of bel^+ and bel^- comes from the non-classical interpretation of formulas. Recall that in BD logic, $|\varphi|^- = |\neg\varphi|^+$, therefore, we only study updating with the positive interpretation of a formula.

Conditioning Belief as Lower Measure. If we look at belief as the lower approximation of the "real" probability function, then we know that the "real" probability function is in the set $\mathcal{M}_{\mathtt{Bel}}$. Therefore, to update the belief after learning that φ is the case, one can compute the Bayesian update of every probability in $\mathcal{M}_{\mathtt{Bel}}$. In that framework, this boils down to ignoring information states (that is, the elements $s \in S$ not supporting φ. If $(\mathcal{M}_{\mathtt{Bel}})^*(|\varphi|^+) = 1 - \mathtt{Bel}(\overline{|\varphi|^+}) > 0$, then one can define the conditional belief on φ as follows: for every $X \in \mathcal{P}(S)$, $\mathtt{Bel}_{|\varphi|^+}(X) = ((\mathcal{M}_{\mathtt{Bel}})_{|\varphi|^+})_*(X)$ which gives us the following conditional belief function on formulas, for each $\psi \in \mathscr{L}_{\mathtt{BD}}$, $\mathtt{bel}^+_{|\varphi|^+}(\psi) = \mathtt{Bel}_{|\varphi|^+}(|\psi|^+) = ((\mathcal{M}_{\mathtt{Bel}})_{|\varphi|^+})_*(|\psi|^+)$. In what follows, for sake of readability, we will write $\mathtt{bel}^+_{|\varphi|}$ and $\mathtt{Bel}_{|\varphi|}$ instead of $\mathtt{Bel}_{|\varphi|^+}$ and $\mathtt{bel}^+_{|\varphi|^+}$. Based on Theorem 2, we have the following explicit formula

$$
\mathtt{bel}^+_{|\varphi|}(\psi) = \begin{cases} 1 & \text{if } \mathtt{Bel}(\overline{|\varphi|^+} \cup |\psi|^+) = 1, \\ \dfrac{\mathtt{Bel}(|\psi|^+ \cap |\varphi|^+)}{1 + \mathtt{Bel}(|\psi|^+ \cap |\varphi|^+) - \mathtt{Bel}(\overline{|\varphi|^+} \cup |\psi|^+)} & \text{if } \mathtt{Bel}(\overline{|\varphi|^+} \cup |\psi|^+) < 1. \end{cases}
$$

Notice that since the update is performed on $\mathtt{Bel}$, both $\mathtt{bel}^+$ and $\mathtt{bel}^-$ are affected by the update. In addition, $\mathtt{bel}^+_{|\varphi|}(\varphi) = 1$ as expected. However, in general, $\mathtt{bel}^+_{|\varphi|}(\neg\varphi) \neq 0$, because $|\varphi|^+ \cap |\neg\varphi|^+ \neq \emptyset$.

Conditioning Belief via Mass Functions. If we interpret belief as representing the information coming from pieces of evidence, then one can also update the belief function $\mathtt{Bel}$ via its associated mass function $\mathtt{m}_{\mathtt{Bel}}$ and Dempster-Shafer combination rule. We call that method DS conditioning. A piece of evidence fully supporting exactly φ is usually represented by the mass function $\mathtt{m}_{|\varphi|^+} : \mathcal{P}(S) \to [0,1]$ such that $\mathtt{m}_{|\varphi|^+}(|\varphi|^+) = 1$ and $\mathtt{m}_{|\varphi|^+}(X) = 0$ otherwise. Therefore, the updating of $\mathtt{Bel}$ by finding positive information about φ, denoted $(\mathtt{Bel})^{|\varphi|}$, is the belief function associated to the mass function $\mathtt{m}_{\mathtt{Bel}} \oplus \mathtt{m}_{|\varphi|^+}$. Then, based on Proposition 2, we have:

Proposition 3. *The belief function* $(\mathtt{bel}^+)^{|\varphi|}$ *is defined if* $1 - \mathtt{Bel}(\overline{|\varphi|^+}) > 0$, *and, for every* $\psi \in \mathcal{L}_{\mathtt{BD}}$,

$$
(\mathtt{bel}^+)^{|\varphi|}(\psi) = (\mathtt{Bel})^{|\varphi|}(|\psi|^+) = \frac{\mathtt{Bel}(|\psi|^+ \cup \overline{|\varphi|^+}) - \mathtt{Bel}(\overline{|\varphi|^+})}{1 - \mathtt{Bel}(\overline{|\varphi|^+})}.
$$

It is well-known that DS combination rule is associative and commutative [8], therefore DS conditioning of belief functions is commutative and associative as well. Notice that, for every $\varphi, \psi \in \mathcal{L}_{\mathtt{BD}}$, we have $\mathtt{m}_{|\varphi|^+} \oplus \mathtt{m}_{|\psi|^+} = \mathtt{m}_{|\psi|^+} \oplus \mathtt{m}_{|\varphi|^+} = \mathtt{m}_{|\varphi|^+ \cap |\psi|^+} = \mathtt{m}_{|\varphi \wedge \psi|^+}$, which implies that $((\mathtt{bel}^+)^{|\varphi|})^{|\psi|} = ((\mathtt{bel}^+)^{|\psi|})^{|\varphi|} = ((\mathtt{bel}^+)^{|\varphi \wedge \psi|})$. This means that, with DS conditioning, finding both a piece of information supporting φ and a piece of information supporting $\neg\varphi$ is equivalent to finding a contradictory piece of information about φ. Here again, notice that

$(\mathtt{bel}^+)^{|\varphi|}(\neg\varphi)$ can be different than 0 because some states $s \in S$ can support both φ and $\neg\varphi$. In addition, it is worth noticing, that $\mathtt{Bel}_{|\varphi|}(X) \leq \mathtt{Bel}^{|\varphi|}(X)$ for every $X \subseteq S$ (see [6, Theorem 3.8.6]). Therefore, $\mathtt{bel}^+_\varphi(X) \leq (\mathtt{bel}^+)^{|\varphi|}(X)$ and $\mathtt{bel}^-_\varphi(X) \leq (\mathtt{bel}^-)^{|\varphi|}(X)$.

3.3 Updating Plausibility

Recall that the interpretation of φ and $\neg\varphi$ are independent, therefore, when we consider a mass function $\mathtt{m}$ and its associated belief and plausibility functions $\mathtt{bel}_\mathtt{m}$ and $\mathtt{pl}_\mathtt{m}$, it is often the case that $\mathtt{bel}_\mathtt{m}(\varphi) \not\leq \mathtt{pl}_\mathtt{m}(\varphi) = 1 - \mathtt{bel}_\mathtt{m}(\neg\varphi)$. Hence, if one wants to reason with a belief function and a plausibility function that provide an interval that contains the probability of φ, one needs to consider a belief and a plausibility function that are not associated to the same mass function. In [3, Section 3.3.3], we discuss the interpretation of these mass functions. Then, the question of updating the plausibility function directly, without going through its associated belief function arises. To do so, we introduce $\mathsf{DS}_{\mathtt{pl}}$ models: $\mathcal{M} = (S, \mathcal{P}(S), \mathtt{Bel}, \mathtt{Pl}, v^+, v^-)$. As mentioned above, we can focus on updating based on positive information about a formula φ.

Conditioning Plausibility as Upper Measure. Based on Proposition 1, $\mathtt{Pl}$ is an upper probability, that is, $\mathtt{Pl} = (\mathcal{M}_{\mathtt{Bel}_{\mathtt{Pl}}})^*$. So again, one can define conditioning on a formula φ when $\mathtt{Pl}(\varphi) > 0$ as follows: $\mathtt{pl}^+_{|\varphi|}(\psi) = ((\mathcal{M}_{\mathtt{Bel}_{\mathtt{Pl}}})_{|\varphi|^+})^*(|\psi|^+)$, for every $\psi \in \mathscr{L}_{\mathsf{BD}}$. From Lemma 2, we have the following explicit formula for updating plausibilities.

Proposition 4. *Let φ be a formula such that $\mathtt{Pl}(|\varphi|^+) > 0$,*

$$\mathtt{pl}^+_{|\varphi|}(\psi) = \frac{\mathtt{Pl}(|\psi|^+ \cap |\varphi|^+)}{\mathtt{Pl}(|\psi|^+ \cap |\varphi|^+) + \mathtt{Pl}(\overline{|\psi|^+} \cap |\varphi|^+)}$$

Notice that, as expected, $\mathtt{pl}^+_{|\varphi|}(\varphi) = 1$ and, since $|\psi|^- = |\neg\psi|^+$, we have:

$$\mathtt{pl}^-_{|\varphi|}(\psi) = ((\mathcal{M}_{\mathtt{Bel}_{\mathtt{Pl}}})_{|\varphi|^+})^*(|\psi|^-) = ((\mathcal{M}_{\mathtt{Bel}_{\mathtt{Pl}}})_{|\varphi|^+})^*(|\neg\psi|^+) = \mathtt{pl}^+_{|\varphi|}(\neg\psi).$$

Conditioning Plausibility via Mass Function. DS conditioning can be applied to plausibility functions via their associated mass functions $\mathtt{m}_{\mathtt{pl}}$ (see Lemma 3). The mass function associated to the update of $\mathtt{Pl}$ based on some piece of information positively supporting φ is computed via Dempster-Shafer combination rule as follows: $\mathtt{m}_{\mathtt{pl}} \oplus \mathtt{m}_{|\varphi|^+}$. Based on Proposition 2, we get the following formula for the corresponding plausibility function $(\mathtt{pl}^+)^{|\varphi|}$ over formulas.

Proposition 5. *The function $(\mathtt{pl}^+)^{|\varphi|}$ is defined if $\mathtt{Pl}(|\varphi|^+) > 0$, and, for every $\psi \in \mathscr{L}_{\mathsf{BD}}$, we have $(\mathtt{pl}^+)^{|\varphi|}(\psi) = \frac{\mathtt{Pl}(|\psi|^+ \cap |\varphi|^+)}{\mathtt{Pl}(|\varphi|^+)}$.*

4   Further Work and Conclusion

This article presents methods to update belief and plausibility functions within the framework of BD logic. Recall that BD logic was introduced to reason about incomplete and contradictory information. In DS models, even though the underlying logic is non-classical, namely BD logic, the belief and plausibility functions are defined over the powerset of states. Therefore, we can import the techniques for updating belief functions from the classical logic literature. This work is in fact a first step. Indeed, we wish to look at belief and plausibility functions over De Morgan algebras to get a better understanding of the implication of combining belief functions and BD logic. Recall that De Morgan algebras provide the algebraic semantics for BD logic, and that there is no duality between BD models and De Morgan algebras. Therefore, there is no way to directly import our results on frames to De Morgan algebras. A natural first step will be to study the mathematical properties of belief and plausibility functions over De Morgan algebras, and to establish whether they can be represented as lower and upper probabilities over sets of non-standard probabilities. This would open various options to update the belief. Indeed, [7] presents different ways to update non-standard probabilities, among which two ways that generalise Bayesian update. In addition, Dempster-Shafer combination rule can straight forwardly be transferred to De Morgan algebras (see [3]) which provides a natural way to update belief functions over De Morgan algebras. However, it remains to be checked whether this method is equivalent to DS conditioning on DS models.

References

1. Barthélemy, J.-P.: Monotone functions on finite lattices: an ordinal approach to capacities, belief and necessity functions. In: Fodor, J., De Baets, B., Perny, P. (eds.) Preferences and Decisions Under Incomplete Knowledge, pp. 195–208. Physica Verlag (2000)
2. Belnap, N.D.: How a computer should think. In: Omori, H., Wansing, H. (eds.) New Essays on Belnap-Dunn Logic. SL, vol. 418, pp. 35–53. Springer, Cham (2019). https://doi.org/10.1007/978-3-030-31136-0_4
3. Bílková, M., Frittella, S., Kozhemiachenko, D., Majer, O., Nazari, S.: Reasoning with belief functions over belnap-dunn logic. Preprint arXiv:2203.01060 (2022)
4. Frittella, S., Manoorkar, K., Palmigiano, A., Tzimoulis, A., Wijnberg, N.: Toward a Dempster-Shafer theory of concepts. Int. J. Approx. Reason. **125**, 14–25 (2020)
5. Grabisch, M.: Belief functions on lattices. Int. J. Intell. Syst. **24**(1), 76–95 (2009)
6. Halpern, J.Y.: Reasoning About Uncertainty, 2nd edn. MIT Press, Cambridge (2017)
7. Klein, D., Majer, O., Rafiee Rad, S.: Probabilities with gaps and gluts. J. Philos. Log. **50**(5), 1107–1141 (2021)
8. Shafer, G.: A Mathematical Theory of Evidence. Princeton University Press, Princeton (1976)
9. Zhou, C.: Belief functions on distributive lattices. Artif. Intell. **201**, 1–31 (2013)

Applications

Real Bird Dataset with Imprecise and Uncertain Values

Constance Thierry[(✉)], Arthur Hoarau, Arnaud Martin,
Jean-Christophe Dubois, and Yolande Le Gall

University Rennes, CNRS, IRISA, DRUID, Rennes, France
constance.thierry@irisa.fr

Abstract. The theory of belief functions allows the fusion of imperfect data from different sources. Unfortunately, few real, imprecise and uncertain datasets exist to test approaches using belief functions. We have built real birds datasets thanks to the collection of numerous human contributions that we make available to the scientific community. The interest of our datasets is that they are made of human contributions, thus the information is therefore naturally uncertain and imprecise. These imperfections are given directly by the persons. This article presents the data and their collection through crowdsourcing and how to obtain belief functions from the data.

Keywords: Datasets · Imprecise · Uncertain

1 Introduction

The theory of belief functions allows for uncertainty and imprecision in the data. However, there are very few real datasets available to consider belief functions. [4] and [1] work on imprecise and uncertain real data that the authors have collected but these data are not made available to the community. Similarly, [3] proposes an MCQ that allows students to give imprecise and uncertain answers that can be modeled with belief functions. However, the experimental data are not reported. It was important to build real datasets to evaluate proposed methods in real context [9]. To do so, we collected human contributions from crowdsourcing campaigns, as human contributions are uncertain and imprecise information.

Crowdsourcing is the outsourcing of tasks to a crowd of contributors on a platform dedicated to the domain [5]. The tasks that can be achieved through crowdsourcing are very diverse. In this paper, we presented to the contributors a picture of a bird and asked them to identify the bird from a list of proposed names. We use interfaces that allow us to collect imprecise and/or uncertain responses. We conducted six crowdsourcing campaigns for bird photo annotation. For all these campaigns the contributor had to give his certainty in his answer. Two of them are only composed of precise contributions. For the four other campaigns, the contributor can be imprecise and choose more bird names. For two of the imprecise campaigns, after the contributor has given his answer he is offered to enlarge or restrict his selection consequently.

The rest of the paper is as follows, Sect. 2 introduces the belief functions. Section 3 reviews the crowdsourcing campaigns and Sect. 4 presents the datasets. We propose examples of modelisation thanks to the belief function Sect. 5. Section 6 concludes the paper.

2   Belief Functions

The theory of belief functions, also called Dempster-Shafer theory [2,8], is used in this study in order to model both data imprecision and uncertainty.

One considers $\Omega = \{r_1, \ldots, r_M\}$ the frame of discernment for M exclusive and exhaustive hypotheses. In this paper, Ω represents all possible bird species of a given photo among M bird species. The power set 2^Ω is the set of all subsets of Ω. A basic belief assignment is the belief that a source may have about the elements of the power set of Ω, this function assigns a mass to each element of this power set such that the sum of all masses is equal to 1.

$$m : 2^\Omega \rightarrow [0,1]$$
$$\sum_{A \in 2^\Omega} m(A) = 1 \tag{1}$$

Focal element:	An element of 2^Ω with a non-null mass.
Simple support mass function:	Only has two focal elements, and one of them is the frame of discernment Ω.
Consonant mass function:	Each focal element is nested.

3   Crowdsourcing Campaign

The main objective for these campaigns is always the same, a photo of a bird is presented to the contributor with a set of species names (including the good answer) and he has to select the right answer. The Wirk platform (Crowdpanel[1]) is used to realize the crowdsourcing campaigns. As the users of the platform live in France, the birds used for the campaigns are all of species visible in metropolitan France. For all the campaigns, the contributor has to give his answer, then specify his certainty according to the following Likert scale: "Totally uncertain","Uncertain", "Rather uncertain", "Neutral", "Rather certain", "Certain", "Totally certain". We explained to the contributors that there is no penalty for being uncertain and/or imprecise in their answers. After having given his answer and his certainty, he can validate his contribution in order to move on to the next question.

---

[1] https://crowdpanel.io/ (15/04/2022).

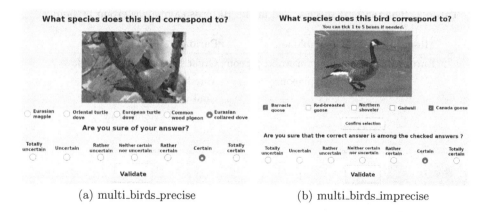

(a) multi_birds_precise   (b) multi_birds_imprecise

Fig. 1. Interfaces used for crowdsourcing campaigns multi_birds

3.1   Campaigns Multi_birds

For these two campaigns, five bird names are proposed to the contributor. The names change from one question to another and a bird species is presented only once. We have tried to introduce different levels of difficulty in the questions. For example, for a difficult question, a photo of an eagle is presented to the contributor and the five answer items are different species of eagles. Conversely, for a simpler question, a photo of a gull is presented to the contributor and the four other answers are names of duck species. For a single photo, responses were presented in random order to each contributor to avoid selection bias. In addition, the questions were also asked in a random order, so that when a contributor c_1 answers a question q_i, c_2 answers q_j. These crowdsourcing campaigns include 3 attention questions for which the contributor is asked to give the same answer as the one given in the previous question.

Multi_birds_precise. The interface used for this task is given Fig. 1(a). Participants have to provide a precise answer by selecting a single bird name, and a self-assessment of their certainty in this answer.

Multi_birds_imprecise. For this task the contributor can be imprecise and select up to all of the bird names offered. The interface is given Fig. 1(b). The contributors first must give his answers, validate it and then he is asked to give his certainty in this answer.

For both campaigns the crowds are composed of 100 contributors, each one must annotate 50 photos, for a total of 5000 contributions for each campaign. A contributor allowed to do the first campaign cannot participate in the second.

3.2   Campaigns 10_birds

For these campaigns, ten bird species are selected and proposed as response elements to the contributors. In order to observe the contributor's ability to be imprecise in case of hesitation, the ten birds presented are composed of subgroups

Table 1. The ten bird species used in the 10_birds campaigns group by family

Muscicapidae	Columbidae	Paridae	Corvidae
European robin	Common wood pigeon	Great tit	European jackdaw
	Rock pigeon	Marsh tit	Carrion crow
		Coal tit	Common raven
			Rook crow

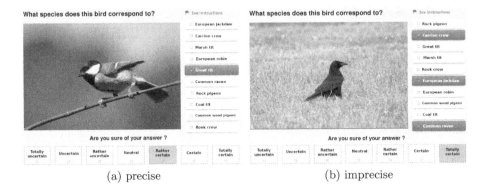

(a) precise   (b) imprecise

Fig. 2. Interfaces used for crowdsourcing campaigns 10_birds precise and imprecise

from the same bird family given Table 1. The bird names are presented to each contributor in a different order to avoid selection bias. This ordering of names is nevertheless fixed for a contributor throughout the campaign. Such as the campaigns multi_birds, the questions are asked in a random order. The same scale is used for certainty and 3 attention questions are also asked. The contributor is no longer required to validate his answer before he can give his certainty.

10_birds_precise. The contributor should select from the interface Fig. 2(a) a unique bird name and then give his certainty about it.

10_birds_imprecise. The contributor can choose thanks to the interface Fig. 2(b) 1 to a maximum of 5 answers from the ten provided bird names. We impose a maximum number of answers to 5 because we admit that if the contributor hesitates it is between names of birds of the same family. He should not hesitate between a pigeon and a chickadee for example. We have chosen to offer the crowd a maximum selection of 5 names because we do not want to introduce a bias and encourage him to choose exactly the 4 corvidaes in case of hesitation.

10_birds_iterative. This campaign is called iterative because the contributor is asked to expand or refine the contribution they have entered. To do so, in a first step the contributor answer the question as shown Fig. 3(a) and then:

– If he is precise but not "totally certain" of his answer, he is offered to expand his selection if he feels the need. In this case, the first selected answer is kept in step 2 and he can complete it by selecting new names.

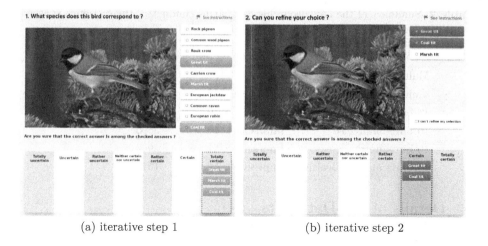

(a) iterative step 1   (b) iterative step 2

Fig. 3. Interfaces used for the campaigns 10_birds iterative and machine learning

– If he is imprecise in his contribution, he is asked in a second step if he is able to restrict his choice of answer while giving his new certainty as in the example Fig. 3(b). When he is offered to restrict his selection, only the previously chosen answer elements are proposed again.

These interactions with the contributor increased the number of responses collected, and therefore the time spent soliciting the contributor.

10_birds_machine_learning. For classification problems, a larger number of observations is often required. This campaign provides imperfectly labeled observations that can be used for classification problems and more generally in machine learning. This is a similar campaign to 10_birds_iterative with more photos per class and fewer responses per photo. A number of 20 photos per bird species is used instead of 5. A total of 200 photos separated into 10 species are then labelled. Contributors are only invited to give 20 answers, 2 randomly selected per species of bird. No attention questions are asked in this campaign.

Each crowdsourcing campaign required a crowd of 50 contributors. As with the other campaigns, a contributor who has participated in one experiment cannot participate in another. For each of the ten bird that make up the proposed answer set, a contributor is presented with 5 photos of a bird, so that he answers 50 questions. Thus 2500 data are collected for the experiments 10_birds_precise, with precise answers, and 10_birds_imprecise, for which the contributor can choose up to five answers. There is 2990 data collected for 10_birds_iterative because for this experiment, as there are 2550 first step answers and 440 s step answers. Finally 1515 data are collected for 10_birds_machine_learning, with 1040 first step answers and 475 s step answers.

All crowdsourcing campaigns are summarized in Table 2 which includes the name of the campaign, whether the data collected is accurate or inaccurate, the number of contributors and the total number of contributions collected.

Table 2. Summary of the crowdsourcing campaigns conducted, the certainty is asked for all the answers

Campaign	Answers	Crowd size	Number of answer
Multi_birds_precise	Precise	100	5000
Multi_birds_imprecise	Imprecise	100	5000
10_birds_precise	Precise	50	2500
10_birds_imprecise	Imprecise	50	2500
10_birds_iterative	Imprecise	51	2990
10_birds_machine_learning	Imprecise	52	1515

4   Details of the Data Sets

The datasets are made available to the community on the INRIA git account https://gitlab.inria.fr/cthierry/imprecise_uncertain_dataset. The repository is structured as follows: a folder is associated with each crowdsourcing campaign presented above, there is also a csv file named "answers_multi_birds.csv". This file is used as the answer set propose to the contributor of the multi_birds campaign. It includes the following variable:

- photo: the number of the bird photo display by the interface
- goodanswer: the true bird name of the photo
- answ1, answ2, answ3, answ4: other bird names propose as answer elements to the contributor in addition of the true bird name
- difficulty: an hypothesis of the difficulty of the question according to the author. Values range from 1 (easy question) to 5 (difficult question). In fact the difficulty observed is not correlated to those one supposed by the author.

Thus, it is possible to use the file "answers_multi_birds.csv" to construct a frame of discernment $\Omega_q = \{goodanswer, answ1, answ2, answ3, answ4\}$ for each question q of the multi_birds campaigns. In the files named after the crowdsourcing campaigns there are several files providing different information.

Data. This csv file includes the contribution from the crowd for the bird annotation. The contribution includes therefore a selection of bird names and a certainty associated. The file includes:

- log_id: indicates the line of the file
- user: unique user ID for a contributor
- currenttrial: number of the question asked to the contributor
- img: number of the photos shown to the contributor
- goodanswer: the true name of the bird to identify
- answer: the set of bird names selected by the contributor (a unique answer for the precise crowdsourcing campaign)
- answer: initial response from the contributor
- answerhistory: set of values checked/unchecked by the contributor to answer the question

- isgoodanswer: boolean indicated if the true bird name is include in the answer set given by the contributor
- certitude: certainty given by the contributor to express his confidence in his bird names selection
- certitudehistory: history of the certainty values for the answer
- timestamp: time recorded by the interface
- time: time of the contributor's answer to the question

Some variables such as user are common to several files. For all the files, the contributor certainty ranges from 1 (totally uncertain) to 7 (totally certain).

For the campaign multi_birds_imprecise 57.04% of the data includes in the csv file are imprecise. And for the 10_birds campaigns we have the following results: imprecise 55.64%, iterative 45.32% and machine_learning 58.22%. Contributors have made good use of the opportunity to be imprecise when possible. On average, when contributors are imprecise they choose two answers.

Attention. In crowdsourcing campaigns, attention questions are asked to the contributor in order to ensure its seriousness. This csv file includes the answers to the three attention questions. These questions consist in asking the contributor a previous question in order to get him to give exactly the same contribution. The variables included in the file are the following:

- log_id, user, currenttrial, answer, certitude, timestamp, time
- attention_answer: contributor's answer to the attention question
- answerhistory: set of values checked/unchecked by the contributor to answer the attention question
- certainty: certainty of the first selected answer set
- certaintyhistory: history of checked/unchecked certainty values
- issamecertitude: boolean indicating if the certainty given to the attention question is identical to the certainty given to the initial question
- issameanswer: boolean indicating if the set of bird names selected at the attention question is identical to the set initially selected

Event. This file records the principal events of the platform named event_type: the connection to the platform (start), the beginning of the crowdsourcing campaign (start_xp), the ending question (questions) and the end of the campaign (finish). It is possible that some contributors have started a crowdsourcing campaign without having finished it, we have only a part of the answers for them. To sort out the contributors (users) to be selected we recommend using the event.csv or question.csv files described below to select the data of users who have reached the final question phase and/or the finish event. When we talk about the number of responses, it is only for contributors who have completed the entire campaign. This file also includes the variables log_id, user and time which gives the date and time when the event took place.

Queries. At the end of the different crowdsourcing campaigns, a questionnaire is sent to the contributors to get their feedback. This questionnaire varies between campaigns. These files include the answers at the end of the campaign (Table 3).

Table 3. Number of precise and then imprecise contributions ($|X_1| = 1$ and $|X_2| > 1$) and imprecise then less imprecise ($|X_1| > |X_2|$)

Campaigns	Size of the dataset	Data subset	Size of the data subset
10_birds_iterative	440	$|X_1| = 1$ and $|X_2| > 1$	88
		$|X_1| > 1$ and $|X_1| > |X_2|$	352
10_birds_machine_learning	475	$|X_1| = 1$ and $|X_2| > 1$	57
		$|X_1| > 1$ and $|X_1| > |X_2|$	418

Iteration. This csv file is present in the 10_birds_machine_learning and 10_birds_iterative folders because it includes the contributors' answers when they expand or specify their answer in the second stage of questioning of these campaigns. The next values are included into the file:

- log_id, user
- trial: equals to currenttrial value of the same contributor (user) in data.csv
- new_answer: new answer given by the contributor
- new_certitude: new certainty given by the contributor
- cant_answer: boolean that takes the value 1 if the contributor cannot modify (refine or enlarge) his answer
- isImprecis: boolean which takes the value 1 if in his first answer the contributor is imprecise (*i.e.* he chooses several bird names)
- aHistory: equivalent to answerhistory
- cHistory: equivalent to certaintyhistory

The file for the campaign 10_birds_iterative includes 1527 rows but for the majority of them, the contributor did not modify his answer (cant_answer = 1). Indeed, for this campaign only 440 responses were modified which represents 17% of the first step answers. More contributors edited their answer for the 10_birds_machine_learning campaign, 475 responses were modified i.e. 46% of the dataset. Thanks to the joint use of this file and data.csv it is possible to build 440 consonant mass functions.

For campaigns with iteration we call X_1 the first set of names given for a photo. The values of X_1 (answer) are present in the data.csv file with the associated certainty. When the contributor is proposed to modify his contribution, the new name selection X_2 (new_answer) and the new certainty (new_certitude) are registered in the iteration.csv file.

For the campaign 10_birds_iterative there are 461 entries in the data.csv file for which the contributor first selected a single answer $|X_1| = 1$, and then was offered to expand his selection so that $|X_2| > 1$. Of these 461 times when the contributor is offered to be imprecise, there are only 88 times when a second answer X_2 is given. Similarly, there are a total of 1066 times when the contributor fills in an imprecise answer, $|X_1| > 1$, and is offered to narrow his selection so that $|X_1| > |X_2|$, a total of 352 contributions report a change in answer.

During the 10_birds_machine_learning campaign contributors also tend to give a second answer more precise than the first one, with 418 responses on 475 iterations, against 57 precise answers at firt step and less precise at second step. Furthermore, 389 responses were listed as totally certain and 186 were listed as totally certain and precise. Among those contributors who were certain and precise, 91% hold the real answer. Of all the answers, 33 were listed as inconsistent, which means that a contributor gave an answer including different bird families, these 33 answers were generated by 15 different contributors.

The following section presents examples of modeling with mass functions.

5   Belief Functions from the Data

We propose a data modeling by simple support and consonant mass functions.

Simple Support Mass Function. This function can be computed for the answer values and certainty of the six campaigns. Given a question q, the set of answers associated to q compose the framework of discernment $\Omega_q = \{r_1, \ldots, r_K\}$. The question being closed, we consider the closed world. The contributor c answers the question q by the contribution $X \in 2^{\Omega_q}$, which can be imprecise, and to which he associates a certainty of value $certainty \in [1, 7]$ which is transformed into a mass $\omega_{cq} \in [0, 1]$ according to the equation:

$$\omega_{cq} = \frac{certainty - 1}{certainty_{max} - 1} \tag{2}$$

For the crowdsourcing campaign introduce in this paper $certainty_{max} = 7$. A mass function with simple support $(X^{\omega_{cq}})$ can be obtained from the contribution:

$$\begin{cases} m_{cq}^{\Omega_q}(X) = \omega_{cq} \text{ with } X \in 2^{\Omega_q} \setminus \Omega_q \\ m_{cq}^{\Omega_q}(\Omega_q) = 1 - \omega_{cq} \end{cases} \tag{3}$$

Consonant Mass Function. During the 10_birds_machine_learning and 10_birds_iterative campaigns, the same question q can be asked twice to the contributor c who can then enlarge or specify his first answer X_1 by a second answer X_2 if he wishes. Let Ω_q be the set of proposed answers and $X_1, X_2 \in 2^{\Omega_q}$. If the first answer of the contributor X_1 is precise and he widens his second answer X_2 then $X_1 \subset X_2$, and conversely if X_1 is more imprecise than X_2 then $X_2 \subset X_1$. At the time of his first selection X_1, the contributor informs a degree of certainty of numerical value $\omega_{cq1} \in [0, 1]$ compute thanks to Eq. (2). If he chooses to fill in a second answer X_2 he must indicate his new certainty whose numerical value is noted $\omega_{cq2} \in [0, 1]$. If the contributor is not asked to modify his selection or if he does not wish to do so, the contribution is modeled by a simple support mass function. In the case where the contributor changes its response X_1 to the response X_2, with $X_1 \subset X_2$, then the contribution can be modeled by a consonant mass function:

$$\begin{cases} m_{cq}^{\Omega_q}(X_1) = \delta_1 * \omega_{cq1} \\ m_{cq}^{\Omega_q}(X_2) = \delta_2 * \omega_{cq2} \\ m_{cq}^{\Omega_q}(\Omega) = 1 - \delta_1 * \omega_{cq1} - \delta_2 * \omega_{cq2} \end{cases} \tag{4}$$

In Eq. (4), the coefficients δ_1 and δ_2 ensure that the mass function belongs to the interval $[0, 1]$, thus: $\delta_1 + \delta_2 = 1$. If we want to give more importance to the first contribution X_1 rather than to the second contribution X_2 then we must choose δ_1 such that $\delta_1 > \delta_2$. Another way to combine the two mass functions from the two iterative responses is to use a combination rule that does not require the assumption of source independence.

We have proposed a modeling of some data by belief functions but it is possible to go further by using them for example to estimate the expertise of the contributor as do [7]. The data can also be used to compare a probabilistic approach to belief functions [6].

6   Conclusion

This paper presents some real credal datasets created through crowdsourcing campaigns for bird photo annotation. To constitute these datasets six crowdsourcing campaigns have been realized. In these six campaigns, the contributor is asked to give his certainty in his answer. For two campaigns the contributor is forced to choose a single bird name as an answer, these data are therefore precise and potentially uncertain. For the other four campaigns the contributor had the possibility to be imprecise in case of hesitation on his answer, these data are imprecise and/or uncertain. For these six crowdsourcing campaigns it is possible to model the contributions by simple support mass functions. Finally, for two of the four imprecise campaigns, the contributor is asked to modify the answer already given by clarifying or expanding it. Thanks to these two campaigns it is possible to model the contributions by consonant mass functions.

References

1. Abassi, L., Boukhris, I.: A worker clustering-based approach of label aggregation under the belief function theory. Appl. Intell. **49**(1), 1–10 (2018). https://doi.org/10.1007/s10489-018-1209-z
2. Dempster, A.P.: Upper and lower probabilities induced by a multivalued mapping. Ann. Math. Stat. **38**(2), 325–339 (1967). https://doi.org/10.1007/978-3-540-44792-4_3
3. Diaz, J., Rifqi, M., Bouchon-Meunier, B., Jhean-Larose, S., Denhiére, G.: Imperfect answers in multiple choice questionnaires. In: Dillenbourg, P., Specht, M. (eds.) EC-TEL 2008. LNCS, vol. 5192, pp. 144–154. Springer, Heidelberg (2008). https://doi.org/10.1007/978-3-540-87605-2_17
4. Dubois, J.-C., et al.: Measuring the expertise of workers for crowdsourcing applications. In: Pinaud, B., Guillet, F., Gandon, F., Largeron, C. (eds.) Advances in Knowledge Discovery and Management. SCI, vol. 834, pp. 139–157. Springer, Cham (2019). https://doi.org/10.1007/978-3-030-18129-1_7
5. Howe, J.: The rise of crowdsourcing. Wired Magazine (2006)
6. Koulougli, D., Hadjali, A., Rassoul, I.: Handling query answering in crowdsourcing systems: a belief function-based approach. In: Fuzzy Information Processing Society (NAFIPS), 2016 Annual Conference of the North American, pp. 1–6. IEEE (2016)

7. Ben Rjab, A., Kharoune, M., Miklos, Z., Martin, A.: Characterization of experts in crowdsourcing platforms. In: Vejnarová, J., Kratochvíl, V. (eds.) BELIEF 2016. LNCS (LNAI), vol. 9861, pp. 97–104. Springer, Cham (2016). https://doi.org/10.1007/978-3-319-45559-4_10
8. Shafer, G.: A Mathematical Theory of Evidence. Princeton University Press (1976)
9. Thierry, C., Martin, A., Dubois, J.-C., Gall, Y.L.: Validation of smets' hypothesis in the crowdsourcing environment. In: Denœux, T., Lefèvre, E., Liu, Z., Pichon, F. (eds.) BELIEF 2021. LNCS (LNAI), vol. 12915, pp. 259–268. Springer, Cham (2021). https://doi.org/10.1007/978-3-030-88601-1_26

Addressing Ambiguity in Randomized Reinsurance Contracts Using Belief Functions

Davide Petturiti[1] (ID), Gabriele Stabile[2] (ID), and Barbara Vantaggi[2(✉)] (ID)

[1] Department of Economics, University of Perugia, Perugia, Italy
davide.petturiti@unipg.it
[2] Department MEMOTEF, "La Sapienza" University of Rome, Rome, Italy
{gabriele.stabile,barbara.vantaggi}@uniroma1.it

Abstract. Motivated by a randomized reinsurance model we consider the lower envelope of the set of bivariate joint probability distributions having a precise discrete marginal and an ambiguous Bernoulli marginal. Under an independence assumption, since the lower envelope fails 2-monotonicity, inner/outer Dempster-Shafer approximations are provided to model the lower expected insurer's annual profit under reinsurance.

Keywords: Belief function · Dempster-Shafer approximation · Reinsurance contract

1 Introduction

Uncertainty is usually modelled through a probability measure, however a demand for more flexible models arises in different fields with the aim to provide tools able to manage partially specified information (imprecision) through a class of compatible probability measures. This work is essentially motivated by an application related to reinsurance and contains some preliminary results incorporating ambiguity in a simple model (see [1]) of reinsurance contract whose outcome depends on both the (unknown) parameter p of an independent Bernoulli distribution and the retention level d in a stop-loss treaty. Since the parameter p of the Bernoulli distribution is partially specified, the aim of present paper is to investigate the effect of ambiguity on the choice of the optimal retention level. A random vector (X, Y) is considered with X having a discrete distribution P_X, whereas the distribution of Y belongs to a class $\mathcal{P}_Y$ of Bernoulli distributions, where p ranges in a closed interval. Under the hypothesis of independence of the two variables, meaning that under P_X and any $P_Y \in \mathcal{P}_Y$ the two variables are independent, we prove that the class of joint probability distributions generates a class of probability measures (credal set) $\mathcal{P}$ that is closed and convex, but whose lower envelope $\underline{P}$ is generally not 2-monotone. Moreover, the core of $\underline{P}$ (i.e., the set of all probability measures dominating $\underline{P}$) strictly contains $\mathcal{P}$.

Following the approach of [5–7], we first look for an inner Dempster-Shafer approximation Bel^i of $\underline{P}$ showing that between the core of Bel^i and $\mathcal{P}$ no

S. Le Hégarat-Mascle et al. (Eds.): BELIEF 2022, LNAI 13506, pp. 286–296, 2022.
https://doi.org/10.1007/978-3-031-17801-6_27

containment relationship may hold. For this, we move towards an outer Dempster-Shafer approximation that preserves the marginal probability distribution of X, namely Bel^{oo}. Both Bel^i and Bel^{oo} allow the computation of lower expectations (with respect to their core) as Choquet expectations and are used to model the lower expected insurer's annual profit under reinsurance. We provide a numerical example showing the implications on the motivating reinsurance application: even taking the outer approximation the optimization problem leads to the same optimal retention level, that is greater than their respective values without modelling ambiguity. We notice that, besides the Choquet integral, other expectation operators could be considered inside Dempster-Shafer theory [10]: this will be addressed in future research.

2   Preliminaries

Let $\Omega = \{\omega_1, \ldots, \omega_n\}$ be a finite non-empty set and denote by 2^Ω its power set. A function $\underline{P} : 2^\Omega \to [0,1]$ such that $\underline{P}(\emptyset) = 0$ and $\underline{P}(\Omega) = 1$ is called a:

- *(coherent) lower probability* if there exists a closed set $\mathcal{P}$ of probability measures on 2^Ω such that, for every $A \in 2^\Omega$,

$$\underline{P}(A) = \min_{P \in \mathcal{P}} P(A);$$

- *k-monotone lower probability* with $k \geq 2$ if for every $A_1, \ldots, A_k \in 2^\Omega$,

$$\underline{P}\left(\bigcup_{i=1}^{k} A_i\right) \geq \sum_{\emptyset \neq I \subseteq \{1,\ldots,k\}} (-1)^{|I|-1} \underline{P}\left(\bigcap_{i \in I} A_i\right).$$

A lower probability which is k-monotone for every $k \geq 2$ is called a *belief function* and is denoted as Bel [2,9].

Every lower probability $\underline{P}$ induces the closed (in the product topology) convex set of probability measures on 2^Ω, called *core*, defined as

$$\mathbf{core}(\underline{P}) = \{P : P \text{ is a probability measure on } 2^\Omega, P \geq \underline{P}\}. \tag{1}$$

A belief function Bel is completely determined [4,9] by its Möbius inverse $m : 2^\Omega \to [0,1]$ with $m(\emptyset) = 0$, summing up to 1 and, for all $A \in 2^\Omega$,

$$Bel(A) = \sum_{B \subseteq A} m(B). \tag{2}$$

Denoting by $\mathbb{R}^\Omega$ the set of all random variables on Ω, the issue of introducing a notion of expectation with respect to a closed set of probability measures $\mathcal{P}$ can be faced in two different manners: either referring to the Choquet integral with respect to the lower probability $\underline{P}$ or to the lower expectation functional with respect to $\mathcal{P}$. Given $\underline{P}$ and $X \in \mathbb{R}^\Omega$, the *Choquet expectation* of X with respect to $\underline{P}$ (see, e.g., [3,4]) is defined through the Choquet integral

$$\mathbb{C}_{\underline{P}}[X] = \sum_{i=1}^{n}(X(\omega_{\sigma(i)}) - X(\omega_{\sigma(i+1)}))\underline{P}(E_i^{\sigma}), \tag{3}$$

where σ is a permutation of Ω such that $X(\omega_{\sigma(1)}) \geq \dots \geq X(\omega_{\sigma(n)})$, $E_i^{\sigma} = \{\omega_{\sigma(1)}, \dots, \omega_{\sigma(i)}\}$ for $i = 1, \dots, n$, and $X(\omega_{\sigma(n+1)}) = 0$.

In particular, if $\underline{P}$ reduces to a probability measure P, then $\mathbb{C}_P[X] = \mathbb{E}_P[X]$, where $\mathbb{E}_P$ denotes the usual expectation operator with respect to P. On the other hand, given $\mathcal{P}$, the corresponding *lower expectation* of $X \in \mathbb{R}^{\Omega}$ is

$$\underline{\mathbb{E}}_{\mathcal{P}}[X] = \min_{P \in \mathcal{P}} \mathbb{E}_P[X]. \tag{4}$$

In general [11,12], we have that $\mathbb{C}_{\underline{P}}[X] \leq \underline{\mathbb{E}}_{\mathbf{core}(\underline{P})}[X] \leq \underline{\mathbb{E}}_{\mathcal{P}}[X]$, where the two inequalities can be strict. Nevertheless, in the particular case $\underline{P}$ is (at least) 2-monotone (see, e.g., [3,4]), then $\mathbb{C}_{\underline{P}}[X] = \underline{\mathbb{E}}_{\mathbf{core}(\underline{P})}[X]$.

3   DS-Approximation of Joint Lower Distributions with an Independent Ambiguous Bernoulli Marginal

Let X, Y be discrete random variables taking values in $\mathcal{X} = \{x_1, \dots, x_m\}$ and $\mathcal{Y} = \{0, 1\}$. Assume that no logical relations (structural zeros) are present between X and Y, therefore, we can simply identify X and Y with the projection maps on the product measurable space $(\mathcal{X} \times \mathcal{Y}, 2^{\mathcal{X} \times \mathcal{Y}})$. We also denote by $2^{\widetilde{\mathcal{X}}}$ and $2^{\widetilde{\mathcal{Y}}}$ the sub-algebras of $2^{\mathcal{X} \times \mathcal{Y}}$ isomorphic to $2^{\mathcal{X}}$ and $2^{\mathcal{Y}}$, respectively.

Let $P_X : \mathcal{X} \to [0, 1]$ be a probability mass function for X and $\mathcal{P}_Y = \{P_Y^p : \mathcal{Y} \to [0, 1] : p \in [p_1, p_2]\}$ be a family of probability mass functions for Y, where

$$P_Y^p(1) = p \quad \text{and} \quad P_Y^p(0) = 1 - p, \quad \text{with } 0 \leq p_1 < p_2 \leq 1.$$

Suppose that, for every $P_Y^p \in \mathcal{P}_Y$, the random variables X, Y are stochastically independent and the joint probability distribution $P^p : 2^{\mathcal{X} \times \mathcal{Y}} \to [0, 1]$ of the vector (X, Y) is obtained extending by additivity the assessment

$$P^p(\{(x, y)\}) = P_X(x) \cdot P_Y^p(y), \quad \text{for all } (x, y) \in \mathcal{X} \times \mathcal{Y}. \tag{5}$$

Therefore, we get the family of joint distributions

$$\mathcal{P} = \{P^p : P^p \text{ is a joint distribution of } (X, Y) \text{ given by (5)}, p \in [p_1, p_2]\}. \tag{6}$$

Proposition 1. *The set $\mathcal{P}$ is a closed and convex subset of $[0, 1]^{2^{\mathcal{X} \times \mathcal{Y}}}$ endowed with the product topology, and its extreme points are $\mathbf{ext}(\mathcal{P}) = \{P^{p_1}, P^{p_2}\}$.*

Proof. The set $\mathcal{P}_Y$ of marginal probability mass functions for Y is a closed and convex subset of $[0, 1]^{\mathcal{Y}}$ endowed with the product topology, and $P^p(\mathcal{X} \times \{y\}) = P_Y^p(y)$, for all $p \in [p_1, p_2]$. In turn, this implies that every sequence $\{P^{p_n}\}_{n \in \mathbb{N}}$ in $\mathcal{P}$ converging pointwise on $2^{\mathcal{X} \times \mathcal{Y}}$ has a limit $P = P^p \in \mathcal{P}$, the convex combination of $P^p, P^{p'} \in \mathcal{P}$ with $\alpha \in [0, 1]$ is such that $P = \alpha P^p + (1 - \alpha)P^{p'} \in \mathcal{P}$, and $\mathbf{ext}(\mathcal{P}) = \{P^{p_1}, P^{p_2}\}$.

Let $\underline{P} = \min \mathcal{P}$ be the lower envelope of $\mathcal{P}$. The following example shows that $\underline{P}$ is generally not 2-monotone and we also have that $\mathcal{P}$ is strictly contained in $\mathbf{core}(\underline{P})$.

Example 1. For $\mathcal{X} = \{x_1, x_2\}$, denote

$$\mathcal{X} \times \mathcal{Y} = \{\underbrace{(x_1,1)}_{=a_1}, \underbrace{(x_1,0)}_{=a_2}, \underbrace{(x_2,1)}_{=a_3}, \underbrace{(x_2,0)}_{=a_4}\},$$

and let $A_i = \{a_i\}$, $A_{ij} = \{a_i, a_j\}$, $A_{ijk} = \{a_i, a_j, a_k\}$ and $A_{1234} = \mathcal{X} \times \mathcal{Y}$. Take the marginal probability mass functions $P_X(x_1) = \frac{3}{4}$, $P_X(x_2) = \frac{1}{4}$, $P_Y^p(1) = p$, $P_Y^p(0) = 1 - p$, where $p \in \left[\frac{1}{4}, \frac{3}{4}\right]$. The family $\mathcal{P}$ of joint probability distributions for (X, Y) has extreme points and lower envelope reported below

$2^{\mathcal{X}\times\mathcal{Y}}$	∅	A_1	A_2	A_3	A_4	A_{12}	A_{13}	A_{14}	A_{23}	A_{24}	A_{34}	A_{123}	A_{124}	A_{134}	A_{234}	A_{1234}
P^{p_1}	0	$\frac{3}{16}$	$\frac{9}{16}$	$\frac{1}{16}$	$\frac{3}{16}$	$\frac{12}{16}$	$\frac{4}{16}$	$\frac{6}{16}$	$\frac{10}{16}$	$\frac{12}{16}$	$\frac{4}{16}$	$\frac{13}{16}$	$\frac{15}{16}$	$\frac{7}{16}$	$\frac{13}{16}$	1
P^{p_2}	0	$\frac{9}{16}$	$\frac{3}{16}$	$\frac{3}{16}$	$\frac{1}{16}$	$\frac{12}{16}$	$\frac{12}{16}$	$\frac{10}{16}$	$\frac{6}{16}$	$\frac{4}{16}$	$\frac{4}{16}$	$\frac{15}{16}$	$\frac{13}{16}$	$\frac{13}{16}$	$\frac{7}{16}$	1
$\underline{P}$	0	$\frac{3}{16}$	$\frac{3}{16}$	$\frac{1}{16}$	$\frac{1}{16}$	$\frac{12}{16}$	$\frac{4}{16}$	$\frac{6}{16}$	$\frac{6}{16}$	$\frac{4}{16}$	$\frac{4}{16}$	$\frac{13}{16}$	$\frac{13}{16}$	$\frac{7}{16}$	$\frac{7}{16}$	1

The lower envelope is easily seen not to be 2-monotone since

$$\underline{P}(A_{123}) = \frac{13}{16} < \frac{15}{16} = \underline{P}(A_{12}) + \underline{P}(A_{23}) - \underline{P}(A_2).$$

We also have that $\mathcal{P} \subset \mathbf{core}(\underline{P})$ since $\mathbf{ext}(\mathbf{core}(\underline{P})) = \{P_1, P_2, P_3, P_4\}$, where, identifying each probability distribution on $2^{\mathcal{X}\times\mathcal{Y}}$ with the vector of its values on the atoms of $2^{\mathcal{X}\times\mathcal{Y}}$ we have

$$P_1 = P^{p_1} \equiv \left(\tfrac{3}{16}, \tfrac{9}{16}, \tfrac{1}{16}, \tfrac{3}{16}\right), \qquad P_2 = P^{p_2} \equiv \left(\tfrac{9}{16}, \tfrac{3}{16}, \tfrac{3}{16}, \tfrac{1}{16}\right),$$

$$P_3 \equiv \left(\tfrac{7}{16}, \tfrac{5}{16}, \tfrac{1}{16}, \tfrac{3}{16}\right), \qquad P_4 \equiv \left(\tfrac{5}{16}, \tfrac{7}{16}, \tfrac{3}{16}, \tfrac{1}{16}\right).$$

We now investigate the approximation of $\underline{P}$ with a belief function, referred to as *DS-approximation* (where "DS" stands for Dempster and Shafer), by following [5–7]. Searching for an *inner DS-approximation* means to look for a belief function Bel^i that dominates $\underline{P}$, i.e., $Bel^i \geq \underline{P}$ pointwise on $2^{\mathcal{X}\times\mathcal{Y}}$, and is as close as possible to $\underline{P}$ according to the squared Euclidean distance D_2 defined over the set of lower probabilities on $2^{\mathcal{X}\times\mathcal{Y}}$:

$$\text{minimize } D_2(\underline{P}, Bel)$$

subject to:

$$\begin{cases} \sum_{B \subseteq A} m(B) \geq \underline{P}(A), & \text{for all } A \in 2^{\mathcal{X}\times\mathcal{Y}}, \\ \sum_{B \subseteq \mathcal{X}\times\mathcal{Y}} m(B) = 1, & \\ m(B) \geq 0, & \text{for all } B \in 2^{\mathcal{X}\times\mathcal{Y}}, \\ m(\emptyset) = 0. & \end{cases} \qquad (7)$$

We have that D_2, besides assuring uniqueness of the inner DS-approximation, has a justification in terms of a penalty coherence condition for belief functions [8]. Other distances besides D_2 can be considered (see [5–7]).

It trivially holds that there are infinitely many inner DS-approximations of $\underline{P}$, as every P^p will work, so problem (7) is always feasible. The following Example 2 shows that the D_2-optimal inner DS-approximation Bel^i of $\underline{P}$ is generally non-additive, nevertheless, the same example shows that, even though $\mathbf{core}(Bel^i) \subset \mathbf{core}(\underline{P})$, we have that $\mathbf{core}(Bel^i) \not\subseteq \mathcal{P}$ and $\mathcal{P} \not\subseteq \mathbf{core}(Bel^i)$. This last fact has important consequences when computing lower expectations, since no dominance relation can be established between the lower expectation computed with respect to $\mathcal{P}$ and the Choquet integral with respect to Bel^i.

Example 2. Let $\underline{P}$ be as in Example 1. The D_2-optimal inner DS-approximation Bel^i of $\underline{P}$ and its Möbius inverse m^i are reported below

$2^{X \times Y}$	$\emptyset$	A_1	A_2	A_3	A_4	A_{12}	A_{13}	A_{14}	A_{23}	A_{24}	A_{34}	A_{123}	A_{124}	A_{134}	A_{234}	A_{1234}
m^i	0	$\frac{4}{16}$	$\frac{4}{16}$	$\frac{2}{16}$	$\frac{2}{16}$	$\frac{4}{16}$	0	0	0	0	0	0	0	0	0	0
Bel^i	0	$\frac{4}{16}$	$\frac{4}{16}$	$\frac{2}{16}$	$\frac{2}{16}$	$\frac{12}{16}$	$\frac{6}{16}$	$\frac{6}{16}$	$\frac{6}{16}$	$\frac{6}{16}$	$\frac{4}{16}$	$\frac{14}{16}$	$\frac{14}{16}$	$\frac{8}{16}$	$\frac{8}{16}$	1

We have that $\mathbf{ext}(\mathbf{core}(Bel^i)) = \{Q_1, Q_2\}$ where

$$Q_1 \equiv \left(\tfrac{4}{16}, \tfrac{8}{16}, \tfrac{2}{16}, \tfrac{2}{16}\right) \quad \text{and} \quad Q_2 \equiv \left(\tfrac{8}{16}, \tfrac{4}{16}, \tfrac{2}{16}, \tfrac{2}{16}\right).$$

Though $\mathbf{core}(Bel^i) \subset \mathbf{core}(\underline{P})$, since none between Q_1, Q_2 can be expressed as the convex combination of P^{p_1}, P^{p_2} and vice versa, it follows that between $\mathbf{core}(Bel^i)$ and $\mathcal{P}$ no containment relationship holds.

The previous example suggests to move towards an outer DS-approximation. In this case, we search for a belief function Bel^o that is dominated by $\underline{P}$, i.e., $Bel^o \leq \underline{P}$ pointwise on $2^{X \times Y}$, and is as close as possible to $\underline{P}$ according to D_2. We notice that $\underline{P}_{|\widetilde{2^X}}$ coincides with the probability distribution of X and this property is inherited by any inner approximation Bel^i but generally not by an outer approximation Bel^o. Thus, we search for an outer approximation Bel^{oo} such that $Bel^{oo}_{|\widetilde{2^X}}$ coincides with the probability distribution of X. Such an outer DS-approximation will be called X-*preserving* and can be found solving the following optimization problem

$$\text{minimize } D_2(\underline{P}, Bel)$$

subject to:

$$\begin{cases} \sum_{B \subseteq A} m(B) = \underline{P}(A), \text{ for all } A \in \widetilde{2^X}, \\ \sum_{B \subseteq A} m(B) \leq \underline{P}(A), \text{ for all } A \in 2^{X \times Y} \setminus \widetilde{2^X}, \\ \sum_{B \subseteq X \times Y} m(B) = 1, \\ m(B) \geq 0, \qquad \text{for all } B \in 2^{X \times Y}, \\ m(\emptyset) = 0. \end{cases} \qquad (8)$$

Denoting by P_{X_*} the inner measure induced by P_X on $2^{\mathcal{X} \times \mathcal{Y}}$, defined as

$$P_{X_*}(A) = \sup \left\{ \sum_{x \in B} P_X(x) : B \times \mathcal{Y} \subseteq A, B \in 2^{\mathcal{X}} \right\}, \quad \text{for all } A \in 2^{\mathcal{X} \times \mathcal{Y}}, \tag{9}$$

P_{X_*} is an X-preserving outer DS-approximation of $\underline{P}$. In turn, this implies that problem (8) is always feasible and taking D_2 it admits a unique optimal solution.

Example 3. Let $\underline{P}$ be as in Example 1. The D_2-optimal X-preserving outer DS-approximation Bel^{oo} of $\underline{P}$ and its Möbius inverse m^{oo} are reported below

$2^{\mathcal{X} \times \mathcal{Y}}$	$\emptyset$	A_1	A_2	A_3	A_4	A_{12}	A_{13}	A_{14}	A_{23}	A_{24}	A_{34}	A_{123}	A_{124}	A_{134}	A_{234}	A_{1234}
m^{oo}	0	$\frac{3}{16}$	$\frac{3}{16}$	$\frac{1}{16}$	$\frac{1}{16}$	$\frac{6}{16}$	0	0	0	0	$\frac{2}{16}$	0	0	0	0	0
Bel^{oo}	0	$\frac{3}{16}$	$\frac{3}{16}$	$\frac{1}{16}$	$\frac{1}{16}$	$\frac{12}{16}$	$\frac{4}{16}$	$\frac{4}{16}$	$\frac{4}{16}$	$\frac{4}{16}$	$\frac{4}{16}$	$\frac{13}{16}$	$\frac{13}{16}$	$\frac{7}{16}$	$\frac{7}{16}$	1

We have that $\mathbf{ext}(\mathbf{core}(Bel^{oo})) = \{Q_1, Q_2, Q_3, Q_4\}$ where

$$Q_1 = P^{p_1} \equiv \left(\tfrac{3}{16}, \tfrac{9}{16}, \tfrac{1}{16}, \tfrac{3}{16}\right), \qquad Q_2 = P^{p_2} \equiv \left(\tfrac{9}{16}, \tfrac{3}{16}, \tfrac{3}{16}, \tfrac{1}{16}\right),$$

$$Q_3 \equiv \left(\tfrac{3}{16}, \tfrac{9}{16}, \tfrac{3}{16}, \tfrac{1}{16}\right), \qquad Q_4 \equiv \left(\tfrac{9}{16}, \tfrac{3}{16}, \tfrac{1}{16}, \tfrac{3}{16}\right).$$

Both Bel^i and Bel^{oo} allow the following decomposition of the corresponding Choquet expectation.

Proposition 2. *For every $f : \mathcal{X} \to \mathbb{R}$ and $g : \mathcal{X} \times \mathcal{Y} \to \mathbb{R}$ we have:*

(i) $\mathbb{C}_{Bel^i}[f(X) + g(X,Y)] = \mathbb{E}_{P_X}[f(X)] + \mathbb{C}_{Bel^i}[g(X,Y)]$;
(ii) $\mathbb{C}_{Bel^{oo}}[f(X) + g(X,Y)] = \mathbb{E}_{P_X}[f(X)] + \mathbb{C}_{Bel^{oo}}[g(X,Y)]$;

where $\mathbb{E}_{P_X}$ denotes the expectation with respect to the marginal P_X.

Proof. We only prove *(i)* since the proof of *(ii)* is analogous. Every $P \in \mathbf{core}(Bel^i)$ is such that $P_{|2^{\widetilde{\mathcal{X}}}}$ coincides with the marginal distribution of X, therefore

$$
\begin{aligned}
\mathbb{C}_{Bel^i}[f(X) + g(X,Y)] &= \min_{P \in \mathbf{core}(Bel^i)} \mathbb{E}_P[f(X) + g(X,Y)] \\
&= \min_{P \in \mathbf{core}(Bel^i)} \left(\mathbb{E}_{P_X}[f(X)] + \mathbb{E}_P[g(X,Y)]\right) \\
&= \mathbb{E}_{P_X}[f(X)] + \mathbb{C}_{Bel^i}[g(X,Y)].
\end{aligned}
$$

4   Ambiguous Randomized Reinsurance Contracts

Referring to X, Y of Sect. 3, here variable X denotes the (non-negative) aggregate loss of an insurer over one year, while Y is an ambiguous Bernoulli random variable independent of X, indicating reinsurance. Let P_X be the marginal probability mass function of X and let $\mathcal{P}_Y$ be the family of marginal probability mass functions of Y. Consider the set $\mathcal{P}$ of joint distributions of (X, Y) given by (6).

Following [1], we consider a reinsurance contract in which the retained loss of the insurer is singled out by the random variable

$$r(X, Y, d) = \begin{cases} \min(X, d) & \text{if } Y = 1, \\ X & \text{if } Y = 0, \end{cases} \tag{10}$$

where $d \geq 0$ denotes the retention in a stop-loss treaty. For every $P^p \in \mathcal{P}$

$$\mathbb{E}_{P^{p2}}[r(X, Y, d)] \leq \mathbb{E}_{P^p}[r(X, Y, d)] \leq \mathbb{E}_{P^{p1}}[r(X, Y, d)]. \tag{11}$$

Let $\pi(X)$ and $\pi_R(d)$ be, respectively, the total premium the insurer receives from the policyholders for the aggregate loss X and the premium required from the reinsurer. By adopting the expected value principle with safety loading $\theta > 0$, and assuming a pessimistic attitude towards ambiguity we set

$$\pi_R(d) = (1 + \theta)\mathbb{E}_{\mathcal{P}}[X - r(X, Y, d)] = (1 + \theta)\left(\mathbb{E}_{P_X}[X] + \underline{\mathbb{E}}_{\mathcal{P}}[-r(X, Y, d)]\right). \tag{12}$$

For $\alpha \in (0, 1)$, we further define the risk measure associated to $r(X, Y, d)$ as

$$\underline{\text{VaR}}_\alpha(r(X, Y, d)) := \inf\{x : \underline{F}_{r(X,Y,d)}(x) \geq \alpha\}, \tag{13}$$

where $\underline{F}_{r(X,Y,d)}(x) := \underline{P}(r(X, Y, d) \leq x)$.

The insurer's annual profit under reinsurance is

$$Z(X, Y, d) = \frac{\pi(X) - \pi_R(d)}{1 - r_{coc}} - r(X, Y, d) - \frac{r_{coc}}{1 - r_{coc}}\underline{\text{VaR}}_\alpha(r(X, Y, d)), \tag{14}$$

where r_{coc} denotes the cost of capital rate. Due to the translation invariance of the lower expectation operator we get that

$$\underline{\mathbb{E}}_{\mathcal{P}}[Z(X, Y, d)] = \frac{\pi(X) - \pi_R(d)}{1 - r_{coc}} + \underline{\mathbb{E}}_{\mathcal{P}}[-r(X, Y, d)]$$
$$- \frac{r_{coc}}{1 - r_{coc}}\underline{\text{VaR}}_\alpha(r(X, Y, d)). \tag{15}$$

Under this pessimistic attitude towards ambiguity, the issue is to maximize $\underline{\mathbb{E}}_{\mathcal{P}}[Z(X, Y, d)]$ seen as a function of d.

Example 4. Take X ranging in $\mathcal{X} = \{0, 100, 1000\}$ with probability mass function $P_X(0) = \frac{9}{10}, P_X(100) = \frac{6}{100}, P_X(1000) = \frac{4}{100}$, and Y be an ambiguous Bernoulli random variable with probability mass function $P_Y(1) = p, P_Y(0) = 1 - p$ and $p \in \left[\frac{8}{10}, \frac{9}{10}\right]$. Let $\mathcal{P}$ be defined as in (6) and $\underline{P} = \min \mathcal{P}$ pointwise on $2^{\mathcal{X} \times \mathcal{Y}}$. Take $\alpha = 0.99$, $\theta = 0.1$, $r_{coc} = 0.07$, $\pi(X) = (1 + 0.1)\mathbb{E}_{P_X}[X] = 50.6$. Let $r(X, Y, d)$ be defined as in (10):

$$\underline{\mathbb{E}}_{\mathcal{P}}[-r(X, Y, d)] = \begin{cases} -0.08d - 9.2 & \text{if } 0 \leq d < 100, \\ -0.032d - 14 & \text{if } 100 \leq d < 1000, \\ -46 & \text{if } d \geq 1000, \end{cases}$$

therefore, we get

$$\pi_R(d) = \begin{cases} -0.088d + 40.48 & \text{if } 0 \le d < 100, \\ -0.0352d + 35.2 & \text{if } 100 \le d < 1000, \\ 0 & \text{if } d \ge 1000. \end{cases}$$

Moreover, referring to the different definitions of $\underline{F}_{r(X,Y,d)}(x)$, according to the value of d, we have that

$$\underline{\text{VaR}}_\alpha[r(X,Y,d)] = \begin{cases} 100 & \text{if } 0 \le d < 100, \\ d & \text{if } 100 \le d < 1000, \\ 1000 & \text{if } d \ge 1000. \end{cases}$$

Thus we get that

$$\mathbb{E}_P[Z(X,Y,d)] = \begin{cases} \left(\frac{0.088}{0.93} - 0.08\right)d + \left(\frac{3.12}{0.93} - 9.2\right) & \text{if } 0 \le d < 100, \\ \left(-\frac{0.0348}{0.93} - 0.032\right)d + \left(\frac{15.4}{0.93} - 14\right) & \text{if } 100 \le d < 1000, \\ -\frac{19.4}{0.93} - 46 & \text{if } d \ge 1000. \end{cases}$$

It is immediate to verify that $\mathbb{E}_P[Z(X,Y,d)]$, seen as a function of d, has a global maximum at $d^* = 100$. In the precise case the optimal retention level d^* is 100 for $p \in [0.8, 0.9)$ and 0 for $p = 0.9$. Thus, the optimal retention level in the imprecise case is greater than or equal to that in the precise case.

Despite using $\mathcal{P}$ and the associated lower expectation functional $\mathbb{E}_P$, we can consider the D_2-optimal X-preserving outer DS-approximation Bel^{oo} of P together with the corresponding Choquet expectation functional $\mathbb{C}_{Bel^{oo}}$. By virtue of Proposition 2 the premium is

$$\pi_R^{oo}(d) = (1 + \theta)\left(\mathbb{E}_{P_X}[X] + \mathbb{C}_{Bel^{oo}}[-r(X,Y,d)]\right), \tag{16}$$

and the risk measure becomes

$$\underline{\text{VaR}}_\alpha^{oo}(r(X,Y,d)) := \inf\{x : \underline{F}_{r(X,Y,d)}^{oo}(x) \ge \alpha\}, \tag{17}$$

where $\underline{F}_{r(X,Y,d)}^{oo}(x) := Bel^{oo}(r(X,Y,d) \le x)$.

The insurer's annual profit under reinsurance is then changed in

$$Z^{oo}(X,Y,d) = \frac{\pi(X) - \pi_R^{oo}(d)}{1 - r_{coc}} - r(X,Y,d) - \frac{r_{coc}}{1 - r_{coc}}\underline{\text{VaR}}_\alpha^{oo}(r(X,Y,d)), \tag{18}$$

thus we get

$$\mathbb{C}_{Bel^{oo}}[Z^{oo}(X,Y,d)] = \frac{\pi(X) - \pi_R^{oo}(d)}{1 - r_{coc}} + \mathbb{C}_{Bel^{oo}}[-r(X,Y,d)]$$
$$- \frac{r_{coc}}{1 - r_{coc}}\underline{\text{VaR}}_\alpha^{oo}(r(X,Y,d)). \tag{19}$$

The issue is to maximize $\mathbb{C}_{Bel^{oo}}[Z^{oo}(X,Y,d)]$ seen as a function of d. Analogously, we define $\pi_R^i(d)$, $\underline{\text{VaR}}_\alpha^i$ and $Z^i(X,Y,d)$ when we use Bel^i and $\mathbb{C}_{Bel^i}$.

Example 5. Let X, Y, $\mathcal{P}$, α, θ, r_{coc}, $\pi(X)$ as in Example 4. Denote

$$\mathcal{X} \times \mathcal{Y} = \{\underbrace{(0,1)}_{=a_1}, \underbrace{(0,0)}_{=a_2}, \underbrace{(100,1)}_{=a_3}, \underbrace{(100,0)}_{=a_4}, \underbrace{(1000,1)}_{=a_5}, \underbrace{(1000,0)}_{=a_6}\},$$

and let $A_{i_1\cdots i_k} = \{a_{i_1}, \ldots, a_{i_k}\}$.

The D_2-optimal X-preserving outer DS-approximation Bel^{oo} of $\underline{P}$ has Möbius inverse m^{oo} such that

$$m^{oo}(A_1) = \tfrac{720}{1000},\ m^{oo}(A_2) = \tfrac{90}{1000},\ m^{oo}(A_{12}) = \tfrac{90}{1000},$$
$$m^{oo}(A_3) = \tfrac{48}{1000},\ m^{oo}(A_4) = \tfrac{6}{1000},\ m^{oo}(A_{34}) = \tfrac{6}{1000},$$
$$m^{oo}(A_5) = \tfrac{32}{1000},\ m^{oo}(A_6) = \tfrac{4}{1000},\ m^{oo}(A_{56}) = \tfrac{4}{1000},$$

and zero elsewhere. A straightforward computation shows that

$$\mathbb{C}_{Bel^{oo}}[-r(X,Y,d)] = \begin{cases} -0.08d - 9.2 & \text{if } 0 \le d < 100, \\ -0.032d - 14 & \text{if } 100 \le d < 1000, \\ -46 & \text{if } d \ge 1000, \end{cases}$$

therefore, we get that $\pi_R^{oo}(d) = \pi_R(d)$. Moreover, it is easy to verify that $\underline{F}^{oo}_{r(X,Y,d)}(x)$ has the same definitions of $\underline{F}_{r(X,Y,d)}(x)$, according to the value of d. Hence, it follows that $\underline{\text{VaR}}^{oo}_\alpha[r(X,Y,d)] = \underline{\text{VaR}}_\alpha[r(X,Y,d)]$ and $Z^{oo}(X,Y,d) = Z(X,Y,d)$. Finally, we obtain that $\mathbb{C}_{Bel^{oo}}[Z^{oo}(X,Y,d)] = \mathbb{E}_{\underline{P}}[Z(X,Y,d)]$ that has $d^* = 100$ as unique maximizer.

The D_2-optimal inner DS-approximation Bel^i of $\underline{P}$ has Möbius inverse m^i such that

$$m^i(A_1) = \tfrac{725}{1000},\ m^i(A_2) = \tfrac{95}{1000},\ m^i(A_{12}) = \tfrac{80}{1000},$$
$$m^i(A_3) = \tfrac{51}{1000},\ m^i(A_4) = \tfrac{9}{1000},$$
$$m^i(A_5) = \tfrac{34}{1000},\ m^i(A_6) = \tfrac{6}{1000},$$

and zero elsewhere. A straightforward computation shows that

$$\mathbb{C}_{Bel^i}[-r(X,Y,d)] = \begin{cases} -0.085d - 6.9 & \text{if } 0 \le d < 100, \\ -0.034d - 12 & \text{if } 100 \le d < 1000, \\ -46 & \text{if } d \ge 1000, \end{cases}$$

therefore, we get

$$\pi_R^i(d) = \begin{cases} -0.0935d + 43.01 & \text{if } 0 \le d < 100, \\ -0.0374d + 37.4 & \text{if } 100 \le d < 1000, \\ 0 & \text{if } d \ge 1000, \end{cases}$$

and, though the definitions of $\underline{F}^i_{r(X,Y,d)}(x)$ differ from $\underline{F}^{oo}_{r(X,Y,d)}(x)$, it holds that $\underline{\text{VaR}}^i_\alpha[r(X,Y,d)] = \underline{\text{VaR}}^{oo}_\alpha[r(X,Y,d)] = \underline{\text{VaR}}_\alpha[r(X,Y,d)]$. Thus we get that

$$\mathbb{C}_{Bel^i}[Z^i(X,Y,d)] = \begin{cases} \left(\tfrac{0.0935}{0.93} - 0.085\right)d + \left(\tfrac{0.59}{0.93} - 6.9\right) & \text{if } 0 \le d < 100, \\ \left(-\tfrac{0.0326}{0.93} - 0.034\right)d + \left(\tfrac{13.2}{0.93} - 12\right) & \text{if } 100 \le d < 1000, \\ -\tfrac{19.4}{0.93} - 46 & \text{if } d \ge 1000. \end{cases}$$

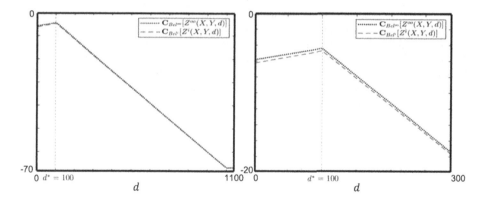

Fig. 1. Graph of $\mathbb{C}_{Bel^{oo}}[Z^{oo}(X,Y,d)] = \mathbb{E}_{\mathcal{P}}[Z(X,Y,d)]$ and $\mathbb{C}_{Bel^i}[Z^i(X,Y,d)]$ as functions of d (left plot for $d \in [0,1100]$, right plot for $d \in [0,300]$)

From Fig. 1 it is easily seen that both $\mathbb{C}_{Bel^{oo}}[Z^{oo}(X,Y,d)] = \mathbb{E}_{\mathcal{P}}[Z(X,Y,d)]$ and $\mathbb{C}_{Bel^i}[Z^i(X,Y,d)]$ have a global maximum at $d^* = 100$.

5   Conclusion

The present paper showed preliminary results on addressing ambiguity in a reinsurance model, inside Dempster-Shafer theory. Future research will focus on: *(i)* characterizing the optimal retention level d^* under the different approximations for a generic X; *(ii)* comparing the effect of the different approximations on d^*.

References

1. Albrecher, H., Cani, A.: On randomized reinsurance contracts. Insur. Math. Econ. **84**, 67–78 (2019)
2. Dempster, A.: Upper and lower probabilities induced by a multivalued mapping. Ann. Math. Stat. **38**(2), 325–339 (1967)
3. Denneberg, D.: Non-Additive Measure and Integral. Kluwer Acad, Pub (1994)
4. Grabisch, M.: Set functions, capacities and games. In: Set Functions, Games and Capacities in Decision Making. TDLC, vol. 46, pp. 25–144. Springer, Cham (2016). https://doi.org/10.1007/978-3-319-30690-2_2
5. Miranda, E., Montes, I., Vicig, P.: On the selection of an optimal outer approximation of a coherent lower probability. Fuzzy Sets Syst. **424**, 1–36 (2021)
6. Montes, I., Miranda, E., Vicig, P.: 2-monotone outer approximations of coherent lower probabilities. Int. J. Approx. Reas. **101**, 181–205 (2018)
7. Montes, I., Miranda, E., Vicig, P.: Outer approximating coherent lower probabilities with belief functions. Int. J. Approx. Reas. **110**, 1–30 (2019)
8. Petturiti, D., Vantaggi, B.: How to assess coherent beliefs: a comparison of different notions of coherence in Dempster-Shafer theory of evidence. In: Augustin, T., et al. (eds.) Reflections on the Foundations of Probability and Statistics: Essays in Honor of Teddy Seidenfeld. Theory and Decision Library A, vol. 54. Springer (2022). https://doi.org/10.1007/978-3-031-15436-2_8

9. Shafer, G.: A Mathematical Theory of Evidence. Princeton Univ, Press (1976)
10. Shenoy, P.: An expectation operator for belief functions in the Dempster-Shafer theory. Int. J. Gen. Syst. **49**(1), 112–141 (2020)
11. Troffaes, M., de Cooman, G.: Lower Previsions. Wiley, Hoboken (2014)
12. Walley, P.: Statistical Reasoning with Imprecise Probabilities. Chapman and Hall (1991)

Evidential Filtering and Spatio-Temporal Gradient for Micro-movements Analysis in the Context of Bedsores Prevention

Nicolas Sutton-Charani[1]([✉]) [iD], Francis Faux[2] [iD], Didier Delignières[1] [iD],
Willy Fagard[3], Arnaud Dupeyron[3] [iD], and Marie Nourrisson[2]

[1] EuroMov Digital Health in Motion, Univ Montpellier, IMT Mines Ales,
6 Avenue de Claviéres, 30100 Alés, France
`nicolas.sutton-charani@mines-ales.fr`, `didier.delignieres@umontpellier.fr`
[2] ISIS Engineer School, Castres, France
`francis.faux@irit.fr`, `marie.nourrisson@etud.univ-jfc.fr`
[3] CHU de Nîmes, Nîmes, France
`willy.fagard@chu.nimes.fr`, `arnaud.dupeyron@umontpellier.fr`

Abstract. In the context of pressure ulcer prevention, this article deals with the problem of detecting and analysing micro-movements in the sacral area of bedridden patients on mattresses equipped with a network of pressure sensors. The study is based on a series of pressure measurements carried out on a cohort of patients lying on two types of mattress at the Nîmes university hospital (France). A spatio-temporal model considers first the local information of measurements from the array of sensors using an evidential filter in order to remove spatial uncertainties or measurement noise before micro-movement analysis. With a Detrended Fluctuation Analysis (DFA), the complexity level of the time series coming from the micro-movement model is finally estimated for different noise filters.

Keywords: Bedsores prevention · Micromovements analysis · Spatio-temporal gradient · Pressure sensors · Detrended fluctuation analysis

1   Introduction

Pressure ulcers, also known as 'bedsores', refer to a type of injury that necroses the skin and underlying tissue because of large amount of pressure applied to an area of skin over a short period. Due to their lack of muscle, elderly people and individuals with severely compromised states of health are especially vulnerable to pressure sores [1]. Prevention of pressure ulcer formation is primarily focused on minimizing episodes of prolonged pressure either by placing appropriate padding at pressure points or by frequent patient repositioning [10,12].

In 2015, the laboratory of Bio-Statistics, Clinical Epidemiology, Public Health, Innovation and Methodology (BCPIM) attached to the Nîmes hospital (France) carried out a series of measurements using a connected pressure

S. Le Hégarat-Mascle et al. (Eds.): BELIEF 2022, LNAI 13506, pp. 297–306, 2022.
https://doi.org/10.1007/978-3-031-17801-6_28

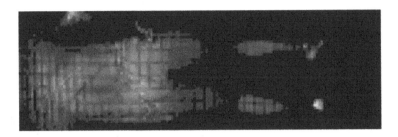

Fig. 1. Pressure image (patient 5, mattress Softform)

48 × 145 uniform sensor array sited on top of mattresses, over a period of 1 month, in order to evaluate the levels of pressure exerted by the mattress on bed ridden patients. The pressure measurements lie in [0 mmHg, 100 mmHg] with a precision of 0.1 mmHg. This study based on the statistical analysis of pressure peaks has highlighted high variability and non-reproducibility between measurements [13]. The measures have been carried out on a cohort of patients spread over 3 centers located in Occitanie (France): CHU Nîmes, EHPAD9 La Chimotaie, EHPAD Notre Dame des Pins. Each center realised measurements on patients lying on two different mattress models: Softform or Airsoft. For each measurement of 30 min, the software recorded the pressure values coming from the sensor network at a frequency of one per second. Hence, we can observe the profile of the patient lying on the bed every second (see Fig. 1). In this paper we pursue this study by considering, instead of the pressure peaks, the spatio-temporal distribution of pressure variations assumed to be caused by micro-movements in the sacral region. These micro-movements are supposed to result from many different processes such as blood pressure, muscular and neural impulsions, or mobility of the tissues when the patients remain immobile in hospital beds. It turns out that the specificity of micro-movements is difficult to observe. According to Descartes a movement refers to the action by which a body moves from one place to another [3]. The concept of *movement* thus generally implies some shape preservation. In our context, the shape in movement is poorly known, not rigid and deformable. Hence, classical image motion analysis methods (optical flow) appears not applicable. Another difficulty raises in the differentiation between micro-movements and disturbing noise due to sensors uncertainties (imprecision, reliability). In the domain of image processing, many approaches have been proposed to search for efficient image denoising [2]. In the framework of Dempster-Shafer theory, this problem has been discussed by Weeraddan et al. in [16] in order to process temporally and spatially distributed multi modality sensor data. The authors extended Temporal Evidence Filtering [6], that allows to merge multi-modalities evidence and to infer in the frequency domain but they do not address the problem of noise buried in the information clutter. In a study close to our problematic, carried out by Li et al. [11] in the domain of visual tracking, the authors incorporate Dempster-Shafer information fusion into the entire image sequence partitioned into spatially and temporally

adjacent sub-sequences. However these models do not respond to our specific context of sparse data, i.e., grey scale images containing many null values. The model of micro-movements proposed in this paper integrates the uncertainty in motion detection using a spatio-temporal gradient method and an evidential filter necessary to smooth, in an adequate way, pictures of pressure levels. Finally a Detrended Fluctuation Analysis (DFA) is proposed to compare in term of complexity the proposed micro-movements model in regards to classic filtering techniques.

The organisation of the rest paper is as follows. In Sect. 2 we recall the necessary basis of the belief function theory and we briefly recall the Detrended Fluctuation Analysis (DFA) for time series complexity estimation. Section 3 presents the evidential filter based on the EKNN model [5] and the micro-movements quantification method. Experiments and results are detailed in Sect. 4 and finally we present our concluding remarks and perspectives.

2   Necessary Background

2.1   Belief Function Theory

In Belief function theory [14], uncertainty regarding the value of a variable ω defined on a finite set of possible values, called the frame of discernment $\Omega = \{\omega_1, \ldots, \omega_N\}$ is represented by a basic belief assignment (BBA) or mass function m defined as a mapping $m : 2^\Omega \to [0, 1]$ verifying $\sum_{A \subseteq \Omega} m(A) = 1$.

The conjunctive combination of BBA's m_j derived from J distinct sources, denoted by $m_{\bigcirc\!\!\!\!\cap}$ is expressed by $m_{\bigcirc\!\!\!\!\cap}(A) = \sum_{A_1 \cap \ldots \cap A_J = A} \left(\prod_{j=1}^J m_j(A_j) \right)$. Dempster's rule, denoted by $\oplus$, is a normalized version of the conjunctive combination rule and is defined such that: $m_\oplus(\emptyset) = 0$ and $m_\oplus(A) = K \cdot m_{\bigcirc\!\!\!\!\cap}$ for $A \neq \emptyset$. The normalization factor K is of the form $(1 - c(m_1, \ldots m_J))^{-1}$ where $c(m_1, \ldots m_J) = \sum_{A_1 \cap \ldots \cap A_J = \emptyset} \left(\prod_{j=1}^J m_j(A_j) \right)$ represents the amount of conflict between the sources. These two combination rules are commutative, associative, and often used to combine BBAs from distinct sources.

The pignistic transformation [15] distributes the ignorance equally among each singleton element in Ω such that:

$$BetP(\omega) = \frac{1}{\left(1 - m(\emptyset)\right)} \sum_{W \subseteq \Omega : W \ni \omega} \frac{m(W)}{|W|}$$

where $|W|$ is the number of singletons in W, we have $\sum_{\omega \in \Omega} BetP(\omega) = 1$.

2.2   Detrended Fluctuation Analysis (DFA)

The complexity of biological signals has given rise to many research works [4, 7, 17]. Among several alternatives (e.g. entropy), Detrended Fluctuation Analysis

(DFA) provides fractality level estimators of any signal, even non-stationary ones, and has shown to efficiently discriminate healthy and ill subjects [8,9].

The DFA algorithm works as follow, for a series $x(i)$ of length N. The series is first integrated, by computing for each i the accumulated departure from the mean of the whole series: $X(i) = \sum_{j=1}^{i} \left(x(j) - \bar{x} \right)$ where $\bar{x} = \sum_{i=1}^{N} x(i)$.

This integrated series is then divided into k non-overlapping intervals of length n. The last $N - kn$ data points are excluded from analysis. Within each interval, a least squares line is fitted to the data. The series $X(i)$ is then locally detrended by subtracting the theoretical values $X^{Th}(i)$ given by the regression. For a given interval of length n, the characteristic size of fluctuation for this integrated and detrended series is calculated by:

$$F(n) = \sqrt{\frac{1}{N - kn} \sum_{i=1}^{N-kn} \left(X(i) - X^{Th}(i) \right)^2}.$$

This computation is repeated over a large range of interval lengths. In the present experiment, from $n = 10$ to $n = 300$, by steps of 1. A power law is expected between n and the average fluctuation size $F(n)$, as $F(n) \propto n^\alpha$.

The exponent α is estimated as the slope of the double logarithmic plot of $F(n)$, as a linear function of n. Typically, $\alpha = 1$ corresponds to an optimal complexity level, generally encountered in healthy, and perennial systems. $\alpha = 0.5$ reveals uncorrelated white noise series, and denotes a complete loss of complexity in the underlying system. At the opposite, $\alpha \approx 1.5$ for quasi-deterministic systems having almost no adaptability.

3   Method

In this section we present an evidential extension of the median (or mean) filter, we propose a micro-movement measure and we explain how the filtering method can impact the micro-movement complexity signal.

3.1   Evidential Filter

The median filter is one of the most famous approach to remove noise from images. It is a non-linear filter which has the property of preserving edges while removing noise. However it is not appropriate for sparse data, e.g. grey scale images containing many null values. Indeed, in that context, the mean filter seems to give better results (see Fig. 2) probably due to higher number of possible estimate, especially for data with frequent values (0 or 1 often arise with pressure sensors). In the bedsores problem, a grid pattern which comes from structural sensors array defect, is visible on all pressure images. This type of perturbation makes necessary some filtering upstream of any analysis.

The idea of our proposal is to use the local information of the sensors array in order to denoise the images before micro-movement analysis and to asses to filter's choice impact in terms of micro-movements and associated complexity. The belief function framework is used in order to take advantage of its uncertainty modelling and fusion abilities. Our model is based on the spatial uncertainty model proposed by Denœux in [5] which extends the K-NN algorithm to uncertain labels where neighbours are considered as information sources with a reliability level that depends on their similarity to the prediction example. The corresponding spatial uncertainty model is recalled hereafter:

$$\begin{cases} m^{s,i}(C_q) = \alpha \\ m^{s,i}(C) = 1 - \alpha \end{cases} \text{ with } \alpha = \alpha_0 e^{-\gamma_q d^\beta} \tag{1}$$

where $m^{s,i}$ is the mass function representing the reliability of neighbour x_i, in regards to a new example x_s to classify, y_i is the label of x_i, which is supposed to belong to class C_q, C is the set of all possible classes and d stands for the distance between x_s and x_i. The $(\alpha_0, \gamma_q, \beta)$ coefficients are tuning hyper-parameters.

With pressure images, as for median and mean filters we slide windows and replace the center of the window by a certain value which is neither the median nor the mean value of the window but is computed as the pignistic expectation of the conjunctive fusion of the window sensors. For a given pixel window, for all pressure measurements a mass function is defined according to the $EKNN$ spatial uncertainty model (1) extended to the numerical context by replacing the categorical focal elements C_q by numerical ones w_i which stands for the pressure measurement of a pixel i. In our approach we chose to redefine the frame of discernment $\Omega = \{w_1, ..., w_K\} \subseteq \{p_1, ..., p_N\} \subset R^+$ for each window of N pixels as the set of K different observed pressure values. By definition, $\forall_i \neq j : w_i \neq w_j$ but the observed pressures can be identical for several pixels of a window, we have $K \leq N$.

The adaptation of the $EKNN$ spatial uncertainty model to the pressure sensors context is given in Eq. (2).
$\forall_i = 1, \cdots, N :$

$$\begin{cases} m_i(p_i) = \alpha_i \\ m_i(\Omega) = 1 - \alpha_i \end{cases} \tag{2}$$

with $\alpha_i = \alpha_0 e^{-\gamma d_i^\beta}$. The distance d_i is computed between the ith pixel having for pressure value p_i and the center of the window.

The conjonctive combination of all the window pixels mass functions leads to a final mass function m^* having for focal elements all the window pixel single values and Ω.

$$\begin{cases} \forall k = 1, ..., K : \\ \\ m^*\left(\{w_k\}\right) = \dfrac{\displaystyle\prod_{\substack{i=1 \\ p_i=w_k}}^{N} \alpha_i \times \prod_{\substack{j=1 \\ p_j \neq w_k}}^{N} (1-\alpha_j)}{1-\kappa} \\ \\ m^*\left(\{w_1, ..., w_K\}\right) = \dfrac{\displaystyle\prod_{i=1}^{N} (1-\alpha_i)}{1-\kappa} \end{cases} \tag{3}$$

with

$$\kappa = 1 - \sum_{k=1}^{K} \left[\prod_{\substack{i=1 \\ p_i = w_k}}^{N} \alpha_i \times \prod_{\substack{j=1 \\ p_j \neq w_k}}^{N} (1 - \alpha_j) + \prod_{i=1}^{N}(1 - \alpha_i) \right]$$

As a first approach we propose to compute the pignistic transform $BetP^*$ of m^* and to replace the central pixel s of the window by the expectation of an unknown pressure value W according to the $BetP^*$ probability.

Results of the pignistic transform applied to m^* and the corresponding expectation of an uncertain pressure value W are given in Eqs. (4) and (5).

$\forall k = 1, ..., K,$

$$BetP^* \left(\{W = w_k\} \right) = m^* \left(\{w_k\} \right) + \frac{1}{K} m^* \left(\Omega \right) \qquad (4)$$

$$E_{BetP^*} \left[W \right] = \sum_{k=1}^{K} w_k * BetP^* \left(\{W = w_k\} \right) \qquad (5)$$

The choice of redefining the frame of discernment Ω for each pixels window comes from the local nature of the belief filter, we only take into account the of pixels neighbors to smooth images. The fact that we consider discrete spaces Ω is counterbalanced by the pignistic transform which can result in any value between the elements of Ω (see Eq. (5)).

The belief filter we propose smooth the pressure images with a sliding window approach. For all pixel windows the central pixel is replaced by the pignistic expectation of the (uncertain) pressure values of all its neighbors and their corresponding bbas which are defined according their distance to the central pixel (see Eq. (3)). In the following subsection we propose a micro-movement indicator which can be computed on any sets of pressure images, smoothed or not.

3.2 Micro-movements Quantification

Once all images have been smoothed with our belief filter, we quantify the amount of micro-movement as the level of deformation enhanced by variations of pressure on the sensor network. Indeed, all the information available about these movements is provided by pressure sensors which provide grey intensity images at each timestamp.

A sequence of pressure measurements images can be considered as a 3-dimensional discrete spatio-temporal field: $W_t^{(x,y)}$, where x and y denote the coordinates of an individual sensor in the sensor network, and t is the time index. The gradient vector $\vec{\nabla}(W_t^{(x,y)})$ is defined such that $\vec{\nabla} = [\partial/(\partial x), (\partial/\partial y)]^T$. The Sobel operator allows to calculate approximations of the horizontal and vertical derivatives using two 3×3 kernels which are convolved with the original source image $W_t^{(x,y)}$. The Sobel operator allows to calculate two images

$$S_x(t) = \begin{pmatrix} 1 & 0 & -1 \\ 2 & 0 & -2 \\ 1 & 0 & -1 \end{pmatrix} \star W_t^{(x,y)} \text{ and } S_y(t) = \begin{pmatrix} 1 & 2 & 1 \\ 0 & 0 & 0 \\ -1 & -2 & -1 \end{pmatrix} \star W_t^{(x,y)} \text{ which contain}$$

respectively the horizontal and vertical approximations of the gradient vector for each point (x, y). The 2-dimensional signal processing convolution operation is denoted by $\star$.

The micro-movement measure μ_{mvt} is computed as the absolute value of the spatio-temporal gradient of the videos corresponding to the set of pressure images of one measurement of 30 min:

$$\mu_{mvt}(t, x, y) = \left| \frac{\partial}{\partial t} S(t, x, y) \right| \tag{6}$$

where $S(t, x, y) = \sqrt{S_x(t)^2 + S_y(t)^2}$.

It is noticeable that micro-movements quantification can be computed independently of the considered filter, even on raw image (without any filtering). In the case of filtered images, the Sobel values $S(t, x, y)$ are computed on images that have been smothed according to Eqs. (3) and (5).

4   Experiment and Results

In this section the evaluation criteria considered in this study are detailed and finally results are presented in terms of filtering, micro-movement and DFA.

The evidential filtering proposed in this work is first visually evaluated. The micro-movement and complexity time series computed on raw and filtered images are compared. This comparison is realised with the raw signals as reference, *i.e.*, the best denoising filters corresponding to the closest micro-movement and complexity signals to the ones computed on raw images (without filtering).

Visual Evaluation of Filtering (Fig. 2). The median filtered images are very smoothed which is not an optimal context for micro-movement analysis since micro-movements are of very low intensity by definition. The belief filter gives a less smoothed image than the mean filter (the grid has almost disappeared), which is suitable for the analysis of of micro-movements that could become undetectable with too strong smoothing.

Micro-movement Quantification
The micro-movement signal computed as in Eq. (6) contains large peaks which do not correspond to micro-movement but are rather due to real movements when the patient lies down on his bed. In order to extract those real movements from the signal, we removed upstream all micro-movement data that were higher than the 99th signal centile. The obtained micro-movement time series seems stationary and rough enough for complexity computations.

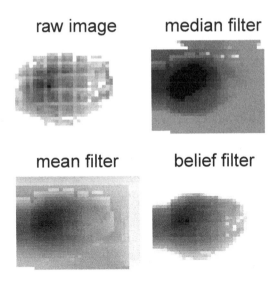

Fig. 2. Sacrum denoising

In terms of DFA

For three measurements, the evolution of complexity is presented in Fig. 3, nevertheless all the following comments have been made on the set of 18 measurements.

We first observe that for all three filters (median, mean and belief) the same curve shapes are observed. This means that the filter choice has no impact on the complexity evolution shape. However, the filtered level of complexity computations are generally biased from the ones computed on raw images and these biases tend to decrease with time. On the set of 18 measurements we note that the raw complexity signals lie almost always between filtered signals, except for 2 or 3 cases when the raw complexity signal has especially large variations probably due to some real movement track perturbations. In terms of DFA, this seems to attenuate the filter choice importance and even the importance of filtering itself which is sometimes a delicate pre-processing question for analysts and researchers.

Finally, in terms of filtering, mean and belief filters curves are always very close. However, significance test comparison have been made on the final levels of complexity in order to compare the complexity estimation absolute errors (in regards to the final complexity level computed on raw images). On the set of 18 measurement the median filters complexity estimator has the lowest error: 0.0154, the belief and mean filter having errors of 0.0161 and 0.0380. With a risk of 5% there is no significant (Wilcoxon test) difference of complexity estimation errors between median and belief filters ($p = 0.290$) whereas belief filter estimation errors are significantly lower than mean filter one ($p = 0.009$).

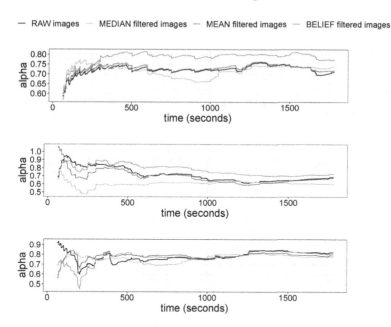

Fig. 3. Filtering vs complexity

5   Conclusion and Perspectives

In this paper a spatio-temporal model of micro-movement located in the sacral area of bedridden patients has been proposed as weel as a study of the evolution of complexity during the measurement (30 min). This preliminary work lays the foundations for a micro-movement analysis integrating uncertainty/imprecision management using an evidential filter. Results are encouraging.

In standard DFA, the complexity parameter α is computed on the whole signal, *i.e.* once the measurement is complete. In our case we monitored the evolution of complexity over time in order to estimate the necessary measurement duration to get a relatively stable level of sacrum micro-movement complexity (α coefficient). Concerning DFA, the measurement noise does not seem to impact the estimates and it does not seem necessary to filter for complexity analysis.

In an industrial perspective this type of estimation could be of precious value for connected mattresses and bedsores risk notifications (when the amount of micro-movement is too low or the level of complexity seems abnormal).

The future works concern the search of a similarity measure able to discriminate patients and mattresses from micro-movement and complexity signals based on machine learning methods. In the evidential perspective, uncertainty modeling should be improved with more complex fusion and decision rules and finally it would be interesting to consider uncertainties at the complexity level in the DFA model.

References

1. Bhattacharya, S., Mishra, R.: Pressure ulcers: current understanding and newer modalities of treatment. Indian J. Plast. Surg. **48**(01), 004–016 (2015)
2. Buades, A., Coll, B., Morel, J.M.: A review of image denoising algorithms, with a new one. Multisc. Model. Simul. **4**(2), 490–530 (2005)
3. Clavelin, M.: Galilée et Descartes sur la conservation du mouvement acquis, vol. 242[1]. Dix-septiéme siécle (2009)
4. Delignieres, D., Lemoine, L., Ramdani, S., Torre, K., Fortes, M., Ninot, G.: Fractal analyses for "short" time series: a re-assessment of classical methods. J. Math. Psychol. **50**, 525–544 (2000)
5. Denæux, T.: A k-nearest neighbor classification rule based on Dempster-Shafer theory. IEEE Trans. Syst. Man Cybern. **219**, 804–813 (1995)
6. Dewasurendra, D.A., Bauer, P.H., Premaratne, K.: Evidence filtering. IEEE Trans. Signal Process. **55**(12), 5796–5805 (2007)
7. Eke, A., Herman, P., Kocsis, L., Kozak, L.: Fractal characterization of complexity in physiological temporal signals. Physiol. Meas. **23**, 31–38 (2002)
8. Goldberger, A.L., Amaral, L.A.N., Hausdorff, J.M., Ivanov, P.C., Peng, C.K., Stanle, H.E.: Fractal dynamics in physiology: alterations with disease and aging. Proc. Natl. Acad. Sci. **99**(suppl_1), 2466–2472 (2002)
9. Hausdorff, J., et al.: Altered fractal dynamics of gait: reduced stride-interval correlations with aging and Huntington's disease. J. Appl. Physiol. (Bethesda, Md. : 1985) **82**, 262–9 (1997)
10. Kim, S.Y., Shin, Y.S.: A comparative study of 2-hour interface pressure in different angles of laterally inclined, supine, and fowler's position. Int. J. Environ. Res. Public Health **18**(19), 9992 (2021)
11. Li, X., Dick, A., Shen, C., Zhang, Z., van den Hengel, A., Wang, H.: Visual tracking with spatio-temporal Dempster-Shafer information fusion. IEEE Trans. Image Process. **22**(8), 3028–3040 (2013)
12. Mervis, J.S., Phillips, T.J.: Pressure ulcers: Prevention and management. J. Am. Acad. Dermatol. **81**(4), 893–902 (2019)
13. Neverov, C., et al.: Modélisation incertaine à partir de mesures non-reproductibles. Application à la comparaison de pression exercée par des matelas. In: LFA 2019– 28ème Rencontres Francophones sur la Logique Floue et ses Applications. Alès, France (2019)
14. Shafer, G.: A mathematical theory of evidence. In: A Mathematical Theory of Evidence. Princeton University Press (1976)
15. Smets, P.: Decision making in the TBM: the necessity of the pignistic transformation. Int. J. Approx. Reason. **38**(2), 133–147 (2005)
16. Weeraddana, D.M., Kulasekere, C., Walgama, K.S.: Dempster-Shafer information filtering framework: temporal and spatio-temporal evidence filtering. IEEE Sens. J. **15**(10), 5576–5583 (2015)
17. West, B.J.: Fractal Physiology and Chaos in Medicine, 2nd edn. World Scientific, Singapore (1990). https://doi.org/10.1142/8577

Hybrid Artificial Immune Recognition System with Improved Belief Classification Process

Rihab Abdelkhalek[(✉)] and Zied Elouedi

LARODEC, Institut Supérieur de Gestion de Tunis, Université de Tunis,
Tunis, Tunisia
rihab.abdelkhalek@gmail.com, zied.elouedi@gmx.com

Abstract. In the past few years, a high worldwide interest in Machine learning (ML) algorithms, involving Artificial Immune Recognition System (AIRS) have increased rapidly. In fact, AIRS is a supervised learning technique inspired by the main concepts and methods of the human immune system. It has shown a great success on broad range of classification problems. AIRS includes different powerful versions for solving complex problems and making appropriate decisions. Despite the high efficiency of these AIRS versions, they suffer from some weaknesses, which could affect the classification accuracy. Actually, some of these versions are incapable to deal with uncertainty presented during the classification procedure. This is treated as one of the most important challenges in real-life classification difficulties. In this paper, we aim to overcome this issue by handling uncertainty under AIRS using the belief function theory. In addition, we aim to improve existing AIRS versions by proposing a new hybrid AIRS approach. This hybrid approach combines a feature selection process as well as various optimization techniques such as Genetic Algorithm (GA) and Gradient Descent. Furthermore, this novel proposed AIRS includes a new enhanced belief classifier in order to increase the classification accuracy. A comparison between diverse AIRS versions and the suggested approach on real world data sets has been performed showing the high efficiency of the proposed hybrid AIRS.

Keywords: Machine learning · Artificial immune recognition systems · Uncertainty · Belief function theory · Genetic algorithm · Feature selection

1 Introduction

Artificial Immune Systems (AIS) constitute one of the new and motivated fields of biologically inspired computing and natural computing. Artificial Immune Recognition System (AIRS) [6] represents a powerful and robust AIS approach, which has been successfully exploited as an efficient solution to different challenges in computer science and engineering. It has been widely applied in diverse fields

S. Le Hégarat-Mascle et al. (Eds.): BELIEF 2022, LNAI 13506, pp. 307–316, 2022.
https://doi.org/10.1007/978-3-031-17801-6_29

including optimization, data mining more specifically anomaly detection, pattern classification and clustering, robotics and malware detection [7]. AIRS introduces various versions such as AIRS2 [1] and AIRS3, which have shown a great success in solving complex problems under a certain environment. However, under an uncertain environment, these versions face some difficulties in dealing with uncertainty that could deeply exist in the classification process. In order to overcome this weakness, researchers have improved the basic AIRS versions by employing various uncertainty theories such as the possibility theory and the fuzzy set [10]. The belief function theory (BFT) is a powerful modeling tool, which has been introduced in 1976 by Shafer [8,9]. It is a rich and popular framework known by its high ability in reasoning under uncertainty and modeling imprecise information. It is also called Dempster-Shafer theory or the evidence theory. Relying on this theory, new evidential versions of AIRS2 and AIRS3 [2] have been proposed. Furthermore, an extension of evidential AIRS3 has been suggested in [3] where the weight of the training patterns is considered during the learning reduction phase. After that, a great optimization of this approach has been achieved in [4] where a significant progress in classification accuracy has been made. Comparing [1] with this optimized AIRS version, a high improvement has been proved. Yet, this latter employs numerous user-predefined parameters which could negatively influence the decision-making process. Moreover, this AIRS model involves a large number of random methods in the several stages of the process which could degrade the approach's performance. So, to find a solution for these issues, a recent research work has been introduced in [5] where authors employ the genetic algorithm during the classification procedure. This approach has shown a success in terms of classification accuracy. However, such approach is unable to select the relevant features, it exploits all the features in the data set without excluding the redundant and the unnecessary ones. This could decrease the efficiency of the AIRS model and produce incorrect predictions. To solve this problem, a recent improved AIRS approach has been proposed where a pre-processing phase was implemented. During this phase, a feature selection is performed and a reduction of data dimension is achieved. Thanks to this improvement, the AIRS model has attain great classification results. Despite the high effectiveness of this approach, it does not take into account the number of classes during the classification process. It depends only on limited parameters such as the number of training instances represented by each pattern and the stimulation level between instances. In such approach, during the classification process, all data sets with different number of classes are treated equally and the number of classes is ignored. In order to alleviate this problem and obtain more accurate and effective classification results, we propose an evidential hybrid AIRS3 approach where we combine the optimized evidential AIRS3 with genetic algorithm and feature selection and at the same time, we exploit the number of classes during the classification phase. Experiments of the new optimized hybrid AIRS3 within the belief function theory, conducted on different public data sets, have proven that the proposed approach outperforms all other AIRS versions in terms of classification accuracy.

This paper is divided as follows: Sect. 2 represents the Artificial Immune Recognition System. Our novel hybrid approach is detailed in Sect. 3. After that, obtained results accomplished on real-world data sets are described in Sect. 4. At the end, Sect. 5 concludes the paper.

2   The Artificial Immune Recognition System

Artificial immune system (AIRS) is a supervised learning algorithm, which employs the human immune system as inspiration [6]. In AIRS, objects are represented as antigens and antibodies sharing the same data representation. AIRS2 [1] is an improved version of AIRS, which starts with a normalization of attributes and an initialization of the Memory cell (MC) pool and the Artificial Recognition Ball(ARB) pool. The selection of the most stimulated cell denoted by mc_match is performed based on the affinity between the test antigen ag and cells having the same class as ag. Then, clones from the mc_match are generated and added to the ARB pool and only the most stimulated ones are maintained. The classification phase in AIRS2 is achieved relying on the basic K-Nearest Neighbors (KNN) [12] algorithm. While, in the AIRS3 version, the classification process is accomplished in a different way. The K value does not represent the number of neighbors but it corresponds to the sum of number of training antigens represented by each memory cell denoted by $numRepAg$. Finally, the unknown antigen is attributed to the class owning the highest sum of $numRepAg$.

3   Optimized Hybrid Evidential AIRS3

In order to enhance the performance of the existing AIRS approaches, we propose an optimized hybrid method based on artificial immune recognition system (AIRS) with a new belief classification process. We call this approach Hybrid-EAIRS-BCKN. Unlike the majority of AIRS versions which employ the basic K-Nearest Neighbors (KNN) [12], our approach adapts a new special belief C*K neighbors classifier denoted by BCKN with C is the number of classes in the whole training data set. The C*K factor represents the sum of $numRepAg$ related to each selected cell in the MC pool. The BCKN is characterized by its simple implementation and its high efficiency in reasoning under uncertainty. Its integration in the hybrid AIRS has highly improved the classification accuracy. The belief framework is known by its great robustness and flexibility in dealing with imperfect information. In addition to that, the optimized hybrid AIRS combines multiple optimization techniques such as the genetic algorithm and the gradient descent in order to build an optimized and a powerful AIRS model. The whole process is illustrated in the following flowchart in Fig. 1.

3.1   Notations

To give a better explanation of our proposed approach, basic notations are defined in what follows:

– $\Theta = \{c_1, c_2, \cdots, c_M\}$: The frame of discernment including a finite set of classes where M refers to the number of classes.

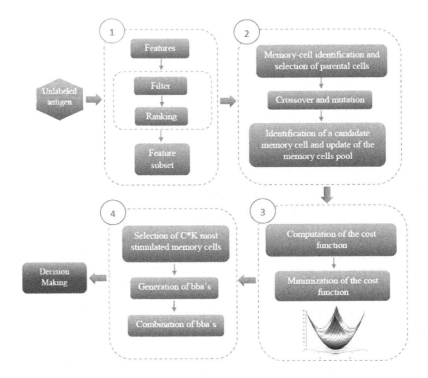

Fig. 1. Optimized hybrid Evidential AIRS3 process

- n: The number of the obtained memory cells after the execution of the BCKN method.
- $I_n = \{I_1, \cdots, I_n\}$: corresponds to the indexes of the n obtained neighbors.
- $MC = \{mc^{(1)}, mc^{(2)}, \cdots, mc^{(n)}\}$: represents the MC pool containing n memory cells (mc).
- $m(\{c_p\}|mc)$: The basic belief mass related to the class c_p.
- ω: represents the inverse of $numRepAg$ corresponding to mc.
- ωd: corresponds to the weighted euclidean distance.

3.2 Stage 1: Data Pre-processing with Feature Selection Method

Feature selection is an important element of data pre-processing which aims to improve the predictive model efficiency. Actually, this method reduces the dimensions of data set by selecting only the most relevant features. It removes noisy and redundant attributes. Among the numerous feature selection techniques, we opt for the Infinite Latent Feature Selection (ILFS) algorithm since it is considered as among the most powerful and robust probabilistic ranking approaches [18].

3.3 Stage 2: Learning Reduction with Genetic Algorithm

This stage of our contribution is divided into three fundamental phases:

- **Memory cell recognition and parents selection**
  In this phase, we select the *mc_match* as the most stimulated memory cell owning an identical class as the antigen *ag*. This chosen cell and the training antigen will be considered as the parental chromosomes. Inspired by genetic algorithm concepts, each cell corresponds to a portion of chromosomes.
- **Crossover and mutation**
  In genetic algorithm, crossover and mutation represents two fundamentals stages. The main goal of the crossover is the generation of new offsprings derived by the combination of the two parental chromosomes (the *mc_match* and the training antigen). In our approach, we opt for the linear combination as one of the simplest and efficient crossover's operators. After the crossover step, the mutation is performed in order to introduce more efficient solutions.
- **Candidate memory cell recognition and refresh of the memory cells pool**
  During this phase, the selection of the appropriate candidate memory cell is performed based on the various computed affinities and the stopping criterion. We compare the affinity of the training antigen and the resulting offspring with the affinity threshold (*AT*). In the case, this latter surpasses the calculated affinity, this antigen will be designated as the candidate memory denoted by *mc_candidate*. In the contrary case, new solutions will be created until meeting the stopping criterion. If the obtained *mc_candidate* is paring the presented antigen better than the actual *mc_match* does, it will join others cells in the MC pool and become a long-lived memory cell. Furthermore, the *mc_match* could be replaced by the *mc_candidate* if the affinity between them is less than the product of affinity threshold scalar and affinity threshold and affinity threshold scalar.

3.4 Stage 3: Belief Parameters Optimization with Gradient Descent

The belief classification phase of AIRS depends on two main parameters which are α and $\gamma = (\gamma_1, \ldots, \gamma_p)$, where $p \in \{1, \ldots, M\}$. In fact, α has the value of 0.95 which has been proved to produce the best classification precision [11]. For the value of the parameter γ, it is computed using an evidential optimization technique. In fact, a cost function is defined based on the computed basic belief assignment(*bba*) is defined as:

$$E(mc^{(l)}) = \sum_{p=1}^{M} (BetP_p^{(l)} - t_p^{(l)})^2 \tag{1}$$

With $BetP^{(l)}$ is the distribution associated to the *bba* $m^{(l)}$ such that:

$$BetP^{(l)} = (BetP^{(l)}(\{c_1\}), \ldots, BetP^{(l)}(\{c_M\}))$$

$mc^{(l)}$ represents a memory cell of the MC pool associated to class c_p. Its related class membership is encoded as a vector $t^{(l)} = (t_1^{(l)}, ..., t_M^{(l)})$ of M binary indicator variables $t_j^{(l)}$ presented by $t_1^{(l)} = 1$ if $j = p$ and $t_1^{(l)} = 0$ otherwise.

In the next step, a minimization of the Mean Squared Error (MSE) is performed using the gradient descent. As a result, the optimum value of the parameter γ is selected. This latter as well as the value of α will be used in Eq. 2.

3.5   Stage 4: Belief Classification with BCKN Algorithm

The classification process of our proposed hybrid AIRS3 approach relies on the Belief C*K neighbors method. We recall that the C*K value in the AIRS3 approach is not the number of nearest selected neighbors but it is the sum of $numRepAg$ of all selected cells as follows $\sum numRepAg = C * K$. The chosen cells in the classification process are those having the highest stimulation values to the unlabeled antigen. This technique has achieved a great success in improving the classification accuracy of the AIRS model. It involves two main steps:

– **Basic belief assignment's generation**
  During this phase, traditional AIRS versions generate K basic belief assignments $(bba's)$ corresponding to the chosen nearest neighbors. Nonetheless, in some cases where the value of K is too large, this technique could be very complicated and it makes performance of the model worse. So, we aim in our work to alleviate such weakness by generating only the $bba's$ corresponding to the C*K selected cells. Thus, the resulting memory cells will be considered as sources of evidence. Relying on these pieces of evidences, the choice of the adequate class will be performed. Then, during the computation process, contrary to the traditional AIRS methods, we do not employ the traditional γ_p as the reciprocal of the mean distance between two cells owning the same class c_p, yet we rely on the optimal value of γ attained during the optimization process. This value is designated by γ_p^*. Therefore, the generation of basic belief assignment from the most stimulated selected antigens is accomplished as follows:

$$m(.|mc^{(i)}) = \begin{cases} m(\{c_p\}|mc^{(i)}) = \alpha \ e^{-(\gamma_p^* \cdot (\omega d_{(i)})^2)} \\ m(\Theta|mc^{(i)}) = 1 - (\alpha \ e^{-(\gamma_p^* \cdot (\omega d_{(i)})^2)}) \end{cases} \quad (2)$$

– **Basic belief assignment's fusion**
  Once the $bba's$ are induced for each selected memory cell, a fusion of these $bba's$ is processed based on the Dempster rule of combination such that:

$$m = m(.|mc^{(i)}) \oplus ... \oplus m(.|mc^{(n)}) \quad (3)$$

  where n represents the number of selected memory cells
  The main objective of the combination process is to obtain a final bba related to the unknown antigen. To attain this goal, the evidence of the n selected

memory cells are aggregated. In fact, this aggregation can be accomplished relying on several rules such as conjunctive rule of combination introduced by Smets [16], Combination With Adapted Conflict (CWAC) rule [17], the Dezert-Smarandache Theory (DSmT) [13], the Dempster's rule of combination [8], etc. In. In our approach, we opt for the Dempster's rule of combination since we are inspired by the evidential K-nearest neighbors [15].

3.6 Stage 5: Decision Making with Pignistic Probability

The final stage of the optimized hybrid evidential AIRS3 is the decision making process. At this level, the assignment of the suitable class to the test antigen is accomplished relying on the pignistic transformation function [14], designated by $BetP$. Based on the calculated $BetP$ values, we assign to the unknown antigen the class owning the highest value of pignistic probability.

4   Experimental Study

In order to evaluate the efficiency of our hybrid AIRS3 approach within the belief function theory, we did numerous experiments where we compare our novel approach with various other AIRS methods namely, AIRS2 denoted by $A1$ [1], Fuzzy AIRS2 [10] denoted by $A2$, Evidential AIRS3 denoted [2] by $A3$, Weighted Evidential AIRS3 [3] denoted by $A4$, Optimized Weighted Evidential AIRS3 [4] denoted by $A5$, and Optimized Evidential AIRS3 with genetic algorithm [5] denoted to $A6$.

4.1   The Framework

The different real data sets employed in our experiments are: Wine (W), Cryotherapy (C), Fertility (F), Pima Indians Diabetes (PID) Hebernal (H) and User Knowledge Modeling (UKM). Characteristics of each data set are presented in Table 1, where # nbInstances is the number of antigens, # nbAttributes corresponds to the number of attributes and # nbclass is referred to the number of class labels of a data set.

Table 1. The characteristics of used data sets

Data set	#nbInstances	#nbAttributes	#nbClass
C	90	6	2
W	178	13	3
F	100	9	2
PID	768	8	2
H	306	3	2
UKM	403	5	4

For all these data sets, we used the following parameter values for all the diverse versions of AIRS employed in our comparison:

Clonal rate = 10, Mutation rate = 0.4, HyperClonal Rate = 2, Number of resources = 200, Stimulation threshold = 0.3, Affinity threshold scalar = 0.2.

Furthermore we have tested with distinct values of K = [1, 2, 3, 5, 7, 10 and 12].

4.2   Evaluation Criteria

For the evaluation of our approach, we rely on four metrics which is the Percent Correct Classification (PCC), the precision (P), the Recall (R) and the F-measure also known by F1 score (F1). In addition, during our tests, we adopt the cross-validation method to determine the effectiveness of our novel approach. In fact, used the 10-fold cross-validation where we average the values derived in all the 10 cases of cross validation.

Table 2. The mean *PCC* (%)

Data sets	A1	A2	A3	A4	A5	A6	Hybrid EAIRS-BCKN
W	65.71	76.51	86.12	86.78	86.92	97	**97.68**
C	66.61	74.58	75.69	75.8	76.16	84.58	**85.72**
PID	63.4	65.75	69.2	69.44	69.6	71.97	**73.66**
F	82.45	84.87	85.26	85.53	85.69	87.24	**87.86**
H	64.52	61.21	64.66	64.81	65.08	68.07	**69.88**
UKM	62.63	63.99	48.23	51.39	62.63	86.37	**86.8**

Table 3. The mean *Recall* (%)

Data sets	A1	A2	A3	A4	A5	A6	Hybrid EAIRS-BCKN
W	64.93	75.41	86.66	87.45	87.45	97.2	**97.52**
C	74.29	67.22	74.79	74.86	75.5	**89.15**	83.72
PID	71.98	71.23	81.06	81.26	81.31	77.4	**79.12**
F	92.39	93.72	94.13	94.33	94.33	97.35	**98.52**
H	77.98	70.13	78.56	78.7	77.01	87.22	**90.57**
UKM	58.75	61.26	46.8	53.47	59.78	81.15	**81.2**

4.3   Experimental Results

To prove the success of our contribution, we compared our proposed approach with nine traditional AIRS versions. Our comparison is based on different popular evaluation measures including *PCC*, *Recall*, *Precision* and *F − Measure* of the numerous used K-values. The different averages of these metrics are represented in Table 2, Table 3, Table 4 and Table 5. As shown in these four tables, our

Table 4. The mean *Precision* (%)

Data sets	A1	A2	A3	A4	A5	A6	Hybrid EAIRS-BCKN
W	61.76	72.72	86.31	87.01	87.01	97.17	**97.84**
C	67.07	78.5	78.78	78.84	79.02	80.63	**87.19**
PID	72.29	75.02	74.06	74.63	75.31	80	**82.09**
F	88.44	90.02	90.13	90.22	90.38	89.39	**89.63**
H	74.64	75.5	75.25	75.31	75.74	**75.8**	74.14
UKM	51,27	62,94	46,75	38,13	64,31	78,8	**79,26**

Table 5. The mean *F − measure* (%)

Data sets	A1	A2	A3	A4	A5	A6	Hybrid EAIRS-BCKN
W	53.06	65.81	81.88	82.57	82.57	97.03	**97.64**
C	67.79	71.02	75.69	76.11	76.11	**84.33**	84,27
PID	71.04	72.71	77.23	77.51	81.81	78.42	**80.45**
F	90.12	91.39	91.74	91.83	91.91	93	**93.79**
H	75.69	72.28	76.4	76.59	76.18	81	**81.43**
UKM	43.43	54	39.72	23.78	57.77	**77.25**	73.43

approach hybrid evidential AIRS3 with BCKN achieve better results than most other versions of AIRS in term of classification accuracy, F-measure, recall and precision for all the employed data sets. Let's take as instance the data set W, our approach attain the best *PCC* with the value 97.68% compared to 65.71% for A1, 76.51% for A2, 86.12 for A3, 86.78 for A4, 86.92 for A5, 97 for A6.

5   Conclusion

In this paper, we have developed a new hybrid evidential AIRS approach with a new classification process named BCKN. It is so powerful that it handles the uncertainty pervaded in the classification phase, optimizes the belief parameter, selects only the relevant features, takes into account the number of represented antigens by each memory cell denoted by *numRepAg* and finally, proceeds on belief classification process based on *numRepAg* and the number of classes of each data set. Experimental results have proved the high performance of our approach in terms of *PCC*, *Precision*, *Recall* and *F − measure* against all the AIRS methods described in this paper under certain and uncertain frameworks.

References

1. Watkins, A., Timmis, J.: Artificial immune recognition system (AIRS): revisions and refinements. In: Congress on Genetic Programming and Evolvable Machines, pp. 173–181 (2002)

2. Abdelkhalek, R., Elouedi, Z.: A belief classification approach based on artificial immune recognition system. In: Lesot, M.-J., et al. (eds.) IPMU 2020. CCIS, vol. 1238, pp. 327–340. Springer, Cham (2020). https://doi.org/10.1007/978-3-030-50143-3_25

3. Abdelkhalek, R., Elouedi, Z.: WE-AIRS: a new weighted evidential artificial immune recognition system. In: International Conference on Robotics and Artificial Intelligence, pp. 874–881 (2020)

4. Abdelkhalek, R., Elouedi, Z.: Parameter optimization and weights assessment for evidential artificial immune recognition system. In: Li, G., Shen, H.T., Yuan, Y., Wang, X., Liu, H., Zhao, X. (eds.) KSEM 2020. LNCS (LNAI), vol. 12275, pp. 27–38. Springer, Cham (2020). https://doi.org/10.1007/978-3-030-55393-7_3

5. Abdelkhalek, R., Elouedi, Z.: An optimized evidential artificial immune recognition system based on genetic algorithm. In: Yin, H., et al. (eds.) IDEAL 2021. LNCS, vol. 13113, pp. 188–195. Springer, Cham (2021). https://doi.org/10.1007/978-3-030-91608-4_19

6. Watkins, A.: A resource limited artificial immune classifier. In: The 2002 Congress on Evolutionary Computation, pp. 926–931 (2002)

7. Castro, L.N., De Castro, L.N., Timmis, J.: Artificial Immune Systems: A New Computational Intelligence Approach. Springer, Heidelberg (2002)

8. Dempster, A.P.: A generalization of Bayesian inference. J. R. Stat. Soc. Ser. B (Methodol.) 205–247 (1968)

9. Shafer, G.: A Mathematical Theory of Evidence. Princeton University Press, Princeton (1976)

10. Chikh, M.A., Saidi, M., Settouti, N.: Diagnosis of diabetes diseases using an artificial immune recognition system2 (AIRS2) with fuzzy k-nearest neighbor. J. Med. Syst. **36**(5), 2721–2729 (2012)

11. Zouhal, L.M., Denoeux, T.: An evidence theoretic kNN rule with parameter optimization. IEEE Trans. Syst. Man Cybern. **28**(2), 263–271 (1998)

12. Dasarathy, B.V.: Nearest Neighbour NN Norms: NN Pattern Classification Techniques. IEEE Computer Society Press (1991)

13. Dezert, J., Tchamova, A.: On the behavior of Dempster's rule of combination (2011)

14. Smets, P., Kennes, R.: The transferable belief model. In: Yager, R.R., Liu, L. (eds.) Classic Works of the Dempster-Shafer Theory of Belief Functions, pp. 693–736. Springer, Heidelberg (2008). https://doi.org/10.1007/978-3-540-44792-4_28

15. Denoeux, T.: A k-nearest neighbor classification rule based on Dempster-Shafer theory. In: Yager, R.R., Liu, L. (eds.) Classic Works of the Dempster-Shafer Theory of Belief Functions, pp. 737–760. Springer, Heidelberg (2008). https://doi.org/10.1007/978-3-540-44792-4_29

16. Smets, P.: The combination of evidence in the transferable belief model. IEEE Trans. Pattern Anal. Mach. Intell. **12**(5), 447–458 (1990)

17. Lefèvre, E., Elouedi, Z.: How to preserve the conflict as an alarm in the combination of belief functions? Decis. Support Syst. **56**, 326–333 (2013)

18. Roffo, G.: Feature selection library (MATLAB toolbox). arXiv preprint arXiv:1607.01327 (2016)

Author Index

Printed in the United States
by Baker & Taylor Publisher Services